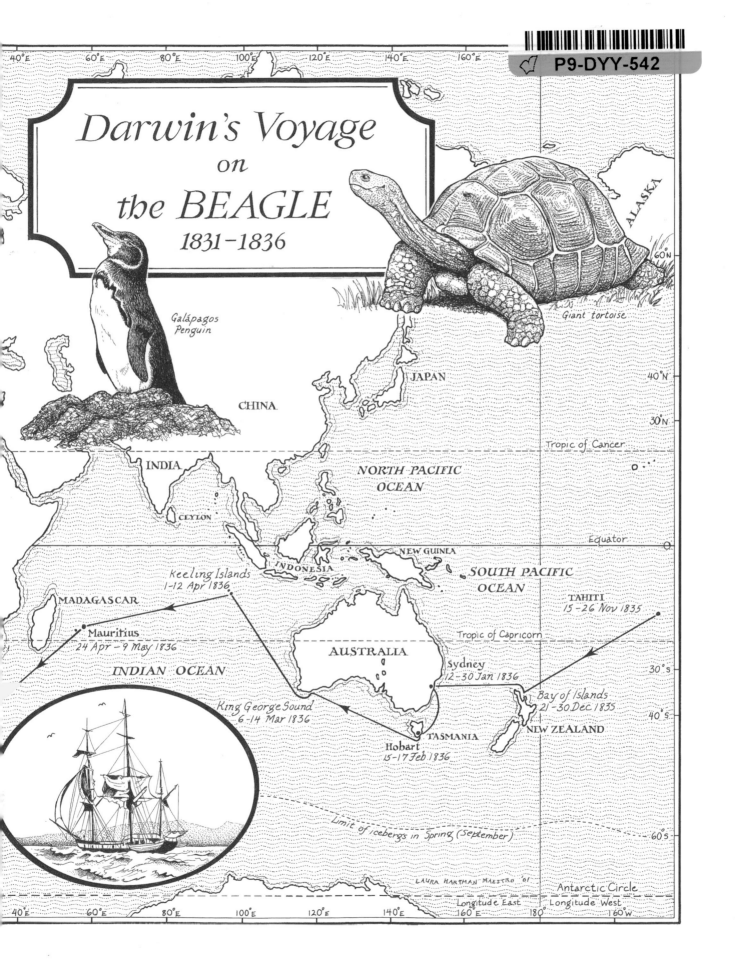

Darwin's Voyage
on
the BEAGLE
1831–1836

Galápagos
Penguin

Giant tortoise

ALASKA

JAPAN

CHINA

INDIA

CEYLON

NORTH PACIFIC
OCEAN

NEW GUINEA

INDONESIA

SOUTH PACIFIC
OCEAN

Keeling Islands
1–12 Apr 1836

TAHITI
15–26 Nov 1835

MADAGASCAR

Mauritius
24 Apr – 9 May 1836

Tropic of Capricorn

AUSTRALIA

Sydney
12–30 Jan 1836

INDIAN OCEAN

King George Sound
6–14 Mar 1836

Bay of Islands
21–30 Dec 1835

NEW ZEALAND

TASMANIA

Hobart
15–17 Feb 1836

Tropic of Cancer

Equator

Limit of icebergs in Spring (September)

LAURA HARTMAN MAESTRO '01

Antarctic Circle

Longitude East
160°E

Longitude West
180°   160°W

40°E   60°E   80°E   100°E   120°E   140°E   160°E

60°N
40°N
30°N
0
30°S
40°S
60°S

evolution

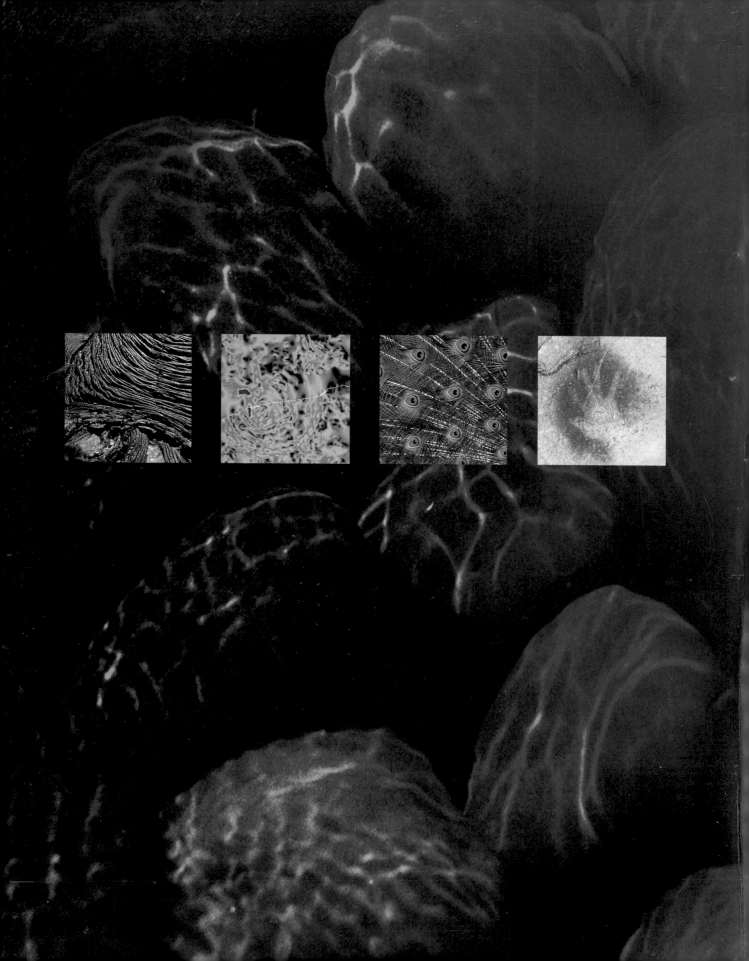

CARL ZIMMER

evolution

THE TRIUMPH OF AN IDEA

Introduction by Stephen Jay Gould

Foreword by Richard Hutton

HarperCollins*Publishers*

HarperCollins books may be purchased for educational, business, or sales promotional use. For information, please write: Special Markets Department, HarperCollins Publishers Inc., 10 East 53rd Street, New York, NY 10022.

ILLUSTRATION CREDITS

ENDPAPERS AND P. 18: Laura Hartman Maestro, adapted from *Charles Darwin* by Janet Browne, Princeton University Press, 1995; p. 18: Deborah Perugi, adapted from *Charles Darwin* by Janet Browne, Princeton University Press; 42: Laura Hartman Maestro, adapted from *At the Water's Edge* by Carl Zimmer, Free Press, 1998; 70–71: Deborah Perugi; 77: Laura Hartman Maestro; 83: Deborah Perugi, adapted from *Evolution* by Mark Ridley, Blackwell Science, 1996; 89: Deborah Perugi, adapted from *Proceedings of the National Academy of Sciences* 96: 5101–6; 102: adapted from *Science* 285: 1025–6; 120: Deborah Perugi; 123, 130: Laura Hartman Maestro; 138: Deborah Perugi, adapted from *At the Water's Edge,* Zimmer; 185: Deborah Perugi, adapted from *Nature* 403: 853–8; 220: adapted from *Science* 287: 607–14; 222: Deborah Perugi; 263: adapted from *Proceedings of the National Academy of Sciences* 96:5077–82; 264: Deborah Perugi, adapted from *The Human Career* by Richard Klein, University of Chicago, 1999; 298: adapted from *Nature* 403: 708–13; 299: Deborah Perugi, adapted with permission from Richard Klein

Additional illustration credits appear on p. 364.

FIRST EDITION

Designed by Lindgren/Fuller Design

Printed in Japan

Library of Congress Cataloging-in-Publication Data is available upon request.

ISBN 0-06-019906-7

01 02 03 04 05 TPN 10 9 8 7 6 5 4 3 2 1

To Grace

CONTENTS

Introduction by Stephen Jay Gould   ix

Foreword by Richard Hutton   xv

Acknowledgments   xix

PART 1   SLOW VICTORY: DARWIN AND THE RISE OF DARWINISM

1 • Darwin and the *Beagle*   3

2 • "Like Confessing a Murder": The Origin of *Origin of Species*   27

3 • Deep Time Discovered: Putting Dates to the History of Life   57

4 • Witnessing Change: Genes, Natural Selection, and Evolution in Action   73

PART 2   CREATION AND DESTRUCTION

5 • Rooting the Tree of Life: From Life's Dawn to the Age of Microbes   101

6 • The Accidental Tool Kit: Chance and Constraints in Animal Evolution   117

7 • Extinction: How Life Ends and Begins Again   143

PART 3   EVOLUTION'S DANCE

8 • Coevolution: Weaving the Web of Life   189

9 • Doctor Darwin: Disease in the Age of Evolutionary Medicine   211

10 • Passion's Logic: The Evolution of Sex   229

PART 4   HUMANITY'S PLACE IN EVOLUTION AND EVOLUTION'S PLACE IN HUMANITY

11 • The Gossiping Ape: The Social Roots of Human Evolution   259

12 • Modern Life, 50,000 B.C.: The Dawn of Us   293

13 • What about God?   313

Further Reading   345

Index   353

INTRODUCTION

A famous legend (perhaps even true) from the early days of Darwinism provides a good organizing theme for understanding the centrality and importance of evolution both in science and for human life in general. A prominent English lady, the wife of a lord or a bishop (yes, they may marry in the Church of England), exclaimed to her husband when she grasped the scary novelty of evolution: "Oh my dear, let us hope that what Mr. Darwin says is not true. But if it is true, let us hope that it will not become generally known!"

Scientists invoke this familiar story to laugh at the recalcitrant stodginess of old belief and breeding—especially the risible image of the upper classes keeping a revolutionary fact of nature in the Pandora's box of their own private learning. Thus, the lady of this anecdote enters history as a quintessential patrician fool. Let me suggest, however, if only to organize the outline of this introduction, that we reconceptualize her as a prophet. For what Mr. Darwin said is clearly true, and it has also not become generally known (or, at least in the United States, albeit uniquely in the Western world, even generally acknowledged). We need to understand the reasons for this exceedingly curious situation.

EVOLUTION AS TRUE

The task of science is twofold: to determine, as best we can, the empirical character of the natural world; and to ascertain why our world operates as it does, rather than in some other conceivable, but unrealized, way—in other words, to specify facts and validate theories. Science, as we professionals always point out, cannot establish absolute truth; thus, our conclusions must always remain tentative. But this healthy skepticism need not be extended to the point of nihilism, and we may surely state that some facts have been ascertained with sufficient confidence that we may designate them as "true" in any legitimate, vernacular meaning of the word. (Perhaps I cannot be absolutely certain that the earth is round rather than flat, but the roughly spherical shape of our planet has been sufficiently well verified that I need not grant the "flat earth society" a platform of equal time, or even any time at all, in my science class-room.) Evolution, the basic organizing concept of all the biological sciences, has been validated to an equally high degree, and may therefore be designated as true or factual.

In discussing the truth of evolution, we should make a distinction, as Darwin explicitly did, between the simple *fact* of evolution—defined as the genealogical connection among all earthly organisms, based on their descent from a common ancestor, and the history of any lineage as a process of descent with modification—and *theories* (like Darwinian natural selection) that have been proposed to explain the causes of evolutionary change.

Three broad categories of evidence best express the factuality of evolution. First, direct evidence of human observation, guided by an explicit theory since Darwin's publication in 1859, but buttressed by data on longer periods of breeding for improved crop plants and domesticated animals, provides hundreds of exquisitely documented examples of the small-scale changes that our theories anticipate over such geologically brief periods of time. These include the familiar cases of changing pigmentation in moth wings as an adaptive response to substrates darkened by industrial soot, altered beak shapes in Galápagos species of Darwin's finches as climates and food resources change, and the development of antibiotic resistance by numerous strains of bacteria. No one—not even among creationists—has denied this overwhelming weight of evidence in the small, but we also need proof that such minor changes can accumulate through geological time into the substantial novelties that build the history of life's expanding diversity.

We must therefore turn to a second category of direct evidence from transitional stages of major alterations found in the fossil record. A common claim, stated often enough to merit the label of "urban legend," holds that no such transitional forms exist and that paleontologists, dogmatically committed to evolution, have either withheld this information from the public or have claimed that the fossil record is too imperfect to preserve the intermediates that must once have existed. In fact, although the fossil record is indeed spotty (a problem with nearly all historical documents, after all), the assiduous work of paleontologists has revealed numerous elegant examples of sequences of intermediary forms (not just single "in between" specimens) joining ancestors in proper temporal order to very different descendants—as in the evolution of whales from terrestrial mammalian ancestors through several intermediate stages, including *Ambulocetus* (literally, the walking whale), the evolution of birds from small running dinosaurs, of mammals from reptilian ancestors, and a threefold increase in brain size during the last 4 million years of human evolution.

Finally, a third major category of more indirect, but ubiquitous, evidence allows us to draw a clear inference of change from a different historical past by observing the quirks and imperfections, present in all modern organisms, that make no sense except as holdovers from an otherwise altered (that is, evolved) ancestral state—that is, except as products of evolution. This principle governs the analysis of all kinds of historical series, not just biological evolution. We can infer that an abandoned railroad line once linked a group of well-spaced and linearly ordered towns (that would have no other reason for such an alignment). We can also identify social change from a more rural past by the etymological evidence of many words now used in very different meanings in our modern industrial world ("broadcast" as a mode of planting by throwing out seeds by the handful; or "pecuniary" advantages, literally measured in cattle, from the Latin *pecus,* or cow). In the same manner, all organisms carry useless remnants of formerly functional structures that make no sense except as holdovers from different ancestral states—the tiny vestiges of leg bones, invisibly embedded in the skin of certain whales, or the nonfunctional nubs of pelvic bones in some snakes, surviving as vestiges of ancestors with legs.

EVOLUTION AS NOT GENERALLY KNOWN OR ACKNOWLEDGED

No scientific revolution can match Darwin's discovery in degree of upset to our previous comforts and certainties. In the only conceivable challenge, Copernicus and Galileo moved our cosmic location from the center of the universe to a small and peripheral body circling a central sun. But this cosmic reorganization only fractured our concept of real estate; Darwinian evolution, on the other (and deeper) hand, revolutionized our view of our own meaning and essence (insofar as science can address such questions at all): Who are we? How did we get here? How are we related to other creatures, and in what manner?

Evolution substituted a naturalistic explanation of cold comfort for our former conviction that a benevolent deity fashioned us directly in his own image, to have dominion over the entire earth and all other creatures—and that all but the first five days of earthly history have been graced by our ruling presence. In evolutionary terms, however, humans represent but one tiny twig on an enormous and luxuriantly branching tree of life, with all twigs interconnected by descent, and the entire tree growing (so far as science can tell) by a natural and lawlike process. Moreover, the unique and minuscule twig of *Homo sapiens* emerged in a geological yesterday, and has flourished for only an eyeblink of cosmic immensity (about 100,000 years for our species and only 6–8 million years for our entire lineage since our branchlet split from the node of our closest living relative, the chimpanzee. By contrast, the oldest bacterial fossils on Earth arose 3,600 million years ago).

We might mitigate the challenge of these basic facts if we could espouse a theory of evolutionary change that remained congenial to our old comforts about human necessity and inherent superiority—as in the common misconception that evolution implies predictable and progressive pathways of change, and that human origins (however belated) may therefore be viewed as both inevitable and culminating. But our best understanding of how evolution operates—that is, our preferred *theory* for the mechanism of evolutionary change (as contrasted with the simple *factuality* of evolution, discussed in the last section)—does not even grant us this ideological comfort. For our favored and well-attested theory, Darwinian natural selection, offers no solace or support for these traditional hopes about human necessity or cosmic importance.

Hence, when I ask myself why evolution, although true by our strongest scientific confidence, has not become generally known or acknowledged in the United States—that is, nearly 150 years after Darwin's publication, and in the most technologically advanced nation on earth—I can only conclude that our misunderstanding of the broader implications of Darwinism, in particular our misreading of his doctrine as doleful, or as subversive to our spiritual hopes and needs, rather than as ethically neutral and intellectually exhilarating, has impeded public acceptance of our best documented biological generality. Hence, I treat the meaning of Darwinism, or the implications of evolutionary theory (rather than the mere understanding of evolution's factuality), as my major theme in trying to explicate why such an evident fact has not become generally known.

Public difficulty in grasping the Darwinian theory of natural selection cannot be attributed to any conceptual complexity—for no great theory ever boasted such a simple structure of three undeniable facts and an almost syllogistic inference therefrom. (In a famous, and true, anecdote, Thomas Henry Huxley, after reading *Origin of Species,* could only say of natural selection: "How extremely stupid not to have thought of that myself.") First, that all organisms produce more offspring than can possibly survive; second, that all organisms within a species vary, one from the other; third, that at least some of this variation is inherited by offspring. From these three facts, we infer the principle of natural selection: since only some offspring can survive, on average the survivors will be those variants that, by good fortune, are better adapted to changing local environments. Since these offspring will inherit the favorable variations of their parents, organisms of the next generation will, on average, become better adapted to local conditions.

The difficulties lie not in this simple mechanism but in the far-reaching and radical philosophical consequences—as Darwin himself well understood—of postulating a causal theory stripped of such conventional comforts as a guarantee of progress, a principle of natural harmony, or any notion of an inherent goal or purpose. Darwin's mechanism can only generate local adaptation to environments that change in a directionless way through time, thus imparting no goal or progressive vector to life's history. (In Darwin's system, an internal parasite, so anatomically degenerate that it has become little more than a bag of ingestive and reproductive tissue within the body of its host, may be just as well adapted, and may enjoy just as much prospect of future success, as the most complex mammalian carnivore, wily, fleet, and adept, living free on the savannas.) Moreover, although organisms may be well designed, and ecosystems harmonious, these broader features of life arise only as consequences of the unconscious struggles of individual organisms for personal reproductive success, and not as direct results of any natural principle operating overtly for such "higher" goods.

Darwin's mechanism may sound bleak at first, but a deeper view should lead us to embrace natural selection (and a variety of other legitimate evolutionary mechanisms from punctuated equilibrium to catastrophic mass extinction) for two basic reasons. First, truthful science is liberating in the practical sense that knowledge of nature's actual mechanisms gives us the potential power to cure and to heal when factual matters cause us harm. When, for example, we know how bacteria and other disease-causing organisms evolve, we can understand, and find means to combat, the development of antibiotic resistance, or the unusual mutability of the AIDS virus. Also, when we recognize how recently our so-called human races diverged from a common African ancestry, and when we measure the minuscule genetic differences that separate our groups as a result, then we can know why racism, the scourge of human relations for so many centuries, can claim no factual foundation in any real differences among human groups.

Second, and more generally, by taking the Darwinian "cold bath," and staring a factual reality in the face, we can finally abandon the cardinal false hope of the ages—that factual nature can specify the meaning of our life by validating our inherent superiority, or by proving that evolution exists to generate us as the summit of life's purpose. In principle, the factual state of the universe, whatever it may be, cannot teach us *how we should live* or *what our lives should mean*—for these ethical questions of value and meaning belong to such different realms of human life as religion, philosophy, and humanistic study. Nature's facts can help us to realize a goal once we have made our ethical decisions on other grounds—as the trivial genetic differences among human groups, for example, can help us to understand human unity once we have agreed on the unalienable rights of all people to life, liberty, and the pursuit of happiness. Facts are just facts, in all their fascination, their pristine beauty, and, sometimes, their unfortunate necessity (bodily decline and mortality, as the obvious example), and ethical rectitude, or spiritual meaning, reside within other domains of human inquiry.

When we thought that factual nature matched our hopes and comforts—all things bright and beautiful, and all things made for our superior selves—

then we easily fell into the trap of equating actuality with righteousness. But when we sense the different fascination of evolution's naturalistic ways, and of life's astonishingly rich diversity and history of change, with *Homo sapiens* as but one contingent twig on the most luxuriant of all trees, then we finally become free to detach our search for ethical truth and spiritual meaning from our scientific quest to understand the facts and mechanisms of nature. Darwin, in defining the factual "grandeur of this view of life" (to quote the last line of *Origin of Species*), liberated us from asking too much of nature, thus leaving us free to comprehend whatever fearful fascination may reside "out there," in full confidence that our quest for decency and meaning cannot be threatened thereby, and can emerge only from our own moral consciousness.

Stephen Jay Gould
Museum of Comparative Zoology
Harvard University

I'm sitting in Down House, watching Charles Darwin teach his 8-year-old daughter, Anne, about barnacles. He moves his crude light microscope in front of her so that she can see the ugly, fleshy thing more closely. As he begins to recite a whole series of facts (after all, he's in the middle of a 10-year study of barnacles that will buttress his theory of evolution), she laughingly plays with the pronunciation of the word. "Barney Ickle," she calls it, after a toy. Darwin has seven children already, and will go on to have ten. But Anne will always be his favorite.

Darwin is obsessed with his scientific ideas, but still very close to his children. As I watch, I realize that he is somehow, improbably, trying hard to bring the two sides of his life together, his work and his family. He is acutely aware of the world around him and knows about the turmoil just outside his door. The traditional English way of life is under siege, as the Industrial Revolution changes how gentry and commoners, bosses and workers, interact. And his theory of evolution, which offers a biological basis for the relationships among plants and animals, including humans, promises to threaten not only the beliefs of those who consider him a friend but the very structure of the society in which he lives. Even his family will not be immune to the danger. Yet this is something he has to do.

Suddenly Alastair Reid, the director of our drama, yells "Cut!" and Darwin becomes the actor Chris Larkin, Anne the actress Eleanor Martha Ogbourne, and I am thrust out of the nineteenth century into the present. We are amidst a tangle of lights, electrical cords, and cameras in a wonderful old home just outside of Bath, watching the rain pelt against the windows.

It is the third day of a 15-day shoot. We are chronicling Charles Darwin's life as a part of our 8-hour television series on evolution. After almost two years of developing the ideas and exploring some wondrous science, this is our final push. During the past year, our crews have visited the deserts of South Africa and Egypt and the hills of Mexico and Hawaii; the forests of Uganda and Tanzania and the tubercular prisons of Russia; caves along the Mediterranean in Turkey, caves in southern France. Authorities escorted one of our teams out of the rain forests of Ecuador to protect them from rebels infiltrating from Colombia, forcing us to return a month later to finish our work. A military government refused us permission to film in the foothills of the Himalayas in Burma—only to have us find a similar story in the jungles of Thailand, just next door.

Telling the story of evolution is almost as daunting as filming it. Cameras record only what is occurring right in front of them, and evolution doesn't happen in a few minutes. Yet we've taken it on, and we find ourselves exploring 3.8 billion years of life on Earth in 8 hours—which comes to about 132,000 years of life per second of television time.

A project like this is long overdue. Far from a quaint notion developed by a Victorian gentleman some 160 years ago, evolution is an idea that ties together all the information that biologists and naturalists have gathered throughout history. It makes sense of observations about the behavior of newts and snakes, fig trees and wasps, birds and tigers, orchids and algae. It ties our emerging knowledge of genes to our understanding of our biological past and weaves a history of common ancestry for all living things. And it explains the range and breadth of the nearly 4-billion-year-old fossil record—both for animals and plants that came before us and for our own heritage.

Oddly, our task as filmmakers parallels Darwin's. The conflict that affected him 150 years ago in many ways still exists. We, too, are trying to reconcile what we have learned with how we see the world around us, how we live. We, too, have access to knowledge of extraordinary range and scope—knowledge that for some people is frightening in all its implications.

Perhaps most important, however, is that evolution affects our daily lives. Evolution matters today, and matters a great deal—in health care, for instance, as it informs our fight against antibiotic resistance, or in agriculture and pesticide resistance, or in the radically changing composition of life on this planet. We are witnessing the likely extinction of 50 percent or more of the species on Earth; understanding evolution gives us both tools to deal with this crisis and a sense of what the world will look like in its aftermath.

Philosopher Daniel Dennett once wrote of the theory of evolution: "If I were to give an award for the single best idea anyone has ever had, I'd give it to Darwin, ahead of Newton and Einstein and everyone else. In a single stroke, the idea of evolution by natural selection unifies the realm of life, meaning, and purpose with the realm of space and time, cause and effect, mechanism and physical law. But it is not just a wonderful idea. It is a dangerous idea."

Dangerous because, for people who interpret the Bible literally, it threatens dearly held religious beliefs about a six-day process of creation. Dangerous because its implications have been historically misused and abused by everyone from the Nazis in Germany to eugenicists trying to "improve" the human race. Dangerous because, writ large, it could be seen as countering the notion that there is some greater meaning in life than simply the here and now.

Translating this dangerous idea to television has been absorbing, illuminating—but, most of all, challenging. Our job is to report on what is known, and not known, from a scientific point of view. And that means examining facts and hypotheses—the accumulation of evidence over the past 150 years; the exploration of ideas that are testable; the explanation of experiments that are repeatable—all to frame the story. Then we offer it to everyone who has a stake—everyone whose life is affected by it, everyone who is curious about it, everyone who is troubled by its implications for who we are and how we came to be—in other words, everyone.

Our project, *Evolution,* is therefore a story of change over time, of who lives, who dies, and who grabs the opportunity to pass on traits to the next generation—and the next, and the next. It's the story of a simple mechanism discovered by Charles Darwin, and how that mechanism has altered the way we view ourselves. And it's the story of why evolution is so hard to accept for so many people. In sum, it is the story, the true story, of all life on Earth and how we are inextricably bound to one another.

In his life, Charles Darwin embodied the idea that a world understood through the lens of science was no less awe-inspiring, no less meaningful than one explained by the mythology of the past. Leave it to Darwin to sense beauty in a barnacle and wonder in the workings of nature.

Richard Hutton
Executive Producer, Evolution
WGBH

ACKNOWLEDGMENTS

This book is part of the Evolution Project, a coproduction of the WGBH/NOVA Science Unit and Clear Blue Sky Productions. The project also includes a seven-part television series, a Web site, a multimedia library, and an educational outreach program. It has been an honor and a pleasure to be involved with so many talented people on such a large enterprise. Special thanks to Paul G. Allen and Jody Patton at Clear Blue Sky Productions for their inspiring vision for and generous support of the Evolution Project, and to Paula Apsell of the WGBH/NOVA Science Unit for guiding its development. I am grateful to them for bringing me aboard.

Richard Hutton, the executive producer of the television series, reined in the unruly science of evolutionary biology to create seven compelling episodes, consulted with dozens of leading scientists, oversaw a production staff traveling around the world, and yet always had time to chat with me about the fine points of the Cretaceous-Tertiary mass extinctions or Charles Darwin's family life. I thank him for his leadership and intellectual companionship. To shape the editorial structure of the series, Richard worked with NOVA senior editor Steve Lyons, series science editor Joe Levine, and associate producer Tina Nguyen. I particularly want to thank Joe for many insightful e-mail exchanges, and Tina for helping me track down various chunks of information, no matter how esoteric.

Many thanks to Caroline Chauncey, publishing manager at WGBH Enterprises. She played many roles during the writing of this book—unofficial editor, diplomat, and cheerleader—and for all of those roles I am grateful. Thanks also go to art consultant Toby Greenberg, who sifted through thousands of pictures for the book's lovely artwork. Thanks also to Betsy Groban, managing director of WGBH Enterprises, WGBH literary agent Doe Coover, and the core administrative staff at WGBH: Lisa Mirowitz, coordinating producer; Karen Carroll Bennett, business manager; Denise Drago, unit manager; Kerrie Iasi, series associate producer; and Cecelia Kelly, production secretary, who all worked so hard to bring the disparate pieces of this project together. At Clear Blue Sky Productions, vice president and general manager Eric Robison;

Bonnie Benjamin-Phariss, director of documentary productions; Jason J. Hunke, director of publicity and marketing; and Pamela Rosenstein, coordinating producer, kindly offered editorial advice and review.

The producers of the individual episodes all took time away from their own frantic schedules to share their research, transcripts, and footage with me and were kind enough to review parts of my book. Thanks to them all: David Espar, senior producer, Linda Garmon, producer of drama, and Susan K. Lewis, producer, "Darwin's Dangerous Idea"; Joel Olicker, producer, and Chris Schmidt, coproducer, "Great Transformations"; Richard Hutton and Kate Churchill, producers, "Extinction!"; Gail Willumsen and Jill Shinefeld, producers, "The Evolutionary Arms Race"; Noel Buckner and Rob Whittlesey, producers, "Why Sex?"; John Heminway and Michelle Nicholasen, producers, "The Mind's Big Bang"; Bill Jersey, producer, and Jamie Stobie, coproducer, "What about God?"

This book is about science, so of course profound thanks must go to the scientists who helped make it possible. The Evolution Project's board of advisers, who offered wisdom and guidance for both the series and the book, included Charles Aquadro, William H. Calvin, Sharon Emerson, Jane Goodall, Sarah Blaffer Hrdy, Don Johanson, Mary-Claire King, Ken Miller, Steven Pinker, Eugenie Scott, and David Wake. Thanks in particular go to Stephen Jay Gould, who not only served as an adviser to the series but graced this book with an introduction.

I want to thank all the scientists who were profiled in the show and this book. I'm also grateful to the experts who graciously explained their research to me or reviewed portions of the manuscript, including Chris Adami, Maydianne Andrade, Connie Barlow, Wouter Bleeker, Ed Brodie Jr., Ed Brodie III, David Burney, Joseph Cain, Sean Carroll, Steve Case, Chris Cheng-DeVries, Robert Cowie, Carla Dantonio, Robin Dunbar, David Dusenbury, Stephen Emlen, Douglas Erwin, Brian Farrell, John Flynn, Beatrice Hahn, Kristen Hawkes, Nicholas Holland, David Inouye, Christine Janis, Kenneth Y. Kaneshiro, Judy Kegl, Richard Klein, Andrew Knoll, Thurston Lacalli, Laura Landweber, Stuart Levy, Michael Lynch, Axel Meyer, Kenneth Miller, Stephen Mojzsis, Anders Møller, Ulrich Mueller, Martin Nowak, Stephen O'Brien, Maureen O'Leary, Norman Pace, Nipam Patel, Marion Petrie, Stuart Pimm, David Resznick, Mark Ridley, Dolph Schluter, Kurt Schwenk, Eugenie Scott, John Thompson, Frans de Waal, Peter Ward, Andrew Whiten, Brad Williamson, and Mark Winston. Thanks go especially to Kevin Padian, who read the entire manuscript and probably used up a whole pack of red pens marking it up.

Gail Winston, my editor at HarperCollins, provided the guidance that helped me transform a companion volume into a book of its own. For that invaluable help, my thanks. Thanks also to my agent, Eric Simonoff.

And finally, the greatest thanks go to my wife, Grace. During a busy year, she always kept things in perspective.

PART ONE

Slow Victory:
Darwin and the
Rise of Darwinism

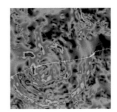

Darwin and
the *Beagle*

IN LATE OCTOBER 1831 a 90-foot coaster named HMS *Beagle* lay docked at Plymouth, England. Its crew scrambled about it like termites in a nest. They were packing the ship as tightly as they could for a voyage around the world, one that would last five years. They rolled barrels of flour and rum into the hold and crammed the deck with wooden boxes that contained experimental clocks resting on beds of sawdust. The *Beagle*'s voyage was a scientific one: its crew would be testing the clocks for the British navy, which depended on precise timekeeping to navigate. Exquisitely detailed maps would be drawn on the voyage as well, so mahogany lockers were installed in the poop cabin and packed with navigational charts. The crew replaced the ship's 10 steel cannons with brass ones so that not even the slightest interference could confuse the *Beagle*'s compasses.

Fossils (such as this relative of the nautilus shown at left) were long believed to be rocks that had spontaneously taken on biological shapes. In fact, fossils form when skeletons, shells, and other organic remains are replaced by minerals.

Amid the flurry of preparations, a 22-year-old man picked his way. He moved awkwardly around the ship, not only because his 6-foot frame was oversized for the cramped quarters, but also because he felt profoundly out of place. He had no official position on the ship, having been invited to keep the captain company during the voyage and act as an unofficial naturalist. It was usually up to a ship's surgeon to act as the naturalist for a voyage, but this awkward young man had no such practical skill. He was a medical school dropout who, for want of any other respectable line of work, was considering a career as a country parson when the voyage was over. Once he had stowed away his preserving jars, his microscope, and the rest of his equipment in the poop cabin, he had nothing more to do. He tried helping the assistant surveyor calibrate some of the timepieces, but he didn't even know enough math to do the most basic calculations.

The name of this awkward young man was Charles Darwin. By the time the *Beagle* returned to England five years later, he would be transformed into one of Britain's most promising young scientists. And out of his experiences on the journey, he would discover the single most important idea in the history of biology, one that would permanently alter humanity's perception of its place in the natural order. From clues that he collected aboard the *Beagle,* Darwin would show that nature had not been created in exactly the form it takes today. Life evolves: it changes gradually but perpetually over vast gulfs of time, driven through those changes thanks to the laws of heredity, without any need of direct divine intervention. And humans, far from being the pinnacle and destiny of God's creation, were but a single species among many, another product of evolution.

Darwin would send Victorian England into a crisis with his theory, but he would offer an alternative view of life that has turned out to have a grandeur of its own. It is clear today that evolution connects us to the dawn of Earth, to showers of comets and death-winds of stars. It produced the crops we eat and now helps insects destroy them. It illuminates the mysteries of medicine, such as how mindless bacteria can outwit the best minds in science. It holds a warning for those who would take from Earth without limits; it reveals how our minds were assembled among lonely bands of apes. We may still struggle with what evolution says about our place in the universe, but that universe is all the more remarkable.

The *Beagle* is remembered today only because of Darwin's experience on board the ship. But if you tried to tell that to the sailors rolling barrels aboard they might have laughed without even a glance at the young man who was pretending to know what he was doing.

"My chief employment," Darwin wrote to his family from Plymouth, "is to go on board the *Beagle* and try to look as much like a sailor as ever I can. I have no evidence of having taken in man, woman or child."

IN SEARCH OF BEETLES AND RESPECTABILITY

Darwin had grown up along the banks of the Severn River in Shropshire, collecting pebbles and birds, completely unaware of the fortunes that made his life pleasant. His mother, Susannah, came from the wealthy Wedgwood family, which made china of the same name. Although his father, Robert, came from less wealthy stock, he built up a fortune of his own by working as a doctor and discreetly lending money to his patients. He eventually became rich enough to build his family a large house, the Mount, on a hillside overlooking the Severn.

Charles and his older brother Erasmus had the close, practically telepathic connection that brothers sometimes have. As teenagers they built themselves a laboratory at the Mount where they would dabble in chemicals and crystals. When Charles was 16, Erasmus went to Edinburgh to study medicine. Their father sent Charles along with him to keep Erasmus company, and ultimately to go to medical school as well. Charles was happy to tag along, for the company of his brother and for the adventure.

Charles Darwin in an 1854 photograph.

When they arrived in Edinburgh, Charles and Erasmus were shocked by the squalor and spectacle of the city. These two boys, raised in the genteel countryside where Jane Austen set her novels, encountered slums for the first time. Politics raged around them as Scottish nationalists, Jacobites, and Calvinists jostled over church and country. At Edinburgh University they faced a rabble of rough students shouting and shooting off pistols in the middle of lectures. Charles and Erasmus recoiled into each other's company, spending their time talking together, walking along the shore, reading newspapers, and going to the theater.

Charles realized very quickly that he hated medicine. The lectures were dreary, the dissected corpses a nightmare, the operations—often amputations without anesthesia—terrifying. He kept himself busy with natural history. But although Charles knew that he could not become a doctor, he had no appetite for standing up to his father. When he came home to the Mount for the summer, he avoided bringing up the matter, spending his days instead shooting birds and learning how to stuff them. He would continue to avoid confrontations for the rest of his life.

Over the summer Robert Darwin decided to send Erasmus to London to continue his studies. Charles returned to Edinburgh alone in October 1826, with only his natural history to distract him from a life he had come to hate.

He became friends with naturalists in Edinburgh, including a zoologist named Robert Grant, who took him under his wing. Grant had been trained as a doctor but had given up his practice to become one of the country's great zoologists, studying sea pens, sponges, and other creatures that scientists of the time still knew almost nothing about. Grant proved a good mentor. "He was dry and formal in manner, but with much enthusiasm beneath this outer crust," Darwin later wrote. He showed Darwin the tricks of zoology: how, for instance, to dissect marine creatures in seawater under a microscope. And Darwin in turn proved to be a bright apprentice; he was the first person ever to see the male and female sex cells of seaweed dance together.

In 1828, at the end of his second year in Edinburgh, Darwin went back home to the Mount. He could no longer avoid his father, and he finally confessed that he couldn't become a doctor. Robert Darwin was furious. He told Charles, "You care for nothing but shooting, dogs and rat-catching and you will be a disgrace, to yourself and all your family."

Robert was not an ogre of a father. His son would become a rich man, and Robert wanted him to be more than idly rich. If Charles wouldn't become a doctor, Robert could imagine only one other respectable profession that was still left open to his youngest son: the clergy. The Darwins were not particularly religious—Robert Darwin even privately doubted whether God existed— but in Britain religion brought security and respectability. Although Darwin had never felt any great passion for the church, he agreed, and the following year he went to Cambridge for a degree in theology.

Darwin did not turn out to be a hardworking student; he was less likely to be studying the Bible than hunting for beetles. He searched for the insects on the heaths and in the forests; to find the rarest species, he hired a laborer to scrape moss off trees and muck out the bottom of barges filled with reeds. And as for the future, Darwin wasn't dreaming of a parsonage but of leaving England altogether.

He read about Alexander von Humboldt's travels through the Brazilian rain forest and over the Andes, and he wanted to travel as well, to discover something about how nature worked. Humboldt had praised the Canary Islands, with their dense lowland jungles and rugged volcanic flanks, and Darwin began scheming an expedition. He found a Cambridge tutor, Marmaduke Ramsay, who was willing to travel to the Canaries with him. He honed his skills at geology by working as an assistant to the Cambridge geologist Adam Sedgwick for several weeks in Wales. When he returned from the expedition, ready to start making serious preparations for his trip to the Canary Islands, he got a message. Marmaduke Ramsay was dead.

Darwin was devastated. He left Cambridge and traveled home to the Mount with no idea what to do. But when he arrived, there was a letter from another of his professors at Cambridge, John Stevens Henslow. Henslow wanted to know whether Darwin cared to take a trip around the world.

THE LONELY CAPTAIN

The offer came from Robert FitzRoy, captain of the *Beagle*. FitzRoy was charged with two missions: to use a new generation of precisely engineered clocks to navigate a trip around the world and to map the coastlines of South America. Argentina and its neighbors had just been freed from Spain's control, and Britain needed to chart the waters as it set up new channels of trade.

Although this would be FitzRoy's second mission as captain of the *Beagle,* he was only 27 years old. He was the product of an aristocratic family with vast estates in England and Ireland, and at the Royal Naval College he had been a sharp student of mathematics and science. He had served in the Mediterranean and in Buenos Aires, and in 1828, at age 23, he became captain of the *Beagle*. The previous captain had gone mad trying to survey the wave-battered islands of Tierra del Fuego, as his crew developed scurvy and his bad maps led him in desolate circles. "The soul of man dies in him," the captain had written in his logbook, and then shot himself.

Robert FitzRoy
(1805–1865), captain
of HMS *Beagle*.

FitzRoy was a jumble of propriety and passion, of aristocratic tradition and modern science, of missionary zeal and solitary desperation. While he was on his first mission as captain of the *Beagle,* surveying Tierra del Fuego, one of his boats was stolen by Indians. FitzRoy retaliated by capturing hostages, most of whom escaped. Those who remained behind, two men and a girl, seemed happy to stay on board, and FitzRoy suddenly decided that he would take them to England, educate them, and bring them back to convert their fellow Indians. On the way home he picked up a fourth Indian, whom he bought with a mother-of-pearl button. Back in England, one of the Indians died of smallpox, but FitzRoy planned to civilize the other three and return them to Tierra del Fuego on his second voyage, along with a missionary who would stay behind to educate the tribes.

FitzRoy had decided that he needed a companion for the coming voyage. Captains did not socialize with their crews, and the enforced solitude could be maddening—the previous captain's suicidal ghost practically haunted the ship. And FitzRoy had an extra worry. His uncle, a politician whose career had disintegrated, had slit his own throat. Perhaps FitzRoy was susceptible to the same dangerous gloom. (His hunch was a good one. Some three decades later, deeply depressed at his own failing naval career, FitzRoy cut his own throat.)

FitzRoy asked the organizer of the expedition, Francis Beaufort, to find him a friend to keep him company. They agreed that this traveling companion would also act as an unofficial naturalist, to document the animals and plants the *Beagle* would encounter. FitzRoy wanted him to be a gentleman as well, someone who could hold a refined conversation, to help him stave off desolation.

Beaufort contacted his friend Henslow at Cambridge. Although the post sounded tempting, he decided he couldn't abandon his wife and child for so long. Henslow offered it to a recent Cambridge graduate, Leonard Jenyns, who went so far as to pack his clothes but then had a change of heart; he had just been appointed to a parish and didn't think it wise to quit so suddenly. So Henslow turned to Darwin. The journey was far beyond Darwin's dream of the Canary Islands, and without a family or post holding him back, he jumped at the chance.

His father was not so eager. He worried about the brutal, filthy conditions on sailing ships; he imagined his son drowning. Besides, the navy was no place for a gentleman to be, and it was disreputable for a future clergyman to be heading off into the wilderness. If Charles went, he might never settle down in a proper life. Charles glumly wrote to Henslow that his father disapproved.

Yet Robert Darwin had not completely made up his own mind. When his son traveled to the Wedgwood estate to distract himself with hunting, Robert sent a note to his brother-in-law, Joseph Wedgwood. He explained his disapproval of the voyage but wrote that "if you think differently from me I shall wish him to follow your advice."

Charles explained the situation to Wedgwood, who bucked up his nephew's spirits. He then wrote a letter to Robert to argue that the pursuit of natural history was very suitable to a clergyman, and that it was a rare opportunity "of seeing men and things as happens to few." After sending off the letter in the early morning, Wedgwood tried to occupy Charles with partridge shooting, but by 10 o'clock the two of them set out for the Mount to argue the case in person. They discovered that Robert Darwin had already read the letter and relented. He gave his son money for the trip; Darwin's sisters gave him new shirts.

Darwin sent a letter to Francis Beaufort, telling him to ignore his previous letter to Henslow: he would be joining the *Beagle*. He began arranging for the voyage, although he hadn't yet actually met FitzRoy. And soon he heard rumors that the captain was having second thoughts. In one of his typical reversals, FitzRoy had started telling people that the position was already taken by a friend of his, and word got back to Darwin.

Darwin was baffled and heartsick, but he kept an appointment with FitzRoy in London despite the rumors. As he stared out the coach window, he worried that this voyage would evaporate as quickly as the first.

When FitzRoy and Darwin met, FitzRoy immediately tried to make out the voyage to be awful—uncomfortable, expensive, and perhaps not even completely around the world. But Darwin would not be deterred. Instead, he charmed FitzRoy with his congenial parson's manner, his ample scientific training, his cultivated Cambridge tone, and his tactical deference. By the end of the meeting FitzRoy was won over. It was agreed: they would sail together.

"Woe unto ye beetles of South America," Darwin declared.

BUILDING THE EARTH

When Darwin arrived at Plymouth in October 1831, with his trunks full of books and scientific equipment, he also brought with him the ideas of his day about Earth and the life that inhabits it. His teachers at Cambridge taught him that by learning about the world, one could learn about God's will. Yet the more British scientists discovered, the harder it was to rely on the Bible as an unerring guide.

British geologists, for example, no longer accepted that the world was only a few thousand years old. It had once been enough to accept the literal word of the Bible, that humanity was created in the first week of creation. In 1658, James Ussher, the archbishop of Armagh, had used the Bible along with historical records to pinpoint the age of the planet. He declared that God had created it on October 22, 4004 B.C. But it soon became clear that Earth had changed since its creation. Geologists could find fossils of shells and other signs of marine life in the layers of rock exposed in cliff faces. Surely God had not placed them there when He created Earth. Early geologists interpreted the fossils as the remains of animals killed during Noah's Flood. When the oceans covered Earth, they were buried in the muck that washed to the bottom. The sediment formed layers of rock on the ocean floor, and when the waters subsided, some of the layers collapsed. In the process, the fossil-strewn cliffs and mountains were created.

By the end of the 1700s, though, most geologists had given up trying to fit Earth's history into a few thousand years, with the only chance for change a single catastrophic flood. Some argued that when the planet had been formed it was covered in a global ocean, which slowly deposited granite and other kinds of rock on top of one another. As the ocean retreated, it exposed parts of the rock, which eroded and formed new layers.

Other geologists argued that the forces creating Earth's surface were coming from within. James Hutton, a Scottish gentleman farmer, envisioned a hot molten core of the planet pushing up granite from below, creating volcanoes in some places and uplifting vast parts of Earth's surface in others. Rains and wind eroded mountains and other raised parts of the planet, and this sediment was carried to the ocean, where it formed new rock that would be raised above sea level later, in a series of global cycles of creation and destruction.

Hutton saw Earth as a finely crafted perpetual-motion machine, always keeping itself habitable for humans.

Hutton discovered evidence for his theory in the rocks of Scotland. He found veins of granite reaching up into sedimentary rocks. He found exposures where the uppermost sedimentary rock layers were arranged in proper horizontal fashion, but then just under them, other layers were tilted practically vertically. The lowest layers, Hutton argued, had been deposited in some ancient body of water, but were then tilted and lifted above sea level by subterranean forces and then gradually eroded by rain. Later, the tilted layers were covered in water, and a new set of horizontal layers of sediment covered them. And finally, the entire sequence had been pushed back up out of the water, producing the outcrops that Hutton saw.

"A question naturally occurs with regard to time," Hutton said when he first presented his new theory to the public. "What had been the space of time necessary for accomplishing this great work?"

His answer was, a lot. In fact he envisioned an "indefinite space of time."

Hutton had discovered a fundamental principle about how Earth changes: that imperceptibly slow forces at work today have been reshaping the planet throughout its history. For that, he's a hero to many geologists today. But in his own time, he was roundly opposed. A few critics complained that his theory went against Genesis. But most geologists disputed him on his assumption that Earth's history had no direction, that it ran through self-sustaining cycles of creation and destruction without beginning or ending. As they looked more closely at the geological record, they saw that the world had not always been the same.

The best evidence came not from the rocks themselves but from the fossils that they encased. In France, for example, a young paleontologist named Georges Cuvier compared the skulls of living elephants to the fossils of elephants that had been unearthed in Siberia, Europe, and North America—all places where no elephants are found today. Cuvier sketched out their massive jaws, their teeth fused together into corrugated slabs. He demonstrated that the fossil elephants (called mammoths) were fundamentally different from those alive today—"as much as, or more than, the dog differs from the jackal and the hyena," he wrote. It would be hard to claim that such a distinctive giant was wandering around without anyone noticing it. For the first time, a naturalist had shown that species had become extinct in the past.

Cuvier went on to demonstrate that many other mammals had gone extinct as well. His discoveries, he wrote, "seem to me to prove the existence of a world previous to ours, destroyed by some kind of catastrophe. But what was this primitive earth? What was this nature that was not subject to man's dominion? And what revolution was able to wipe it out, to the point of leaving no trace of it except some half-decomposed bone?"

Cuvier went even further. There had not been a single catastrophe that had wiped out mammoths and other extinct mammals, but a whole series of them. The fossils of the different ages were so distinct that Cuvier could even use them to identify the rock formation where they had been found. Exactly what had caused the catastrophes Cuvier didn't know for sure. He speculated that revolutions might be caused by a sudden rise of the ocean or a drastic cold snap. Afterward, new animals and plants appeared either by migrating from elsewhere on the planet or by being somehow created. But Cuvier was sure of one thing: revolutions were common in the history of Earth. If Noah's Flood was real, it had been only the most recent in a long succession of catastrophes; each had wiped out a slew of species, and most had taken place long before humans existed.

To work out the true pace of life's history, British geologists such as Adam Sedgwick set out to chart the entire geological record of the planet, layer by layer. The names they gave to the formations, names like the Devonian and the Cambrian, are still used today. But while British geologists in the early 1800s were a long way from a literal reading of the Bible, they still saw their work as a religious calling. They were convinced their science could reveal

Georges Cuvier (1769–1832) made fossils the subject of serious scientific study for the first time in history. He demonstrated that in many cases they belonged to species that had become extinct.

God's work—even His intentions. Sedgwick himself referred to nature as "the reflection of the power, wisdom, and goodness of God."

MANIFESTATIONS OF DESIGN

For British researchers such as Sedgwick, God's goodness was visible not just in the way He created Earth, but in the way He brought life into existence. Each species had been created separately and had been unchanged ever since. Yet species could also be grouped together—into categories such as plants and animals, and within them, into smaller categories such as fishes and mammals. This pattern, British naturalists believed, reflected God's benevolent plan for the world. It was organized as a gradation that started with inanimate objects and the slimy forms of life and reached up toward higher and higher forms—"higher" defined, of course, as being more like humans. Not a single link in this Great Chain of Being could have ever changed, for that would mean that God's creation had been imperfect. As Alexander Pope once wrote, "From Nature's chain whatever link you strike/Tenth, or ten thousandth, breaks the chain alike."

Not only did the Great Chain of Being reveal God's benevolent handiwork, but so did the exquisite designs of individual species—whether you considered a human's eye or a bird's wing. William Paley, an English parson, laid out this argument in books that were mandatory reading for Darwin and other aspiring naturalists and theologians at Cambridge.

Paley's argument centered on a seductive analogy. "In crossing a heath," he wrote, "suppose I pitched my foot against a stone and were asked how the stone came to be there." It might, for all Paley knew, have lain there forever. "But suppose I had found a watch upon the ground, and it should be inquired how the watch happened to be in that place." In that case, Paley argued, he would come to a very different conclusion. Unlike a stone, a watch is made of several parts that work together for a single purpose: to mark the passing of time. And these parts can work only if they work together; half a watch cannot tell time.

The watch therefore must have been created by a designer. Paley could say this even if he didn't know how to make a watch, and even if the watch he found was broken. To say that it was just one of many different possible combinations of bits of metal would be absurd.

When we look at nature, Paley argued, we find countless creations far more intricate than a watch. Telescopes and eyes are built on the same principle, with lenses that bend light in order to create an image. In order to bend light in water, a lens has to be rounder than it would be in air. Lo and behold, the eyes of fishes have rounder lenses than those of land animals. "What plainer manifestation of a design can there be than this difference?" Paley asked.

An oyster, a spoonbill, a kidney: anything Paley examined showed him that nature had a designer. The laws of physics, which astronomers were using at the end of the 1700s to describe the orbits of the planets, may have taken away some of the glory of God. ("Astronomy," Paley admitted, "is not the best medium with which to prove the agency of an intelligent creator.") But life was still fertile theological ground.

From nature, Paley deduced not only the existence of a designer, but that He was benevolent. In the vast majority of cases, he argued, the contrivances of God were beneficial. What little harm there was in the world was just an unfortunate side effect. A person might use his teeth to bite someone, but they were actually designed so that he could eat. If God had wanted us to cause each other harm, He could have designed much better weapons to put in our mouths. These sorts of shadows couldn't distract Paley from the sunshine of life: "It is a happy world after all. The air, the earth, the water teem with delighted existence."

HOW THE GIRAFFE GOT ITS NECK

Darwin was impressed by Paley's rhetoric, but at the same time he was also vaguely aware of some less respectable ideas about how life came to take its current form. Some of them came from his own family. His grandfather, Erasmus Darwin, had died seven years before Charles was born, but even in death he was impossible to ignore. A doctor by trade, he was also a naturalist, an inventor, a botanist, and a best-selling poet. In one of his poems, entitled *The Temple of Nature,* he argued that all animals and plants now living were originally derived from microscopic forms:

Erasmus Darwin
(1731–1802), grandfather
of Charles. A doctor and
botanist, he proposed that
species evolved.

> Organic Life beneath the shoreless waves
> Was born and nurs'd in Ocean's pearly caves;
> First forms minute, unseen by spheric glass,
> Move on the mud, or pierce the watery mass;
> Then as successive generations bloom,
> New powers acquire and larger limbs assume.

Erasmus Darwin's personal life was as scandalous as his scientific views. After the death of his first wife, he began reveling in natural love, fathering two children out of wedlock. "Hail the Deities of Sexual Love!" he declared, "and sex to sex the willing world unite." His son Robert always considered Erasmus something of an embarrassment, and so his grandson

Charles, growing up at the tranquil, proper Mount, did not learn much about him.

But when Charles Darwin traveled to Edinburgh, a city where radical ideas thrived, he discovered that his grandfather had many admirers. One of them was Robert Grant, the zoologist who became Charles's mentor. Grant studied sponges and sea pens not out of idle interest, but because he thought they lay at the root of the animal kingdom. From forms like them, all other animals might have descended. When Grant and Darwin went hunting for specimens in tide pools along the shore, Grant would explain to young Charles his admiration for Erasmus Darwin and his ideas of transmutation, a process by which one species changes into another. And Grant explained to Erasmus's grandson that there were French naturalists who had also dared to contemplate the possibility that life was not fixed—that it evolved.

Grant described a colleague of Cuvier's at the National Museum of Paris named Jean-Baptiste Pierre Antoine de Monet, chevalier de Lamarck. In 1800, Lamarck shocked Cuvier and the rest of Europe by declaring that the fixity of species was an illusion. Species had not all been created in their current form at the dawn of time, Lamarck proposed. Throughout the course of Earth's history, new species formed through spontaneous generation. Each came into existence equipped with a "nervous fluid" that gradually transformed it, over the course of generations, into new forms. As species evolved, they achieved higher and higher levels of complexity. The continual emergence of species and their ongoing transformation created the Great Chain of Being: lower members of the chain had simply started their upward journey later than higher members.

Life could change in another way, Lamarck claimed: a species could adapt to its local environment. Giraffes, for example, live in places where the leaves are far from the ground. The ancestors of today's giraffes might have been short-necked animals that tried to eat the leaves by stretching their neck upward. The more an individual giraffe stretched, the more nervous fluid flowed into its neck. Its neck grew longer as a result, and when it produced baby giraffes, it passed on its longer neck to them. Lamarck suggested that humans might have descended from apes that left the trees, stood upright, and walked out onto the plains. The very effort of trying to walk on two legs would have gradually changed their bodies to our own posture.

Most other naturalists in France and abroad were appalled by Lamarck's ideas. Cuvier led the attack, challenging Lamarck for evidence. The nervous fluid that made evolution possible was a complete conjecture, and the fossil record didn't back him up. If Lamarck were right, the oldest fossils should on the whole be less complex than species are today. After all, they had had less time to rise up the scale of organization. Yet in the oldest rocks that were known in

1800, there were fossils of animals as complex as anything alive today. Cuvier found another opportunity to attack Lamarck when Napoleon's armies invaded Egypt and discovered mummified animals buried in the tombs of the pharaohs. Cuvier argued that the skeleton of the sacred ibis, which was thousands of years old, was no different from the sacred ibis alive in Egypt today.

Most naturalists in Great Britain, steeped in Paley's natural theology, were even more repulsed than Cuvier. Lamarck was reducing mankind and the rest of nature to the product of some unguided, earthly force. Only a few heretics such as Grant admired Lamarck's ideas, and for their heresy they were shut out of Britain's scientific inner circle.

Grant's praise of Lamarck took the young Darwin by surprise. "He one day, when we were walking together, burst forth in high admiration of Lamarck and his views on evolution," Darwin later wrote in his autobiography. "I listened in silent astonishment, and as far as I can judge, without any effect on my mind."

By the time Darwin boarded the *Beagle* four years later, evolution had sunk away from his thoughts altogether. Only after he had returned from his voyage would it rise back again, and in a radically different form.

THE MAKING OF A GEOLOGIST

The voyage started badly. Darwin arrived in Plymouth in October 1831, but it wasn't until December 7, after weeks of repairs and delays and false starts, that the *Beagle* set sail. And as soon as Darwin left shore, he became horribly

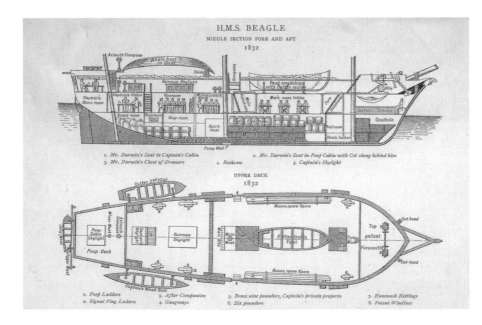

The *Beagle* provided cramped quarters for its crew.

seasick, spewing his meals over the rails. Although Darwin would sail for five years on the *Beagle,* he never managed to get his sea legs.

Darwin found being FitzRoy's companion a tricky job. The captain's temper was sharp and unpredictable, and his navy discipline was a shock to Darwin. On Christmas some of the *Beagle*'s crew got drunk, and FitzRoy had them flogged the following day. Each morning after Darwin emerged from breakfast with FitzRoy, the junior officers would ask, "Has much coffee been spilled this morning?" as a coded way to check on the captain's mood. But Darwin also respected FitzRoy's powerful drive, his dedication to science, and his devotion to Christianity. Every Sunday Darwin attended the captain's sermons.

Darwin yearned for landfall, but it did not come for weeks. At Madeira, the currents were so bad that FitzRoy decided not to anchor there, and at the next port—the Canary Islands—a cholera epidemic was raging. Rather than waste time in a quarantine before going on shore, FitzRoy simply sailed on.

Finally, the *Beagle* stopped for the first time in the Cape Verde Islands. At Saint Jago, Darwin bounded off the ship. He darted about beneath the coconut trees, looking at the rocks, the plants, the animals. He found an octopus that could change colors, from purple to French gray. When he put it in a jar in the hold of the ship, it glowed in the dark.

But it was the geology of the island that Darwin most wanted to see. On the journey from England, Darwin had been engrossed by a new book called *Principles of Geology,* written by an English lawyer named Charles Lyell. It would change the way Darwin viewed the planet, and ultimately lead him to his theory of evolution. Lyell attacked the catastrophe-centered geology that was popular at the time, reviving Hutton's 50-year-old theory of a uniformly changing Earth.

The *Beagle* passes through the Straits of Magellan.

EVOLUTION

Principles of Geology wasn't simply a rehash of Hutton's ideas. Lyell offered a much richer, scientifically detailed vision of how the changes humans witness could have gradually shaped the planet. He offered evidence of volcanic eruptions building islands, of earthquakes having lifted land; he then showed how erosion could grind these exposed features down again. Geological change was happening slowly, imperceptibly, Lyell argued, even over the course of human history. The frontispiece of *Principles* shows the ancient Roman temple of Serapis, with dark bands marking the tops of its pillars, caused by mollusks that at some point had drilled into them. Within the lifetime of the temple it had been completely submerged and then raised from the ocean again. Unlike Hutton, Lyell didn't see Earth going through a grand, global cycle of creation and destruction. The planet changed locally, eroding here, erupting there, in a state of perpetual, directionless flux for an unimaginable span of time.

Darwin was fascinated by *Principles of Geology*. He realized that it offered not only a compelling vision of Earth's history but a method for testing it against the real world. When he landed at Saint Jago he had an opportunity to do just that. He scrambled over the volcanic rock of the island and found clues that the lava had originally poured out underwater, baking coral and shells as it spread. Subterranean forces must have later lifted the rock up to the sea's surface, but they must have then lowered it back down and lifted it up yet again. Some of the rising and falling must have happened only recently, Darwin realized, for in a band of rock in the cliffs he could find fossils of shells that matched those of creatures still alive around the island. Earth was changing in 1832, as it had for eons.

Darwin would write in his autobiography, "The very first place which I examined, namely St. Jago in the Cape de Verde islands, showed me clearly the wonderful superiority of Lyell's manner of treating geology, compared with that of any other author, whose works I had with me or even afterwards read."

He had tried out Lyell's methods and they had worked beautifully. Darwin became a Lyellian on the spot.

Charles Lyell's *Principles of Geology* included a frontispiece of the Temple of Serapis in Italy. The dark bands on the top of the pillars were drilled by mollusks—proof that the pillars had been submerged and then lifted above sea level over the course of recorded history. Lyell argued that such changes could gradually create mountains and all the other physical features of the planet.

"A STRANGE IDEA OF INSECURITY"

At the end of February 1832 the *Beagle* reached South America. FitzRoy kept the ship in Rio de Janeiro for three months, and then sailed south. Although the *Beagle* would travel along the coast of South America for the next three years, Darwin spent most of that time on land. In Brazil, he lived in a cottage in the jungle, overwhelmed by the biological Eden that surrounded him. In Patagonia, he rode horseback through the interior for weeks at a time, always managing to get back to the *Beagle* in time for the next leg of the voyage. He recorded everything he encountered, the fireflies, the mountains, the slaves, and the cowboys. His empty specimen jars began to fill with hundreds of strange new creatures.

On one excursion near Punta Alta on the coast of Argentina, Darwin inspected low cliffs and discovered bones. Prying them out of the gravel and quartz, he found that they were enormous teeth and thigh bones belonging to gigantic extinct mammals. He came back over the course of the next few days and dug out more. At that time, there was only a single fossil of an extinct mammal in all the collections of England, but at Punta Alta Darwin uncovered tons of bones. He didn't know quite what to make of them, guessing that they were giant rhinos and sloths. But Darwin was still just a collector at this point. He simply packed up the fossils and had them sent home.

There was a puzzle to these fossils, the first of many that Darwin would encounter on his voyage. As a good student of Cuvier, Darwin assumed that these were antediluvian monsters, long extinct. But their bones were mixed in with the fossils of shells that were almost identical to those of living species

The voyage of the *Beagle*.

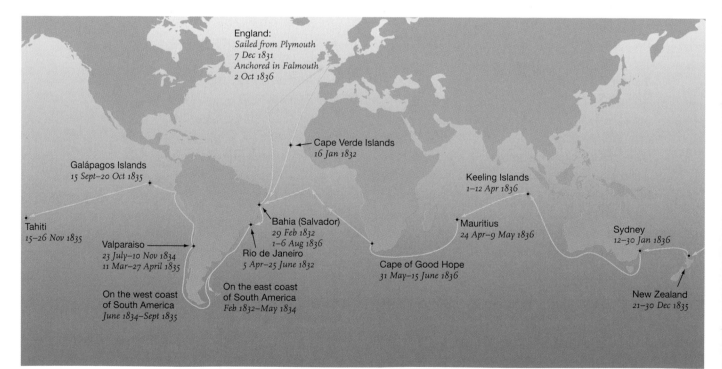

England:
Sailed from Plymouth
7 Dec 1831
Anchored in Falmouth
2 Oct 1836

Cape Verde Islands
16 Jan 1832

Galápagos Islands
15 Sept–20 Oct 1835

Keeling Islands
1–12 Apr 1836

Tahiti
15–26 Nov 1835

Valparaiso
23 July–10 Nov 1834
11 Mar–27 April 1835

Bahia (Salvador)
29 Feb 1832
1–6 Aug 1836

Rio de Janeiro
5 Apr–25 June 1832

Mauritius
24 Apr–9 May 1836

Sydney
12–30 Jan 1836

Cape of Good Hope
31 May–15 June 1836

On the west coast
of South America
June 1834–Sept 1835

On the east coast
of South America
Feb 1832–May 1834

New Zealand
21–30 Dec 1835

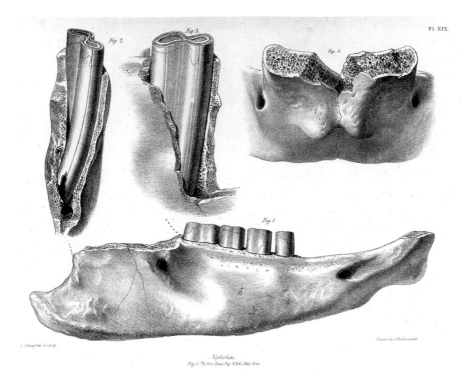

PL. XIX.

Fig. 2 Fig. 3 Fig. 4

Fig. 1

Mastodon
Fig. 1 & 2 Nat. Size. Fig 3 & 4 Nat. Size.

Darwin discovered fossils of giant mammals in South America, such as this jaw of a giant ground sloth, published in his *Zoology of the Voyage of HMS Beagle*. Later they would lead him to formulate his theory of evolution.

on the Argentine coast. The rocks were hinting that the monsters were not as ancient as they might seem.

The *Beagle* rounded Tierra del Fuego in December 1832. This was supposed to be the highlight of the trip for FitzRoy, as he returned his captured Indians to their tribes, bringing with them the powers of civilization. But Tierra del Fuego defeated him again. FitzRoy tried to set up a mission at Wollya Bay, building three wigwams and two gardens. He equipped the mission with the donations of kindly but oblivious London ladies: wineglasses, tea trays, soup tureens, and fine white linens. A few weeks later, when the *Beagle* checked in again at Wollya Bay, the missionary came running to the ship, screaming for his life. The Fuegans had stolen or destroyed everything, and at the ship's return they were amusing themselves by plucking his beard hairs with mussel shells.

FitzRoy gloomily sailed the *Beagle* around Cape Horn and headed for the west coast of South America. Darwin took the opportunity to climb the Andes Mountains. Thinking of Lyell, he tried to picture their tall peaks rising out of the ocean. He rejoined the *Beagle* as it traveled north from Valparaiso, with Mount Osorno looming over them to the east, its peak a perfect cone. As the sailors worked their instruments in the rain, they would stop sometimes to watch clouds of smoke rise from its top. One night in January Osorno exploded, launching boulders and flame. Not even Lyell himself had seen a volcano erupt.

Earth was not finished with its violence, though. The *Beagle* anchored at the town of Valdivia a few weeks later, and on February 20, 1835, the planet danced under Darwin. He had been taking a forest walk near town and had decided to rest. When he lay down, the earth felt as hard and unyielding under him as it ever had, utterly imperturbable. And then it shivered.

"It came on suddenly and lasted two minutes; but the time appeared much longer," Darwin wrote later. Having spent his life in the geological tranquility of England, he had never experienced an earthquake before. "There was no difficulty in standing upright, but the motion made me almost giddy. It was something like the movement of a vessel in a little cross ripple, or still more like that felt by a person skating over thin ice, which bends under the weight of his body."

The trees swayed in the breeze, and the quake stopped. Darwin would not forget the experience. "A bad earthquake at once destroys the oldest associations: the world, the very emblem of all that is solid, has moved beneath our feet like a crust over a fluid; one second of time has conveyed to the mind a strange idea of insecurity, which hours of reflection would never have created."

After the earthquake was over, Darwin hurried back to the town, which he found relatively unharmed. But farther up the coast, the city of Concepción had collapsed into a heap of bricks and timbers. It had been hit by the quake, and by a tidal wave that the quake had triggered. The front of the city's cathedral was sheared off the building as if with a chisel. "It is a bitter & humiliating thing," Darwin wrote in his diary, "to see works which have cost men so much time & labor overthrown in one minute." Cracks had opened in the ground, rocks had shattered. In two minutes, the earthquake had done more damage, Darwin guessed, than a century of ordinary wear and tear.

The earthquake had done something even more profound to the coast, something that was more difficult to appreciate than all the ruined buildings and drowned cattle. Stretches of land that had once been underwater were now up out of the water, covered with dying shellfish. FitzRoy discovered with his surveying instruments that parts of the coast had risen 8 feet during the earthquake. Two months later he returned to Concepción and found that the ground was still elevated.

Darwin realized that Lyell had provided him with the explanation for what he was witnessing. The pressure of molten rock must have triggered the eruption at Osorno and then had enough strength to trigger the earthquake. The injection of fresh molten rock had raised new land out of the ocean. With enough time, it could eventually lift entire mountain ranges into the sky.

A few days later Darwin made his last major inland journey, traveling again into the Andes. At the summits of the mountains that surrounded the Uspallata Pass, Darwin could recognize the same layers of rocks that he had seen months before in the low, flat plains to the east—rocks that had originally been formed by ocean sediments. He found a forest of petrified wood at the pass still standing upright, much like fossils he had found in Patagonia.

"These trees," he wrote to his sister Susan, "are covered by other sandstones and streams of lava to the thickness of several thousand feet. These rocks have been deposited beneath water; yet it is clear the spot where the trees grew must once have been above the level of the sea, so that it is certain the land must have been depressed by at least as many thousand feet as the superincumbent subaqueous deposits are thick."

With earthquakes and volcanoes rumbling through his memory, Darwin concluded that the Andes were a recent creation. Once these 14,000-foot peaks had been as flat as the pampas to the east. The giant mammals whose fossils he had found there had wandered these places as well. And then this land had sunk underwater and then risen up again, as pressure from below had jacked it up. Darwin realized that mountains might be younger than mammals, and that they might still be rising under his feet.

A LITTLE BIRD COLLECTING

With the coastal survey finished, the *Beagle* sailed north to Lima and then westward, leaving South America altogether. After the blasting winds of Tierra del Fuego and the hollow cold of the Andes, Darwin was looking forward to the tropics. The first stop would be at a peculiar cluster of islands called the Galápagos.

The Galápagos Islands have a reputation as the place where Darwin's theory of evolution was born, but Darwin would not realize their significance until nearly two years after his visit. When he arrived at the Galápagos Islands, he was still thinking more about geology than biology, and he looked forward to visiting a place where he could see new land being created as Lyell had proposed.

Darwin first set foot on Chatham Island (now known as San Cristobál), a raw volcanic heap still untamed by plants and soil. Ugly iguanas and countless crabs greeted him. "The natural history of this archipelago is very remarkable," Darwin later wrote. "It seems to be a little world within itself." A world, he meant, unlike the greater one. Here there were enormous tortoises, with shells 7 feet in diameter, that fed on prickly pears and didn't care if Darwin rode on their backs. Here lived not one but two hideous species of iguana—one that stayed on the islands and another that dove into the ocean to eat seaweed. The birds of the Galápagos were so tranquil that Darwin could walk up to them without their taking flight.

Darwin dutifully added the birds to his collection, making only a few notes about them. Some had big beaks, good for crushing large seeds, while others had beaks shaped like needle-nose pliers, for grabbing seeds that were small and hard to reach. Judging from their beaks, Darwin guessed that some were wrens, others finches, warblers, and blackbirds. He did not, however, bother to note which islands the birds came from. He assumed that the birds were South American species that had colonized the islands at some point.

Only after Darwin had finished collecting his animals did he realize that he should have been more careful. Shortly before the *Beagle* left the islands, he met the director of the penal colony on Charles Island (Santa María), an Englishman named Nicholas Lawson. Lawson used tortoise shells for flower pots in his garden, and he had noticed that the tortoises of each island were so distinct from one another that he could tell where they came from by the shape of the flares and flanges on their shells. The tortoises on each island, in other words, were a unique variety, or perhaps even a unique species. The plants of the islands, Darwin learned, were likewise distinct.

Perhaps the birds were as well, but since he hadn't marked where most of his birds had come from, Darwin couldn't know. It would not be until he returned to England that Darwin would finally sort out his birds, and only then would he begin to work out the way in which life changes from one form to another.

The *Beagle* landed on the Galápagos Islands in the eastern Pacific in 1832. Formed by volcanic eruptions, they offer bleak lava-strewn landscapes.

LIFE BUILDS ITSELF

Marine iguanas (left) and blue-footed boobies are among the many species that are found only on the Galápagos Islands.

When the *Beagle* was finished at the Galápagos, it set sail across the glassy Pacific. It traveled quickly, reaching Tahiti in three weeks, New Zealand in another four, and Australia in only two. As it crossed the Indian Ocean, the *Beagle's* objective was to map coral reefs. Coral reefs are living geography, produced by colonies of tiny polyps as they build external skeletons. The polyps can live only near the surface of the ocean because, as marine biologists would later discover, they depend on photosynthetic algae that live inside their tissues. As the *Beagle* passed reefs in the Indian Ocean, Darwin wondered how they formed such perfect circles, sometimes around islands and sometimes around nothing but water. And how was it that the reefs could always be found close to the surface of the water, which was exactly where they needed to be in order to get enough sunlight to grow?

In *Principles of Geology* Darwin had read Lyell's hypothesis on corals: they formed only on the tops of submerged volcanic craters. For once, Darwin thought that Lyell was wrong. The crater hypothesis was ugly and awkward, since it would require that every reef sit precisely on a crater that just happened to be lurking close to the surface of the ocean. Darwin came up with another explanation.

If the Andes were rising, according to Lyell's geology, Darwin reasoned that some other part of the planet must be sinking. That could well be happening in places such as the Indian Ocean. Corals might form in the shallow waters around new islands or along the coasts of mainlands, which then began to sink. As the land began to subside underwater, the corals would sink with them. But they might not be lost, Darwin reasoned, because new corals could grow on top of the reefs as the ground sank. While the old corals died in the darkness, the reef would survive. After a while the island itself might completely erode away, but the reef would have maintained itself near the water's surface.

Every coral reef that the *Beagle* surveyed fit somewhere along this sequence. At Keeling Islands (Cocos Islands), the surveyors on the *Beagle* found that on the seaward edge of the reef there was a sharp drop to the ocean floor. When they scraped off corals from near the bottom, they found that they were dead. It was all as Darwin had predicted.

Darwin was no longer a mere disciple of Lyell, but a mature, independent thinker. He had used Lyell's principles to formulate a better explanation for coral reefs than Lyell's own, and he had worked out how he could test it. Darwin was learning how to study history—in this case, the history of life on Earth—scientifically. He could not replay thousands of years and watch coral reefs grow, but if history had unfolded the way he proposed, he could test his predictions. "We get at one glance an insight into the system by which the surface of the land has been broken up, in a manner somewhat similar but certainly far less perfect, to what a geologist would have done who had lived his 10,000 years, and kept a record of the passing changes," he later wrote.

The planet might look changeless, but Darwin was learning to see it on a scale of millions of years. And from that perspective it was a quivering ball, rising and collapsing, its skin tearing itself apart. Darwin was also learning how life could change on the same scale as well. Given enough time, coral reefs could keep from drowning as the ocean floor fell out from under them. They could build giant fortresses, founded on the skeletons of their ancestors.

It took Darwin six months to travel from the coral atolls back to England, rounding the Cape of Good Hope and passing the Azores, and by then his reputation had preceded him. Henslow, his mentor at Cambridge, had culled some of Darwin's letters and turned the extracts into a scientific paper and a pamphlet. His mammal fossils had made the voyage home safely, and some of them were admired by Britain's leading anatomists. Even Darwin's idol, Lyell, was impatiently waiting to meet him on his return.

Five years after leaving Plymouth, the *Beagle* sailed up the English Channel in a drenching rain. FitzRoy held his final service on October 2, 1836, and later that day, Darwin walked off the ship and headed for home. He would never leave England again; he would barely even leave his own house.

As Darwin set foot on English ground, he knew that he had changed profoundly. There was now no way he could tolerate life as a country parson. He had become a practicing naturalist, and he would spend the rest of his life as one. On top of that, he knew that he would be happy only if he could work as an independent scholar like Lyell, rather than at a university. But if Darwin was to live the life of Lyell, his father would have to give him money to establish himself. As ever, Darwin was nervous about what his father would think.

Darwin stepped off the coach at Shrewsbury late in the evening of October 4. He was eager to see his family again, but was too proper to disturb them in the middle of the night. He slept at an inn, and the next morning, just as his father and sisters were sitting down to breakfast, he walked into the Mount unannounced. His sisters cried out in joy. His father declared, "Why, the shape of his head is quite altered." His dog acted as if Darwin had been gone only a day and was ready for their usual morning walk.

Darwin's fears of his father's anger turned out to be unfounded. While he had been away, his brother Erasmus had abandoned medicine and set himself up as an independent scholar in London. Erasmus had blazed the trail for his younger brother, and their father had not objected. When Robert read Charles's pamphlet, he was filled with pride. He realized that as a naturalist, Charles would not waste his life shooting rabbits. He gave his son stocks and an allowance of 400 pounds a year, enough to establish himself on his own.

Charles Darwin would never again fear his father. But he had inherited Robert's taste for respectability, and whenever possible he avoided unseemly confrontations. He had never been a rebel and would never want to become one. Yet within a few months of returning home, he would terrify himself by beginning a scientific revolution.

I think

Case must be that one generation then should be as many living as now. To do this & to have many species in same genus (as is) requires extinction.

Thus between A & B. immense gap of relation. C & B. the finest gradation, B & D rather greater distinction Thus genera would be formed. — bearing relation

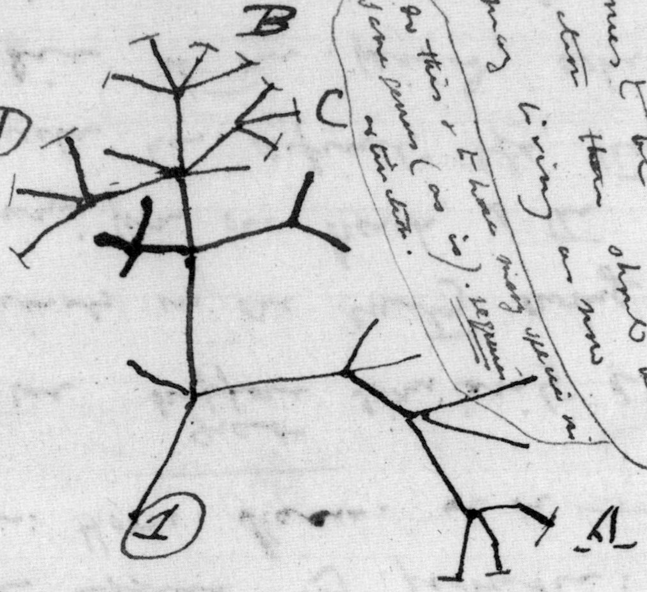

"Like Confessing a Murder"

THE ORIGIN OF *ORIGIN OF SPECIES*

IN LONDON, Darwin discovered that his brother had turned out not to be a very dedicated naturalist. Erasmus was most comfortable at dinner parties and gentlemen's clubs rather than in his laboratory. He introduced Charles into his social circles, and Charles blended in well. But unlike Erasmus, Charles also worked furiously. He wrote papers about geology, put together a book about his travels, and arranged for experts to study his specimens—fossils, plants, birds, flatworms.

Within a few months, Darwin's work had paid off: he had a reputation as one of England's most promising young geologists. But he also began harboring a secret. He would scribble in small, private notebooks, not about geology but biology. He had become obsessed with a disturbing possibility: perhaps his grandfather had been right after all.

In his private notebooks Darwin sketched out a tree of life in 1837 for the first time.

Biology had come far during the five years Darwin had been away. New species were being discovered that challenged the old order, and under microscopes scientists were learning how eggs developed into animals. British naturalists were no longer content with Paley's celebration of God's design on a case-by-case basis. It didn't allow them to answer the profound questions about life. If God had providently designed life, how exactly had He done so? What accounted for how similar some species were, and how dissimilar others? Had all species come into existence at the beginning of Earth, or had God created them as time went by?

For British naturalists, God was no longer a micromanager; instead, He had created laws of nature and had set them in motion. A God who needed to step in at every moment seemed less capable than a God who designed things correctly—and flexibly—at the start. Many British naturalists accepted that over the history of the planet life had changed. Simpler groups of plants and animals had gone extinct, replaced by more complex ones. But they saw it as a stately, divinely guided process, not an earthly evolution like the one that Lamarck had proposed in 1800. And a fresh shudder passed through their ranks in the 1830s as another zoologist at the National Museum in Paris championed a new theory of evolution: Etienne Geoffroy Saint-Hilaire.

ARCHETYPES AND ANCESTORS

Lamarck and Geoffroy had been friends at the museum for decades, but Geoffroy came to accept evolution through his own research, comparing the anatomy of different animals. The conventional wisdom of the day held that animals were similar to each other only when they functioned in similar ways. But Geoffroy was struck by exceptions to this supposed rule. Ostriches have the same bones as flying birds, even though they don't fly. And Geoffroy showed that what looked like unique hallmarks that set off a species from other animals often were not so singular after all. A rhino's horn might seem to make it unique, for example, but it is really just a clump of dense hair.

As Geoffroy struggled to uncover the hidden connections between animals, he found much inspiration in the work of German biologists. They saw science as a transcendentalist quest to discover the hidden unity of life. The poet (and scientist) Goethe argued that the various parts of a plant—from its petals to its thorns—were all variations on one fundamental form: the leaf. For these German biologists, the complexity of life hid certain timeless models, which they called archetypes. Geoffroy set out to find the archetype of all vertebrates.

Every bone in every vertebrate's skeleton, Geoffroy suggested, was a variation on an archetypal vertebra. He then pushed his scheme even further,

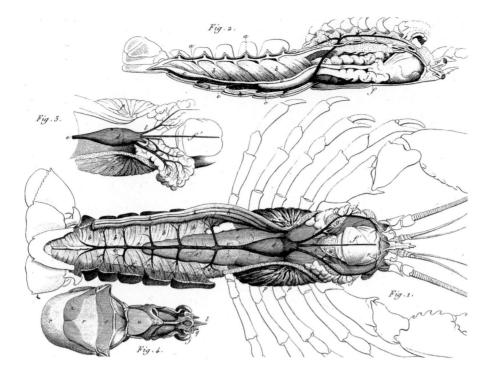

French naturalist Geoffroy Saint-Hilaire demonstrated that animals have many unexpected similarities. In this illustration he demonstrated how a lobster has a digestive system running along its back and a nerve cord running down its ventral side (corresponding to our bellies). Geoffroy pointed out that this was an upside-down version of our own anatomy, with our spinal cord running down our back.

claiming that invertebrates were based on the same transcendental plan. A lobster and a duck, by his reasoning, were variations on the same theme. Lobsters are arthropods, a group that also includes insects, shrimp, and horseshoe crabs. Arthropods bear some faint resemblances to vertebrates: Their bodies are symmetrical along their long axis; they have heads equipped with eyes and a mouth. But the differences are vast. Arthropods build their skeleton—a hard shell—on the outside of their bodies, whereas vertebrates put theirs on the inside. Vertebrates have a spinal cord running along their back and a digestive tract running down the front of their bodies. In a lobster, or any other arthropod, the arrangement is reversed: the gut runs down the back and the nervous system runs along the belly.

This might seem to make the arthropods and vertebrates incomparable, but not to Geoffroy. He claimed that arthropods lived inside a single vertebra. And it was a simple matter to transform belly to back, and thus turn a lobster into a duck. Arthropods had the same design as vertebrates, but it was simply turned upside down. "There is, philosophically speaking, only a single animal," Geoffroy declared.

By the 1830s, Geoffroy had taken his theory a step further. These transformations weren't simply geometrical abstractions, he declared; animals had in fact changed shape over time. Geoffroy was not reviving Lamarck; he didn't accept Lamarck's hypothesis that a trait acquired during an animal's life could be passed down to its offspring. Geoffroy suggested instead that a change in an animal's environment might disturb the way it developed from an egg. Freaks would be born, and would become a new species.

You could see the history of this evolution, Geoffroy claimed, if you looked at how embryos develop today. German scientists were discovering how embryos changed in a matter of days from one bizarre form to another, often bearing no resemblance to the adult forms of the animals. The researchers carefully catalogued their fleeting parts and shapes, and the longer they looked, the more order they claimed to see in the confusion. They were impressed in particular by the way an embryo began as a simple form and gradually grew more complex. They even claimed that each increasingly complex form added a new stage to its development.

Lorenz Oken, one of the German scientists, explained the process this way: "During its development, the animal passes through all stages of the animal kingdom, rising up as it takes on new organs. The foetus is a representation of all animal classes in time." At first it was just a tube, like a worm. Then it developed a liver and a vascular system and became a mollusk. With a heart, and a penis, it became a snail. When it sprouted limbs, it became an insect. When it developed bones, it became a fish; muscles, a reptile; and so on, up to mankind. "Man is the summit, the crown of nature's development," Oken announced.

Geoffroy proposed that embryos didn't just climb the scale of nature; they replayed history. The ancestors of humans really were fish, and the evidence that we sport gill slits at an early stage in our development proved it.

As Geoffroy argued for evolution, European explorers were discovering new species that he claimed fit in perfectly with his theory. The platypus of Australia, for example, was a mammal, but it had a ducklike bill and laid eggs, inspiring Geoffroy to call it a transitional form between mammals and reptiles. In Brazil explorers found lungfish that could breath air through lungs, representing a link between vertebrates in the ocean and on land.

In England, the leading scientists denounced Geoffroy just as they had denounced Lamarck. Adam Sedgwick, the devout Cambridge geologist, declared the work of the two Frenchmen to be "gross (and dare I say it, filthy) views of physiology." But while British scientists generally abhorred evolution, the job of attacking it head-on in the 1830s was left to one man: a brilliant young anatomist named Richard Owen.

Owen was often the first British anatomist to study new species such as the lungfish and the platypus, and he took these opportunities to strike down Geoffroy's claims. Owen showed that platypuses actually secrete milk, a hallmark of mammals. And lungfish, while they might have lungs, didn't appear to have nostrils, which all land vertebrates have. That, for Owen, was enough to relegate them to being ordinary fish.

Yet Owen himself was not content simply to say that God created life, and that its design reflected His goodness. Owen wanted to uncover the natural mechanisms of creation. He abhorred Geoffroy's wild speculations on evolution, but he was too good a naturalist to deny that he was right about some things. The similarities between species, and the ways in which they could be arranged into a series of transformations, were too obvious to be denied.

Owen decided that Geoffroy had just lurched too far in his interpretation of the evidence. Owen knew, for example, that Geoffroy's notions about how embryos formed had been supplanted by new research. A Prussian scientist named Karl von Baer had shown that life was not a simple ladder, with more advanced embryos recapitulating the development of more primitive ones. In the earliest stages of embryos, vertebrates did look like one another, but only because they were just a handful of cells. As time passed they grew more distinctive. Fishes, birds, reptiles, and mammals all have limbs, and they all initially form limb buds as embryos. But in time, those buds turn into fins, hands, hooves, wings, and the other kinds of limbs unique to certain kinds of vertebrates. One kind does not form from another. "A linear arrangement of animals in order of perfection," von Baer wrote, "is impossible."

Sir Richard Owen (1804–1892) examined the fossils that Darwin brought back from his voyage. He would later become Darwin's greatest enemy.

Owen's ambition was to tie together the work of von Baer, Geoffroy, and all the other great biologists of his day into one grand theory of life. He wanted to fight against evolution, but he wanted to do so by finding laws of nature that could account for the evidence of fossils and embryos.

He met Darwin for the first time three weeks after the *Beagle*'s return. They both came to dinner at Lyell's house, where Darwin regaled the party with his stories about the earthquake in Chile. After dinner, Lyell introduced the two young men (Owen was only five years older than Darwin). They got along well, and Darwin recognized in Owen someone famous enough to bring his fossil mammals to national attention. He asked Owen that night if he would examine them. Owen said that he would be happy to do so. They would give him a chance to test his ideas against fossils that no one had seen before.

He had no way of knowing that Darwin would one day render him a fossil as well.

CONFUSION AND HERESY

Four months after the *Beagle*'s return, Darwin began to hear back from his experts about his collection of fossils and carcasses. At first they did nothing but confuse him. Owen had inspected the fossil mammals and announced that they were gigantic variations on the animals that still lived in South

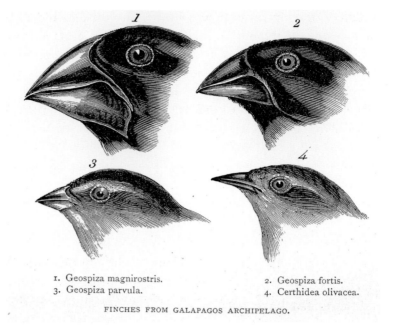

Darwin's *Journal of Researches* included this illustration of four finches from the Galápagos Islands. Each species has a beak specialized for its diet, be it hard seeds or insects.

1. Geospiza magnirostris.
2. Geospiza fortis.
3. Geospiza parvula.
4. Certhidea olivacea.

FINCHES FROM GALAPAGOS ARCHIPELAGO.

America. The rodents were the size of hippos, the anteaters the size of horses. Why, Darwin wondered, was there a continuity between extinct animals and the ones alive on the same spot on Earth? Could the living animals have descended, in a modified form, from the fossil ones?

Darwin had given his Galápagos birds to James Gould, one of Britain's leading ornithologists. He hadn't thought much of them when he had collected them, and when he heard Gould speak about them at a meeting of the Zoological Society, he regretted his carelessness. Judging from their beaks, Darwin had identified many of his birds as finches, wrens, and blackbirds. But Gould announced that they were *all* finches. They simply had wrenlike or blackbird-like beaks, which allowed them to eat particular kinds of foods.

And later, when Darwin paid a visit to Gould's office, Gould showed that he had made an even more grievous mistake. Darwin hadn't noted exactly which island he had gotten most of his birds from, because it seemed unimportant at the time. He had happened to note that three mockingbirds had come from three different islands, and Gould showed him that the mockingbirds belonged to three new—and separate—species.

Darwin wondered why there would be three different species of mockingbirds so close to one another. And were separate species of finches living on separate islands as well? Darwin contacted FitzRoy and asked for some of the birds collected by the *Beagle* crew to be sent to Gould. Fortunately, the crewmen had done a better job than Darwin, jotting down which islands they had shot their birds on. And just like the mockingbirds, the finches of different islands belonged to separate species.

Something, Darwin realized, was very wrong. Why should there be so many unique species on these similar islands? He opened his notebooks and tried to figure out an explanation for the finches of the Galápagos. To the people around him, he seemed unchanged as he went on with his geological work, writing about coral reefs and rising plains and the shape of volcano cones. But in private he was obsessed with an extraordinary thought. Perhaps the finches had not been created in their current form. Perhaps they had evolved.

The land on which species lived, after all, was not eternally unchanging. Darwin's finches now lived on islands that had at some point breached out of the ocean. Once the Galápagos had emerged, an original species of finch might have colonized them from South America, and over time its descendants on each island had changed into the distinct species with their distinct bodies that were now adapted to their way of life. The descendants of the original settlers branched apart into distinct lineages. The same branching might have happened among the mammals of Patagonia. The animals that left behind the gigantic fossils that Darwin had discovered might have given rise to today's mammals, with their smaller bodies.

In his notebook Darwin sketched out a tree, with old species branching into new ones.

Darwin found this idea terrifying. He began to suffer from heart palpitations and stomachaches, to wake up out of strange dreams in the middle of the night. He knew that whatever laws governed finches or anteaters must also govern human beings. He began to think of humans as merely one more species of animal, albeit with some peculiar mental gifts. "It is absurd to talk of one animal being higher than another," he wrote in his notebooks. "People often talk of the wonderful event of intellectual Man appearing—the appearance of insects with other senses is more wonderful.... Who with the face of the earth covered with the most beautiful savannas & forests dare say that intellectuality is the only aim in the world?"

Perhaps humans were the result of evolution, just like the finches. Darwin visited the zoo to look at a newly captured orangutan named Jenny and saw in her face the expressions he could also see in babies. "Man from monkeys?" he wrote.

Although his ideas were still embryonic, Darwin knew that they were dangerous. A public announcement that humans had evolved might alienate him from Lyell and the other naturalists whom he respected, and on whom his career depended. Nonetheless Darwin kept scribbling in his notebooks, working out his theory and gathering facts to support it.

Darwin searched for signs of how traits were handed down from one generation to the next and how they changed in the process. He interrogated gardeners and zookeepers and pigeon fanciers. He interviewed his hairdresser

An 1851 editorial cartoonist imagines an overcrowded London in the future. The British essayist Thomas Malthus raised the possible dangers of overpopulation; his ideas helped Darwin formulate his ideas of natural selection.

about breeding dogs. Although he could see signs that species were not eternal, he still didn't know of any way a species could take on a new form. Lamarck had claimed that an animal could change over its lifetime and pass on its acquired characteristics to its offspring, but there was little evidence that this actually happened. Darwin looked for a different explanation for how evolution could unfold.

He found it in a gloomy book about humanity's inevitable sufferings. In 1798 Thomas Malthus, a country parson, had written *An Essay on the Principle of Population*. In it he pointed out that a country's population, if it is unchecked by starvation or sickness, can explode in a matter of years. If every couple raised four children, the population could easily double in 25 years, and from then on, it would keep doubling. It would rise not arithmetically— by a factor of 3, 4, 5, and so on—but geometrically—by a factor of 4, 8, and 16.

If a country's population did explode this way, Malthus warned that there was no hope that its food supply could keep up. Clearing new land for farming or improving the yields of crops might produce a bigger harvest, but it could only increase arithmetically, not geometrically. Unchecked population growth inevitably brought famine and misery. The only reason that humanity wasn't already in perpetual famine was because its growth was continually checked by forces such as plagues, infanticide, and simply putting off marriage till middle age.

Malthus pointed out that the same forces of fertility and starvation that shaped the human race were also at work on animals and plants. If flies went unchecked in their maggot making, the world would soon be knee-deep in them. Most flies (and most members of every species) must die without having any offspring.

In Malthus's grim essay, Darwin found the engine that could push evolution forward. The fortunate few who got to reproduce themselves wouldn't be determined purely by luck. Some individuals would have traits that would make them better able to survive under certain conditions. They might grow to be big, they might have a particularly slender beak, they might grow thicker coats of fur. Whichever individuals were born with these traits would be more likely to have offspring than weaker members of their species. And because offspring tend to be like their parents, they would pass on those winning traits to their young.

This imbalance would probably be too small to see from one generation to the next. But Darwin was already comfortable with imperceptible geological changes producing mountains. Here was mountain making of a biological sort. If a population of birds ended up on a Galápagos island, the individual birds that were best suited to life on the island would produce the next generation. And with enough time, these changes could produce a new species of bird.

Darwin found a good analogy for this process in the way farmers tend their crops. They breed their plants by comparing how well each stalk or tree turns out. They then use the seeds only from the best ones to plant the next generation. With enough breeding the crops become distinct from other varieties. But in nature there is no farmer. There are only individual animals and plants competing with one another to survive, for light or water or food. They undergo a selection as well, a selection that takes place without a selector. And as a result, Darwin recognized, life's design could come about naturally, with no need for a string of individual acts of creation.

"LIKE CONFESSING A MURDER"

Darwin took a little time away from scribbling heresies to find a wife. Before his voyage he had fallen in love with a woman named Fanny Owen, but shortly after he set sail she married another man. When he got home he wondered if he should get married at all. Ever the methodical scientist, Charles drafted a pro-and-con balance sheet. He wrote "Marry" on the left side, and "Not Marry" on the right, and "This is the question" in the middle. This nuptial Hamlet reasoned that as a single man he would have more time for science and for conversations in men's clubs. He wouldn't have to earn enough to support children. On the other hand, a wife would offer "female chit-chat" and constant companionship in old age. He added up the columns and made his conclusion: "Marry—Marry—Marry. Q.E.D."

Darwin chose his cousin Emma Wedgwood. He had no interest in the sophisticated women he encountered in London. Instead, he looked back to his mother's niece, who had grown up as he did, in the country. Emma had already become interested in Darwin during his occasional visits to the Wedgwood house. She was happy to be courted by him, although he wooed her awkwardly with a series of oblique comments and half-made gestures. She was completely unprepared when he nervously blurted out to her one day that he wanted to marry her. She said yes, but she was so stunned that she promptly went off to teach her Sunday school class.

Soon, though, Emma grew happy with the thought of marrying a man she considered "perfectly sweet-tempered." Darwin, meanwhile, worried that his time on the *Beagle* had made him too unsociable for marriage, but he found hope in the prospect of marrying Emma. "I think you will humanize me," he wrote to her, "and soon teach me there is greater happiness than building theories and accumulating facts in silence and solitude."

Emma's only worry came when Charles talked about nature and the laws that might govern it. Emma, a devout Anglican, could tell that Charles had his doubts about the Bible. "Will you do me a favor?" she wrote to him. She asked him to read part of the Gospel according to John: "A new commandment I give unto you, That ye love one another; as I have loved you, that ye also love one another." If he began with love, Darwin might become a proper Christian.

He promised Emma that he felt "earnest on the subject." But a look in his notebook at the time would have shown he was not being entirely honest. He was wondering if religion was more a matter of instinct than of any love of a real God. It was his love for Emma that kept him from telling her all his thoughts.

After they were married, Charles brought Emma to London, and they settled into comfortable monotony. Emma's anxiety about her husband's soul continued, and she wrote him more letters. In one that she wrote in 1839, she worried that Charles was so consumed with finding the truth in nature that he shut out any other sort of truth—the sort that only religion could reveal. Believing only what could be proved would keep him from accepting "other things which cannot be proved in the same way & which if true are likely to be above our comprehension." She begged Charles not to forget what Jesus had done for him and the rest of the world. Darwin put away her letter without a reply, although he would remember it for the rest of his life.

In 1839 Darwin published *Journal of Researches into the Natural History and Geology of the Countries Visited During the Voyage of HMS* Beagle *Round the World, Under the Command of Captn. FitzRoy, R.N.* It was a huge success in Britain and cemented Darwin's fame as a naturalist. By then Charles and Emma had been married three years and had two children, and they decided it was time to leave London. They were tired of the crime, the coal dust that blackened their clothes, the horse dung that clung to their shoes. They wanted to raise their children in the countryside, where they had grown up. They picked out an estate called Down House, an 18-acre farm in Kent, 16 miles from London. Darwin became a gentleman farmer, planting flowers, buying a horse and a cow. He stopped mingling in the scientific societies altogether. He got whatever information he needed by letter or from carefully selected weekend guests. (Erasmus hated leaving London to visit his brother; he called their house "Down at the Mouth.")

All the while, Darwin continued mulling his theory of evolution in secret. He wrote down an argument for natural selection, and when he finished it, in 1844, he had no idea what to do next. He didn't even know how to talk to anyone about it. To support his theory he had been extracting information from dozens of people, but he had never let any of them know what it was for. The boy who was frightened of telling his father he couldn't become a doctor now had become a man frightened of telling anyone of his dangerous ideas.

But in the end Darwin had to tell someone. He had to find a scientist who could give him a qualified judgment, who might see some fatal flaw he had overlooked. He chose Joseph Hooker, a young botanist who had studied the plants from Darwin's voyage on the *Beagle,* and whom Darwin thought might be open-minded enough not to call him a blasphemer. He wrote to Hooker:

Charles Darwin married his cousin Emma Wedgwood (1808–1896) in 1839.

> I have been now ever since my return engaged in a very presumptuous work, and I know no one individual who would not say a very foolish one. I was so struck with the distribution of the Galapagos organisms, etc., that I determined to collect blindly every sort of fact which could bear any way on what are species. I have read heaps of agricultural and horticultural books, and have never ceased collecting facts. At last gleams of light have come, and I am almost convinced (quite contrary to the opinion I started with) that species are not (it is like confessing a murder) immutable.... I think I have found out (here's presumption!) the simple way by which species become exquisitely adapted to various ends. You will groan, and think to yourself, "on what a man have I been wasting my time and writing to." I should, five years ago, have thought so.

Hooker turned out to be as open-minded as Darwin had hoped. "I shall be delighted to hear how you think this change may have taken place," he wrote back, "as no presently conceived opinions satisfy me on the subject."

Hooker's response gave Darwin enough nerve to show his essay to Emma a few months later. He knew that she might be disturbed, but he wanted her to have the essay published posthumously if he should die too soon. Emma read it. She did not weep or faint. She merely pointed out where the writing got murky. When Darwin wrote that he could imagine natural selection producing something even as complex as an eye, she wrote, "a great assumption."

BACK INTO HIDING

With two people privy to his secret, Darwin was slowly gaining confidence about publishing his essay. But it vanished completely a month later. In October 1844 a book rolled off the presses called *Vestiges of the Natural History of Creation*. Its author, a Scottish journalist named Robert Chambers, chose to be anonymous—going so far as to hide the trail by which his publisher paid him royalties. He was wise to be so cautious.

Vestiges started harmlessly enough, describing the solar system and the neighboring stars, surveying how the laws of physics and chemistry explained Earth's formation out of a gaseous disk. Chambers worked through the geological record as it was then understood, noting the rise of fossils through history. The simple appeared first, and then the complex. As time went by,

In his study at Down House, Darwin wrote *Origin of Species* and many of his other great works.

higher and higher forms of life left their mark. And then Chambers made a scandalizing claim. If people could accept that God assembled the heavenly bodies by natural laws, "what is to hinder our supposing that the organic creation is also a result of natural laws, which are in like manner an expression of his will?" That would make more sense than God stepping in to create every species of shrimp or skink. "Surely this idea is too ridiculous to be for a moment entertained."

As for how those natural laws worked, Chambers offered a mishmash of secondhand chemistry and embryology. He thought that a spark of electricity might turn inanimate matter into simple microbes. After that, life would evolve by altering its development. Chambers relied here on the outdated ideas of German biologists. He pointed out that birth defects often consisted of a failure to carry out all the steps of development—a baby might be born with a heart, for instance, that had only two chambers, like a fish, instead of four. Presumably these defects were the result of "a failure of the power of development in the system of the mother, occasioned by weak health or misery." But if the opposite were the case, a mother might give birth to a child who had passed through a new stage of development. A goose might give birth to a gosling with the body of a rat—producing the first duck-billed platypus. "Thus the new production of new forms, as shown in the pages of the geological record, has never been anything more than a new stage of progress in gestation."

Chambers didn't think his readers should be scandalized that they had descended from fishes. The sequences of events he was proposing were "wonders of the highest kind, for in each of them we have to trace the effect of an Almighty Will, which had arranged the whole in such harmony with external physical circumstances." The middle-class British reader of *Vestiges* could go on with his life as he had before, still guided by the same moral compass. "Thus we give, as is meet, a respectful reception to what is revealed through the medium of nature, at the same time that we fully reserve our reverence for all we have been accustomed to hold sacred, not one tittle of which it may ultimately be found necessary to alter."

Vestiges was a huge hit, selling tens of thousands of copies. For the first time a broad English audience was introduced to the concept of evolution. But the leading scientists of Britain attacked it bitterly. "I believe some woman is the author," wrote Adam Sedgwick, "partly from the utter ignorance the book displays of all sound physical logic." Sedgwick was even more horrified by how such a view of life could undermine decency. If the book were true, he declared, "religion is a lie; human law a mass of folly and a base injustice; morality is moonshine."

The violence of the backlash shocked Darwin and sent him back into scientific hiding. He had not realized just how strongly Sedgwick and his other teachers were opposed to evolution. But he would not abandon his theory. Instead, he would figure out how to avoid Chambers' fate.

Darwin could see that *Vestiges* had clear weaknesses. Chambers had simply read other people's ideas and crushed them together into a shoddy argument. In some ways, Darwin was guilty of the same failing—his ideas were based on things he had read or heard from dozens of people—Lyell, Malthus, even his hairdresser. While he was recognized now as an authority on geology, he worried that he'd be treated as a dilettante when it came to biology. In order to be taken seriously, he had to show himself to be a first-class naturalist, able to wrestle with the complexities of nature.

He turned to the *Beagle* specimens he hadn't yet examined in the eight years that he had been back. In one of his bottles there was a barnacle. Although most people think of barnacles as nothing more than something to scrape off boat hulls, they are actually some of the most unusual creatures of the ocean. Initially zoologists thought they were mollusks, like clams and oysters, with their hard shells cemented to a flat surface. In fact, barnacles are crustaceans, like lobsters and shrimp. It was only in 1830 that their true identity was discovered, when a British army surgeon looked at their larvae and found that they had a resemblance to young shrimp. Once barnacle larvae are released into seawater, they search for a place to land—whether it is a ship's hull or a clamshell—and settle headfirst onto their chosen surface. They then lose most of their crustacean anatomy, developing instead a conical shell, out of which they extend feather-like feet that they use to filter food.

In 1835, off the coast of Chile, Darwin had collected a species of barnacle the size of a pinhead, latched to the inside of a conch shell. Looking at them under his microscope, he now realized that each barnacle was actually two—a large female with a minuscule male attached to it. At the time, scientists were most familiar with hermaphroditic forms of barnacles, equipped with both male and female reproductive organs. The pinhead barnacle was so strange that Darwin was sure it was a new genus.

Darwin was off on another long journey. At first he planned simply to write a short paper describing his discovery. But to do so, he had to figure out where among the many barnacle species he should classify it. He asked Owen to loan him some barnacles and to give him some advice on how to do the work properly.

Owen explained to Darwin that he needed to link his barnacle—no matter how strange it might be—to the basic crustacean archetype. By the 1840s, Owen had decided that archetypes were the key to zoology. Owen himself tried to reconstruct the vertebrate archetype, which he imagined was little more than a spinal column, ribs, and a mouth. This body plan didn't exist in nature, Owen claimed; it was only a blueprint in the mind of God, on which

Barnacles—crustaceans that cement themselves to hard surfaces and filter food from seawater— obsessed Darwin for eight years.

He based more and more elaborate forms. You could see the connection to the archetype if you compared different vertebrates.

Take, for instance, a bat, a manatee, and a bird. The bat has a wing made of a membrane stretched over elongated fingers. A manatee has a paddle for swimming. A bird has a wing as well, but it consists of feathers pinned to hand bones that have been fused together into a hinged rod. Each of these vertebrates has limbs that are adapted to its way of life, but they also correspond to one another precisely, bone for bone. They all have digits connected to marble-like wrist bones, which are connected to two long bones that meet at the elbow with a single long bone. These correspondences (which scientists call homologies) revealed to Owen a common body plan.

Owen urged Darwin to look for homologies between barnacles and other crustaceans. Darwin privately thought that Owen's archetype was nonsense. He thought that the similarities among vertebrates were a sign of their descent from a common ancestor. But to trace the evolution of barnacles from less peculiar crustaceans, Darwin would have to look at a lot of barnacles (1,200 species are known today). He borrowed collections from other naturalists, he studied fossil barnacles, and he even got hold of the British Museum's entire stash. Darwin would end up spending eight years studying barnacles. All the while, his explanation of evolution, an idea as revolutionary as Copernicus's sun-centered cosmology, sat sealed on a shelf.

Why the delay? Fear may have made Darwin procrastinate as he put off the inevitable confrontation with his mentors. Another reason may have been that Darwin was tired. He had taken a grueling five-year voyage, followed by

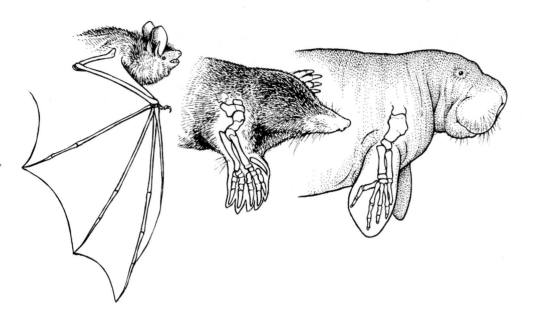

A bat's wing, a mole's foot, and a manatee's fin look very different from one another, but their limbs share underlying similarities. Each has long arm bones, small wrist bones, and five digits. Biologists call this kind of hidden equivalence *homology*.

eight fierce years of writing books and papers. His health had deteriorated after his return to England, to the point that he now was regularly devastated by bouts of vomiting. Darwin was in his midthirties and ready for some peace.

And yet another part of the delay was grief. His favorite daughter, Anne, died of the flu in 1851 at age 10. As he witnessed her undeserved agony, Darwin gave up on the angels. After Anne's death, he couldn't talk to Emma about his crumbling faith. Studying barnacles may have become a way to hide from the pain.

But fear, exhaustion, and grief aside, Darwin was enchanted by his barnacles. They turned out to be a perfect group of animals to study in order to learn how evolution worked. Darwin could see, for instance, how the ancestors of his Chilean barnacles could have descended from hermaphrodites, which had evolved through transitional forms until they began producing males and females. Darwin was also impressed by the variation that he found among the members of individual species of barnacles. No part of the barnacle anatomy was uniform. Here, Darwin realized, was a huge reservoir of raw material for natural selection to work on. Originally he had thought that a species would be subject to natural selection only at certain times—when islands emerged or continents began to sink, for instance. But with so much variation to choose from, natural selection could actually be at work all the time.

None of these thoughts made it explicitly into Darwin's barnacle writings. He published a 1,000-page tome, for which he won praise, awards, and the respect he wanted as a naturalist. By 1854 he was ready to get back to thinking about natural selection.

NATURAL SELECTION UNSEALED

Darwin began by addressing some doubts that Joseph Hooker had raised. One of Darwin's claims was that the plants and animals living on islands had not been created there, that they were modified descendants of colonists. If that were true, then they needed a way to get to the islands in the first place. Hooker, a seasoned botanist, knew that seeds could travel for miles by wind or water, but he doubted that they could travel the enormous distances that Darwin claimed.

Darwin met Hooker's doubts by throwing seeds into tanks of saltwater and found that they could be soaked for four months and still sprout when he planted them in dry ground. He discovered that birds could carry seeds on their feet, and that seeds could even survive being eaten by owls, passing out of their bodies with their droppings. Darwin's theory had generated a hypothesis, and the hypothesis passed a test.

Darwin also went back to his studies of breeding. He drank gin with pigeon breeders as they explained to him how to use tiny variations to produce entirely new forms of birds. Darwin raised pigeons of his own, killing them and boiling the flesh off their skeletons so that he could measure the variability among them. He found that each breed was so distinct that if it were wild, it would be considered its own species. Almost every part of a pigeon's anatomy was distinct from that of other varieties, from their nostrils to their ribs to the size and shape of their eggs. And as far as anyone could tell, every pigeon variety descended from a single kind of wild rock dove.

By 1856 Darwin had found so much more evidence for evolution that he reopened his 1844 essay and began reworking it. It soon puffed up into a monstrous opus, running hundreds of thousands of words long. He threw in everything he had learned over the years—on his voyage, from his reading and conversations, from his studies on barnacles and seeds. He was determined to wear down the opposition with a river of facts.

Since the day that Charles had shown Emma his manuscript in 1844, he had barely spoken of his theory. But now he felt more confident revealing it to a few people—particularly to younger minds more open to new possibilities. For one of his initiates, he chose a brilliant, struggling young zoologist he had recently befriended, by the name of Thomas Huxley.

Huxley could not practice Darwin's brand of gentlemanly science. He had been born above a butcher's shop. His father, a teacher at a failed school and then a director of a failed bank, had no money for Huxley's education. Huxley became a doctor's apprentice at age 13 and three years later followed him down to London, where he was trained as a surgeon, barely getting by on scholarships and paltry loans from in-laws. The only way he could pay off

Thomas Huxley (1825–1895) became known as "Darwin's bulldog" for the way he passionately championed *Origin of Species*.

his debts was to join the crew of HMS *Rattlesnake,* a ship bound for the coasts of New Guinea, as an assistant surgeon. Huxley was developing a taste for zoology, and on the voyage he would be free to gather as many exotic species as he wanted.

After four years, Huxley returned to England in 1850, and like Darwin he had been transformed into a scientist by the journey. Like Darwin, he was preceded home by his reputation, with papers already published by the time he arrived, about bizarre creatures such as the Portuguese man o' war, an animal that is actually a colony of individuals. Huxley arranged a special post with the Royal Navy, with three years' leave with pay, so that he could continue his research. Without any degree, he was elected to the Royal Society at age 26.

The navy ordered Huxley three times to return to active duty, and the third time he refused they struck him from the list. He struggled to find other work in London and eventually ended up working part-time at the School of Mines. By writing columns and reviews on the side, he earned barely enough money for his family. Huxley resented the wealthy men who dominated science simply because they could afford to. But he managed nevertheless to build a reputation for himself, and he wasn't afraid to attack the dean of English biologists, Richard Owen.

Owen at the time was toying with the idea of a kind of divine evolution. Over time, he proposed, God made new species, always referring to His archetypes for their basic design. Owen pictured a stately unfolding of life according to a divine plan, moving from the general to the specialized: an "ordained continuous becoming." To calm Owen's patrons, who still clung to the comfort of natural theology and the fixity of species, he promised that biology was still "connected with the loftiest of moral speculations."

In public lectures and in reviews, Huxley mocked Owen's attempts to make God into a draftsman, and to make the fossil record read like His revisions. Huxley didn't accept evolution in any form, whether divinely guided or simply materialistic. He saw no progress in the history of Earth or of life. But that changed when Darwin invited Huxley to come out to the country one weekend in 1856.

Darwin explained his version of evolution, one that could account for the patterns of nature without resorting to providence or special intervention. He showed Huxley his pigeons and seeds. Soon Huxley was persuaded, and would prove to be Darwin's greatest ally.

Darwin's slow, cautious march toward a public declaration was moving along well, until the mail came to Down House on June 18, 1858. Darwin got a letter from the other side of the world, from a traveling naturalist named Alfred Russel Wallace. Wallace was exploring southeast Asia, collecting animals to pay his way, and looking for evidence of evolution. At 21 he had read *Vestiges* and was seduced by the idea of nature sweeping upward over time. After reading Darwin's tales of his voyages, Wallace decided he would travel as well.

His first journey was to the Amazon in 1848. Later he traveled to what is now Indonesia to look for orangutans and, he hoped, learn about man's ancestry. Along his way he financed his trip by sending back boxes of beetles, bird skins, and other specimens to dealers and patrons in London. Darwin was one of those patrons, receiving the skins of birds for his own research, and the two naturalists began to trade letters.

Darwin encouraged Wallace to think broadly and theoretically about evolution, confiding that he had a theory of his own about how species came to be. Wallace decided to write a letter to Darwin to tell him about his own ideas. When Darwin opened it, his heart sank. Wallace had read Malthus as well, and he had also wondered what effect overpopulation would have on nature. And like Darwin, he concluded that it would change species over time and create new ones.

At the time that Darwin received Wallace's letter, he was planning to write for a few more years before going public. But here was much of his own theory in the handwriting of another scientist. It was not identical—Wallace did not make much of the competition between members of the same species. He proposed simply that the environment culled unfit individuals. But Darwin would not rob Wallace of his proper credit. His sense of honor ran deep, and he would rather have burned his own book than have anyone think he cheated Wallace.

So Darwin arranged with Lyell that the Linnaean Society would hear papers on both his own work and Wallace's at the same time. On June 30, 1858, the society listened to extracts from Darwin's 1844 essay, part of a letter he had written about his idea to Hooker in 1857, and Wallace's paper. Twenty years of cautious research and fretting was suddenly over. The world could now judge.

But no judgment came. Darwin's and Wallace's papers were read during a long, rushed session of the Linnaean Society, and were met with silence. Perhaps the papers were too terse, too polite, for the audience to realize just what Darwin and Wallace were proposing. Darwin decided that he now had to lay out his argument in a paper in a scientific journal.

In the months that followed, he struggled to shrink his gargantuan *Natural Selection* to a brief summary that he could publish. But as he worked on it, the summary swelled to book length again. He simply had too many arguments and pieces of evidence to counter the attacks he knew would come. He contacted John Murray, the publisher of *Journal of Researches,* and asked if he would print a second book. *Journal of Researches* had been a hit, so Murray agreed to publish the new volume, which came to be called *On the Origin of Species by Means of Natural Selection.*

Along the way Darwin's health came under fresh attack. When he received the first finished copy of his book, bound in royal green cloth, in November 1859, he was recovering at a spa in Yorkshire. Soon the complimentary copies arrived, and Darwin sent one to Wallace in Indonesia. Along with the book he sent a note: "God knows what the public will think."

"THERE IS A GRANDEUR IN THIS VIEW OF LIFE"

The argument that appeared in *Origin of Species* had evolved from its ancestral form in 1844. Now it had become something far broader—an all-encompassing explanation for life on Earth.

Darwin chose to start his argument not on the remote Galápagos Islands, not in the murky depths of the oceans among coral reefs, but in ordinary, comfortable English life. He talked about the many forms that animals and plants could assume in the hands of breeders. Pigeon breeders had doubled the normal allotment of feathers on the fantail; they had turned the neck feathers on the Jacobin into a hood. These were the sorts of traits that could set off a bird as a species of its own, and yet the breeders had created them in only a few generations.

Darwin acknowledged that no one really understood how heredity allowed breeders to work these miracles. Breeders simply knew that different traits tended to be inherited with each other. Cats with blue eyes, for instance, are invariably deaf. But while heredity might be a mystery, at least it was clear that parents produced offspring that tended to be like them—although each generation came with a certain amount of variability.

If you came across a fantail and a Jacobin in the wild, you might think they were different species, yet, strangely, they could still mate and produce fledglings. In fact, Darwin pointed out, it is very hard to distinguish between species and varieties in the wild. Biologists argue about whether certain kinds of oak trees belong to the same species. Darwin suggested that the confusion stemmed from the fact that varieties have some of the same characteristics as species. And that is because varieties are often incipient species themselves.

How could an incipient species become a full-blown one? Here Darwin brought Malthus into the argument. Even a slow-breeding species like a

Darwin was impressed by the many different varieties pigeon breeders could create by artificial selection. Nature, he realized, would create new forms of life through a similar process.

human or a condor can double its numbers in 20 or 30 years, easily overrunning the planet in a few millennia. But plants and animals are regularly wiped out in staggering numbers. Darwin recalled how four-fifths of the birds around Down House died one year in a cold snap. The tranquil surface of nature hid a massive slaughter from view.

Some members of a species survived these challenges thanks to luck, while others had certain qualities that made them less likely to die. The survivors would reproduce, while the badly adapted ones would die. Nature, in other words, was a breeder of its own, and a far superior one to humans. A human might breed pigeons for only one trait, such as tail feathers, while nature bred for countless ones—not just traits of flesh and blood, but of instincts as well. "She can act on every internal organ, on every shade of constitutional difference, on the whole machinery of life," Darwin wrote. "Man selects only for his own good; Nature only for that of the being which she tends."

And whereas breeders can do their work only over the course of years or decades, nature has at its disposal a vast expanse of time. "It may be said that natural selection is daily and hourly scrutinizing, through the world, every variation, even the slightest," Darwin wrote. "We see nothing of these slow changes in progress, until the hand of time has marked the long lapses of ages."

If natural selection worked on a variety long enough, it would turn it into a new species of its own. After a thousand generations, a single species of bird made of two varieties might end up as two distinct species. Just as individuals of a given species struggle with one another, they also struggle with members of other species. And the competition between two similar species would be most intense. Eventually one of them might be driven out of existence. This, Darwin argued, accounted for all of the fossils of animals that could no longer be found on Earth. They had not simply disappeared—they had been obliterated by other animals.

To help his readers understand this process, Darwin drew the one illustration that appeared in his book. At the bottom were a few original species, which rose like limbs on a tree, dividing over time into new branches. Most of these branches were nothing more than twigs—varieties or species that became extinct—but some of them branched their way all the way to the top of the page. Life was not a Great Chain of Being, said Darwin, but a bushy tree.

Origin of Species is a deeply defensive book, written by a man who had quietly listened for years to other scientists scoff at evolution, and had imagined them scoffing at him as well. He addressed their objections one by one. If old species gradually turned into new species, then why were animals so distinct from one another? Darwin's answer was that competition between two similar species would tend to drive one of them extinct, so that the animals alive today would be only a scattered selection of all the species that had ever lived.

But shouldn't we be able to see these intermediate forms as fossils? Darwin reminded his readers that fossils, by their nature, could provide only a few fragments of life's history. In order to become a fossil, a carcass had to be properly buried in sediment, turned to rock, and then avoid destruction by volcanoes or earthquakes or erosion. Those chances are abysmally low, and so a species, which once included millions of individual animals, might be known from a single fossil. Gaps in the fossil record shouldn't be a surprise—they should be the rule. "The crust of the earth is a vast museum," Darwin wrote, "but the natural collections have been made only at intervals of time immensely remote."

How could natural selection create complex organs, or entire bodies, that were made of so many interdependent parts? How, for instance, could it make a bat or an eye? Fossils couldn't be expected to tell the whole story. Instead, Darwin turned to living animals as analogies to show at least that such a transformation wasn't impossible. For bats, he pointed to squirrels. Many tree-living squirrels have four ordinary legs and a slender tail. But there are also some species that have flattened tails and loose skin. Then there are squirrels that have broad flaps stretching between their legs and even their tail and can parachute out of trees. Then, Darwin pointed out, there are gliding mammals known as flying lemurs, which have elongated fingers and a membrane that stretches from jaw to tail.

Here was a relatively smooth gradation from an ordinary four-legged mammal to a creature with almost a batlike anatomy. It was possible that the ancestors of bats went through this evolutionary sequence and then went one step farther, evolving the muscles necessary for true flight.

Likewise, there was no need for an eye to pop out of an animal's head all at once. Invertebrates such as flatworms have nothing more than nerves with endings coated in light-sensitive pigments. Some crustaceans have eyes that consist of little more than a layer of pigment coated by a membrane. Over time, this membrane could separate from the pigment and begin to act like a crude lens. With small alterations, such an eye could turn into the precise telescopes that birds and mammals use. Because a little eyesight is better than none at all, each new step along the way would be rewarded by natural selection.

With his discovery of natural selection, Darwin turned back to the ideas of other scientists and showed how they made more sense as parts of his own theory. As a young man Darwin had admired Paley, but now he showed how natural designs could come into being without a designer's direct control. Karl von Baer had demonstrated how embryos of different animals resembled one another early on and grew more particular. For Darwin, this was a sign of the common heritage of animals, and the differences in their development came after their ancestors had diverged.

And Darwin even absorbed Owen's archetype. "I look at Owen's Archetypes as more than ideal, as real representation as far as the most consummate skill & loftiest generalization can represent the parent of the Vertebrata," he once wrote to a colleague. To Owen, the homology between a bat wing and a manatee paddle showed how the mind of God worked. But to Darwin, the homology was a sign of inheritance.

Darwin gingerly avoided writing much of anything about what his theory meant for humanity. "In the distant future I see open fields for far more important researches. Psychology will be based on a new foundation, that of the necessary acquirement of each mental power and capacity by gradation. Light will be thrown on the origin of man and his history."

He was not going to make the mistake Robert Chambers had made in *Vestiges*. He had an argument to make, and he didn't want emotions to interfere. But Darwin did make some attempt to stave off the despair people might feel. "Thus," he wrote in the final lines of his book, "from the war of nature, from famine and death, the most exalted object which we are capable of conceiving, namely the production of the higher animals, directly follows. There is a grandeur in this view of life, with its several powers, having been originally breathed into a few forms or into one; and that, whilst this planet has gone cycling on according to the fixed law of gravity, from so simple a beginning endless forms most beautiful and most wonderful have been, and are being, evolved."

APE VERSUS BISHOP

That winter, as blizzards buried England, thousands of people kept warm by the fire and read Darwin's book. The first printing of 1,250 copies was snapped up in a day, and in January 3,000 more were printed. Huxley sent Darwin words of praise but warned him of the battle to come. "I am sharpening up my claws and beak in readiness," he promised. Newspapers generally offered only brief articles about it, but the reviews, where the literary world debated the great ideas of the nineteenth century, went into more depth. Huxley and other allies of Darwin praised it, but many reviews saw it as blasphemy. *The Quarterly Review* declared that Darwin's theory "contradicted the revealed relation of the creation to its Creator" and was "inconsistent with the fullness of His Glory."

The review that made Darwin angriest appeared in the *Edinburgh Review* in April 1860. It was anonymous, but anyone familiar with Richard Owen knew he had written it. It was stunning in its animosity. Owen called Darwin's book "an abuse of science." He complained that Darwin and his disciples pretended that natural selection was the only possible natural creative law. Owen wasn't actually opposed to evolution; he just didn't like what he considered blind materialism.

Yet Darwin had done what Owen couldn't. Owen had tried to synthesize the discoveries of biology, but he had ended up with murky notions of archetypes and continuous creation. Darwin, on the other hand, could account for the similarities between species with a mechanism that was at work in every living generation.

Owen wrote his review out of anger, both at Darwin and at Huxley. Huxley had been attacking Owen in public lectures with a venom that shocked him. Huxley despised Owen both for the way he curried favor with aristocrats and for what he considered shoddy science. He mocked Owen's theory of continuous creation as absurd. Owen became so angry with his taunts that during one of his public lectures he glared at Huxley, declaring that anyone who didn't see the fossil record as a progressive expression of divine intelligence must have "some, perhaps congenital, defect of the mind."

Their fiercest fights broke out in the years just before *Origin of Species* was published, as Owen tried to prove that humans were distinct from other animals. During the 1850s, orangutans, chimpanzees, and gorillas were beginning to emerge from their jungle obscurity, and Owen dissected their bodies and studied their skeletons. He worked hard to try to find some mark that distinguished humans from them. If we were nothing but a variation on an ape, then what became of morality?

What made humans most distinct from animals, Owen assumed, was our mental capacity: our ability to speak and reason. Owen therefore looked in the brains of apes to discover the anatomy that marked that difference. In 1857 he claimed to have found a key distinction: unlike the brains of apes, the cerebral hemispheres of the human brain extended so far back that they formed a third lobe, with a structure Owen called the hippocampus minor. He declared that its uniqueness warranted putting humans in a subclass of their own. Our brain was as different from a chimp's as a chimp's was from a platypus's.

Huxley suspected that Owen had been misled by studying badly preserved brains. His elaborate classifications were built on a fundamental error. (Huxley liked to say that they stood like "a Corinthian portico in cow dung.") In fact, Huxley argued, the human brain was no more different from a gorilla's than a gorilla's was from a baboon's. "It is not I who seek to base Man's dignity upon his great toe, or insinuate that we are lost if an Ape has a hippocampus minor," Huxley wrote. "On the contrary, I have done my best to sweep away this vanity."

Owen's furious review of Darwin's *Origin of Species* raised the tension between him and Huxley even higher, and finally, a few months later, in

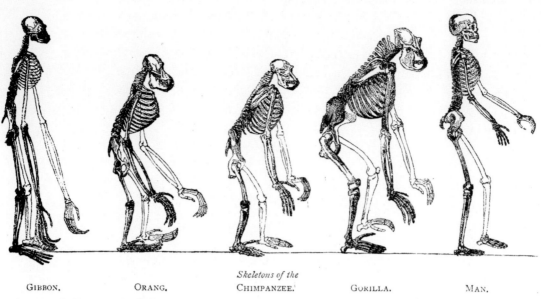

Skeletons of the
GIBBON.   ORANG.   CHIMPANZEE.   GORILLA.   MAN.

Photographically reduced from Diagrams of the natural size (except that of the Gibbon, which was twice as large as nature), drawn by Mr. Waterhouse Hawkins from specimens in the Museum of the Royal College of Surgeons.

The discovery of gorillas and other apes revealed startling similarities between humans and primates, as Huxley demonstrated in this illustration from his 1863 book, *Man's Place in Nature.*

June 1860, their hostility exploded. The British Association for the Advancement of Science held its annual meeting at Oxford, attended by thousands of people. Owen, the association's president, gave a talk on June 28, explaining once again how the human brain was distinct from that of apes. Huxley had an ambush planned. At the end of the talk, Huxley stood and announced that he had just received a letter from a Scottish anatomist who had dissected a fresh chimpanzee brain. The anatomist had discovered that it looked remarkably like a human brain, complete with a hippocampus minor. With a packed audience looking on, Owen had no way to defend himself. Huxley could not have chosen a more public place to humiliate him.

Having won the battle of the brains, Huxley decided to leave the Oxford meeting the next day. But then he bumped into Robert Chambers, the still-anonymous author of *Vestiges*. Chambers was horrified to hear that Huxley was going to depart. Didn't he know what the next day had in store?

Rumors were racing through Oxford that Bishop Samuel Wilberforce was going to attack Darwin. For years, Wilberforce had been a leading religious voice against evolution. He had attacked *Vestiges* in 1844, calling it foul speculation, and now the bishop saw Darwin's book as no different. An American scientist named John William Draper was scheduled to give a talk the next day about "Darwinism" and its implications for society. Wilberforce was going to use the opportunity to denounce Darwin in public, at Britain's most important scientific meeting. Owen was staying at Wilberforce's home during the meeting, and no doubt he was coaching the bishop. Chambers convinced Huxley to stay on for Draper's talk and defend Darwin.

Owen opened the conference the following day. Upward of a thousand people packed the auditorium, and to them he announced, "Let us ever apply ourselves seriously to the task of scientific inquiry, feeling assured that the more we thus exercise, and by exercising improve, our intellectual faculties, the more worthy shall we be, the better shall we be fitted to come nearer to our God."

Draper's talk was entitled "On the Intellectual Development of Europe, Considered with Reference to the Views of Mr. Darwin and Others, That the Progression of Organisms Is Determined by Law." By all accounts it was dull, long, and poorly reasoned. Joseph Hooker was in the audience and described Draper's talk as "flatulent stuff." The hall grew warm, but as woozy as the audience became, no one left. They wanted to hear the bishop.

When Draper was done, Wilberforce stood and spoke. He had recently written a review of Darwin's book, and he essentially repackaged it as a speech. He didn't pretend the Bible should be a test of science, but in his review he had written, "This does not make it the less important to point out on scientific grounds scientific errors, when those errors tend to limit God's glory in creation."

Darwin had made such an error. His book was based on wild assumptions and hardly any evidence. His entire argument hinged on this new idea of natural selection. And yet, Wilberforce wrote, "Has any one such instance ever been discovered? We fearlessly assert not one."

Instead, Wilberforce argued for a loose mix of Paley and Owen. "All creation is the transcript in matter of ideas eternally existing in the mind of the Most High—that order in the utmost perfectness of its relation pervades His works, because it exists as in its centre and highest found-head in Him the Lord of All."

When Wilberforce ended his speech, he looked to Huxley. He asked him, half-jokingly, whether it was on his grandfather's or grandmother's side that he descended from an ape.

Later Huxley would tell Darwin and others that at that moment he turned to a friend seated next to him, struck his hand to his knee, and said, "The Lord hath delivered him into mine hands." He stood and lashed back at Wilberforce. He declared that nothing that the bishop had said was at all new, except his question about Huxley's ancestry. "If then, said I, the question is put to me would I rather have a miserable ape for a grandfather or a man highly endowed by nature and possessed of great means and influence and yet who employs these faculties and that influence for the mere purpose of introducing ridicule into a grave scientific discussion I unhesitatingly affirm my preference for the ape."

The audience broke out in laughter, until a man whom Hooker described as "a gray-haired Roman-nosed elderly gentleman" stood in the center of the audience, quaking in rage. It was Captain FitzRoy.

FitzRoy and Darwin had grown cool to each other over the years. The captain thought that Darwin's book about the *Beagle*'s voyage was self-serving and ignored all the help Darwin had gotten from FitzRoy and his crew. Although FitzRoy had dabbled in Lyell's ancient geology, he had returned to a strict reading of the Bible. In his own book about the voyage, FitzRoy tried to account for all that he and Darwin had encountered with Noah's Flood. He had been appalled to watch Darwin move even further into heresy, not only abandoning the Flood, but even God's work.

The captain had come to Oxford to give a talk about storms, and he happened to get wind of the talk by Draper. After Huxley finished, FitzRoy stood and spoke of how he had been dismayed that Darwin entertained views that contradicted the Bible. He declared that reading *Origin of Species* had brought him "acutest pain." He lifted up both his arms over his head, a Bible clutched in his hands, and asked the audience to believe in God, not man. Whereupon the crowd shouted him down.

Finally it was Joseph Hooker's turn. He climbed to the podium to attack Wilberforce. Later he wrote to Darwin about his speech. "I proceeded to demonstrate that 1) he could never have read your book, 2) he was absolutely ignorant of the rudiments of Botanical Science and the meeting was dissolved forthwith, leaving you master of the field."

If Darwin was master of the field, he was a missing master. Practically a recluse now at age 50, he stayed away from the Oxford meeting. As Hooker and Huxley defended him, he was spending a few weeks in the village of Richmond, where he was being treated yet again for his chronic illness. There he read the letters of his friends, describing their speeches, with an ailing awe. "I would have soon have died as tried to answer the Bishop in such an assembly," he wrote to Hooker.

The Oxford meeting quickly became a legend, and as with all legends, what really happened receded behind a fog bank of embellishment. Each of the players in the drama offered his own version, in which he came off best. Wilberforce was convinced that he had won the debate, while Huxley and Hooker each thought they had delivered the fatal blow to the bishop. To this day, it's not clear what happened, and Darwin himself had a hard time figuring out what took place that warm day in June. Only one thing was certain to him: his 20 years of hiding were over.

In Darwin's own lifetime, he would become recognized as one of the great masters of science. By the 1870s, almost all serious scientists in Britain had accepted evolution, although they might argue with Darwin about how it unfolded. His statue stands at the Natural History Museum in London, and he lays buried in Westminster Abbey, close by Newton's grave.

But the great irony of *Origin of Species* is that only in the twentieth century would its true power be recognized. Only then did paleontologists and geologists work out the chronology of life on Earth. Only then did biologists uncover the molecules that underlie heredity and natural selection. And only then did they begin to truly comprehend how powerfully evolution shapes everything on Earth, from a cold virus to the human brain.

Deep Time Discovered

PUTTING DATES
TO THE HISTORY OF LIFE

GEOLOGISTS HAVE THEIR own mecca, a stretch of the Acasta River deep in the Northwest Territories in Canada. You can get there only by canoeing for days up the river or by flying a float plane north from the town of Yellowknife, over an expanse of half land, half water. The water takes the form of thousands of lakes and ponds, some strung together into blobby rivers, in every shape that Ice Age glaciers could possibly carve. You ski to a landing near a spindly island in the middle of the river. The shore is covered with black spruce, reindeer moss, heather, and lichens. Chirping plovers cut the silence, and blackflies and mosquitoes drill your skin.

A wall of exposed rock tumbles down to the water, and you can clamber down among the boulders. The rocks here are granite, dark gray hunks flecked with bits of feldspar that look pretty much like any other piece of granite you may have encountered. They

The fossil record of animals that lived on land—such as this pterodactyl—reaches back only 360 million years, which represents less than 10 percent of life's history.

The Acasta Shale, a geological formation in the Northwestern Territories, Canada, contains rocks dating back just over 4 billion years. They are the oldest known rocks on Earth.

are exceptional in only one way: some of them are more than 4 billion years old, which makes them the oldest known rocks on Earth. From our planet's infancy the minerals that made them up have held together, as continents have been torn apart and fused back together.

Their age is so vast that it's almost impossible to comprehend. Think of a year as equaling the length of your outstretched arms. To equal the age of the Acasta rocks, you'd have to hold hands in a line of people circling the Earth 200 times. But as hard as it may be to imagine, it is a picture that would have made Darwin very happy.

When Darwin first proposed his theory of evolution, he could not have known the great age of the Acasta rocks. The physics required to determine their antiquity was still 50 years away. Darwin suspected that the world was spectacularly old, which certainly would agree well with his theory of gradual, generation-upon-generation evolution. But it was only during the twentieth century that paleontologists and geologists precisely mapped the terra incognita of Earth's history. They discovered a way to establish not just the order in which new life-forms appeared on Earth, but their actual dates in history, from the earliest signs of life more than 3.85 billion years ago, to the earliest animals 600 million years ago, to the first members of our own species 150,000 years ago.

TOO WARM TO BE OLD

Of all the objections that were raised against Darwin, be they religious, biological, or geological, one of the most troubling to him was over the age of the planet. It came not from a bishop, a biologist, or a geologist, but from an unexpected source: a physicist.

William Thomson (better known as Lord Kelvin) was one of the world's leading physicists when Darwin first published *Origin of Species*. For Kelvin, the universe was a swirl of energy, electricity, and heat. He demonstrated how electricity acted like a fluid, just like water. He also showed how entropy dominates the universe: everything goes from order to disorder unless it receives energy to keep it organized. Burn a candle down to its stump, and the soot, gases, and heat that it releases will never spontaneously join back together into a candle.

As esoteric as Kelvin's work might seem, it made him rich when he applied it to designing the transatlantic cables that joined Europe to North America by telegraph. And sometimes, when Kelvin was bobbing on a cable ship in the middle of the Atlantic, he would think about how old Earth was. Kelvin was a devout man, but he didn't accept that Earth was a few thousand years old simply because someone decided that the Bible said so. He thought that it should be possible instead to put an upper bound on the age of the Earth scientifically, by studying its heat.

Miners, Kelvin knew, found that the deeper they dug, the hotter the rocks became. To account for this heat, Kelvin speculated that Earth formed from the collision of miniature failed planets, and the energy of their impact created a molten blob (a speculation that later proved to be right). Kelvin assumed that once the impacts ended, there would be no way for the planet to receive any new heat, so it would gradually cool like a dying ember. Its surface cooled off the fastest, while its interior remained warm to this day. Only at some point in the distant future would Earth's core become as cold as its surface.

Lord Kelvin (also known as William Thomson, 1824–1907) used the temperature of the Earth to calculate that it was only 20 million years old. He didn't know that radioactivity was skewing his estimates.

Kelvin and other physicists had worked out equations to accurately predict how objects cool, and he applied them to the entire planet. By measuring how fast heat escaped from rocks, and how hot the deepest mine shafts became, he came up with an estimate for the age of the planet. He concluded in 1862 that Earth had been cooling no more than 100 million years.

Kelvin's original motive had been to show how sloppy geology was compared to physics. But after reading *Origin of Species,* he was happy to use his results to attack Darwin. Steeped as Darwin was in Lyell's ancient geology, he assumed the gradual changes of natural selection could take as long as they needed to alter life. But Kelvin's results wouldn't allow him enough time. Kelvin himself wasn't a rabid anti-evolutionist—for all he knew, all of life might have started as a germ—but he saw life today as evidence of design, of the handiwork of God. He used his estimate for Earth's age to cut Darwin down with a single scimitar slice.

Huxley tried to defend Darwin by making a compromise—something Huxley rarely did with critics. He said that biologists had to accept an age for Earth that geologists and physicists decided on and figure out how evolution could work in that span of time. If Earth was only 100 million years old, then evolution must be able to work at high speed. Wallace went further, suggesting that at times evolution could work far faster than it does today. As Earth wobbles on its axis, the planet might experience harsh climates that would make evolution run at high speed.

Darwin wasn't satisfied. "I am greatly troubled at the short duration of the world according to Sir W. Thomson," he wrote in a letter. Kelvin, meanwhile, was getting new reports on the planet's temperature, and kept revising his estimate—each time shortening Earth's life span. By the time he was done, he had winnowed it down to only 20 million years. All the while, Darwin could do nothing but grit his teeth. As he struggled to flesh out his theory of evolution, "then comes Sir W. Thomson like an odious specter."

CLOCKS WITHIN ATOMS

Marie Curie (1867–1934) and her husband, Pierre, discovered that radium released heat as it decayed, a discovery that helped show that Earth was much older than some thought.

Lord Kelvin had based his calculation of Earth's age on a fundamental (and, it would turn out, false) assumption: the planet had no source of heat of its own. But there was a hidden heat inside the planet that Kelvin hadn't counted on. In 1896, 14 years after Darwin died, a French physicist named Henri Becquerel wrapped a piece of uranium salt in a photographic plate. When he developed the plates, he found sharp, bright dots on them. Uranium, he realized, released rays of energy. Seven years later Pierre and Marie Curie showed that a lump of radium released a constant supply of heat.

Becquerel and the Curies had found a source of energy in the basic structure of atoms. Atoms are made out of three building blocks: protons, neutrons, and electrons. Electrons, which carry a negative charge, flit around the edge of atoms, while positively charged protons sit at the center. Each element has a unique number of protons. Hydrogen has one proton, helium has two, and carbon has six. Alongside protons, atoms have neutrally charged particles called neutrons. Two atoms of the same element may have a different number of neutrons. The most common form of carbon on Earth has six protons and six neutrons (known as carbon 12), but there are trace amounts of carbon 13 and carbon 14 as well. These different versions of atoms, known as isotopes, make it possible to tell geological time.

The protons and neutrons in an atom are a bit like piles of oranges at a grocery store: in some arrangements they're perfectly stable, but in others they will sooner or later fall apart. Orange piles are held together by gravity, but protons and neutrons are held together by other forces. When an unstable isotope breaks down, it releases a burst of energy, along with one or more particles (otherwise known as radiation). In the process it may become a different element. Uranium 238, for example, breaks down by releasing a pair of neutrons and a pair of protons, and turns into thorium 234. But thorium 234 is unstable as well and decays into protactinium 234, which also decays. Through a chain of 13 intermediates, uranium 238 finally settles into a stable form, lead 206.

You can't predict exactly when a particular atom will decay, but a large collection of them will obey certain statistical laws. In any given period of time, an atom has a certain probability of decaying. Let's say a pebble has 1 million radioactive isotopes inside it, and this particular kind of isotope has a 50 percent chance of decaying in a year. After the first year, 500,000 of the isotopes will be left. Of those 500,000 isotopes, 50 percent will decay in the second year, leaving 250,000. Year after year, half of the remaining isotopes disappear, until about 20 years later the last isotope disappears. Physicists capture this trailing off in a measurement known as the half-life: how long it takes for half of any given amount of a radioactive element to decay. Uranium 238, for example, has a half-life of 4.47 billion years; other elements have half-lives lasting tens of billions of years, while others have half-lives of only minutes or seconds.

The laws that govern atoms don't submit to any simple intuitive sense, but they work. If they didn't, computers wouldn't crunch numbers and nuclear bombs wouldn't explode. And long before computers and nuclear bombs were invented—in fact, within a few years of the work of Becquerel and the Curies—physicists realized that these laws exposed a fatal flaw in Kelvin's claims of a young Earth. Uranium and other radioactive elements such as thorium and potassium can be found in the Earth, where they decay and give off heat. Kelvin thought that Earth was young because it hadn't cooled down very much from its origin. Radioactivity allows the planet to stay warm far longer.

A physicist named Ernest Rutherford was the bearer of these bad tidings. Rutherford worked out many of the fundamentals of radioactivity, showing that it was a natural alchemy that could transform one element into another. In 1904 he traveled from Montreal, where he taught at McGill University, to England to give a talk about the new discoveries.

I came into the room, which was half dark, and presently spotted Lord Kelvin in the audience and realized that I was in for trouble at the last part of the speech dealing with the age of the earth, where my views conflicted with his. To my relief, Kelvin fell fast asleep, but as I came to the important point, I saw the old bird sit up, open an eye and cock a baleful glance at me! Then a sudden inspiration came, and I said Lord Kelvin had limited the age of the earth, *provided no new source of heat was discovered*. That prophetic utterance refers to what we are now considering tonight, radium! Behold! the old boy beamed upon me.

That may be how Rutherford remembered that day, but Kelvin never publicly retracted his old estimate. Two years after hearing Rutherford's talk, he was writing letters to the London *Times* maintaining there wasn't enough radioactivity in the earth to keep it hot on the inside.

Rutherford realized that radioactivity not only showed that Earth was old, but could show *how* old it was. Any uranium that got trapped inside a cooling rock would decay gradually into lead. And because physicists knew uranium's half-life with great precision, it was possible to use the remaining proportions of lead and uranium to calculate how old a rock was.

With this method, geologists were soon estimating the age of various rocks not in millions of years but in billions. They later learned how to make Rutherford's clock even more accurate. Instead of taking a single measurement of the lead and uranium in a rock, they began to measure their levels in many different parts of it. That allowed them to compare the parts that originally contained very little uranium to others that initially had high levels. If the uranium throughout the rock decayed at a uniform rate, the different samples should all point to the same age. And in many cases they do.

Geologists also learned how to measure time with two clocks at once. In addition to uranium 238, some rocks also contain uranium 235, which decays into a different isotope of lead, lead 207. It also has a different half-life, of only 704 million years. With two independent tests for the age of a rock, geologists can often narrow down the margin of error even more.

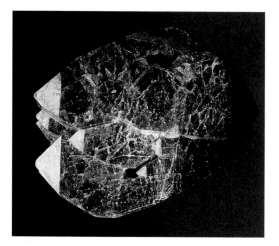

Zircon crystals trap uranium and lead, making them highly accurate geological timekeepers.

They can also eliminate the uncertainty over whether any uranium or lead has crept into a rock after it was formed. As certain types of rocks form, atoms of zirconium and oxygen combine into crystals known as zircons. Zircons act like microscopic prisons: any uranium or lead atoms trapped inside a zircon have a very difficult time escaping, and few new atoms can enter it. Within its zircon cage, the uranium slowly breaks down into lead without any interference from the outside world. The geophysicists who put a date of 4.04 billion years on the rocks of Acasta did so by dating their zircons. They fired a beam of charged particles at the crystals, blasting out tiny clouds of isotopes that they then measured. Thanks to all the different cross-checks they performed, they were able to estimate its age within a margin of error of only 12 million years. Twelve million years may be a vast gulf of time for us, but for the Acasta rocks, it represents a margin of error of less than 0.3 percent.

The rocks at Acasta are the oldest known rocks on the planet, but they formed when Earth was already 500 million years old. Geologists needed a

gift from space to find out the true age of the planet. In the 1940s they began studying the isotopes of lead in meteorites. Most meteorites are jumbles of space junk left behind from the formation of the solar system. In 1953 Claire Patterson, a geologist at the California Institute of Technology, measured the lead and uranium in the meteorite that had carved out the 1.2-kilometer-wide Meteor Crater in Arizona. It had practically no uranium left in it, because most of the atoms had turned into lead. This meteorite had formed at the dawn of the solar system and had circled the sun essentially unchanged ever since.

Meteorites and our planet all formed from the same primordial stuff, but each one ended up with different proportions of elements, including uranium and lead. By comparing the amount of uranium and lead isotopes in rocks from Earth and meteorites such as the one from Meteor Crater, Patterson determined Earth's age. It is 4.55 billion years old.

Why is there a 500-million-year gap between the oldest rocks on Earth and its birth? Thanks in part to their ability to date rocks, geologists have discovered that Earth destroys its crust and creates new rock to take its place. The planet's crust is actually a collection of drifting plates. Magma emerges from the depths of the earth and adds a fresh margin of rock to one side of a plate, and the other side becomes buried underneath its neighboring plate. As the

Meteor Crater in Arizona was created 50,000 years ago by the impact of a meteorite—a relic of the early solar system. By analyzing the meteorite, scientists determined that the Earth is 4.55 billion years old.

sinking edge of the plate plunges into the planet, it warms up until it partially melts. Any fossils it might carry are destroyed with it.

Continents are floating islands of low-density rock that sit on top of the moving plates. When a plate slides under its neighbor, a continent does not get sucked down with it. If a rock is lucky enough to be nestled within a continent, it may be spared Earth's fiery cycle—along with fossils and other clues to life's history it may hold. The rocks of Acasta are geological freaks.

MANY CLOCKS, ONE STORY

Uranium cannot tell the age of all rocks. Zircons, those exquisite clocks, form only in certain kinds of cooling lava. In sedimentary rocks, the uranium-lead system of telling time is almost useless. Another problem lies in the millions of years that uranium requires to turn into measurable levels of lead. On the scale of thousands of years—the scale of human history—uranium cannot tell time. Fortunately, geochemists are not limited to uranium and lead in their choice of clocks. They can turn to dozens of other radioactive elements, depending on the research at hand. To put absolute dates on human history, for example, scientists can turn to an isotope of carbon, carbon 14, which has a half-life of only 5,700 years, making it a good clock for telling time over the past 40,000 years.

Carbon 14 is born when the charged particles that are continually raining down from space slam into nitrogen atoms in the atmosphere. The transformation is only temporary; a carbon 14 atom eventually decays back into a nitrogen atom, shedding a handful of subatomic particles along the way. As long as plants are alive and absorbing fresh carbon dioxide carrying freshly made carbon 14 from the air, they maintain a steady level of the isotope in their tissues; so do the animals that eat them. But as soon as something dies, it can no longer take in any more carbon 14, and its supply starts to dwindle as the isotopes decay into nitrogen. By measuring the carbon 14 still remaining inside the dead tissue of a plant or an animal, you can calculate its age.

Isotopic clocks have allowed paleontologists to organize the history of life against an absolute calendar. Not only did Darwin not know how old Earth was, he didn't know how old any fossils were. The best he and other scientists of his day could say was that a given fossil came from a certain geological period. The oldest period in which fossils had been found was called the Cambrian period, and all rocks that came from older layers were simply labeled Precambrian. For Darwin, the way those Cambrian fossils appeared without any predecessors posed a puzzle as deep as Kelvin's warm Earth.

"If the theory be true," he wrote of evolution by natural selection, "it is indisputable that before the lowest Cambrian stratum was deposited, long periods elapsed . . . and that during these vast periods, the world swarmed with living creatures. . . . To the question why we do not find rich fossiliferous deposits belonging to these assumed earliest periods before the Cambrian system, I can give no satisfactory answer. The case at present must remain inexplicable; and may be truly urged as a valid argument against the view here entertained."

Paleontologists now know that the Precambrian actually did swarm with living creatures, and it was swarming more than 3.85 billion years ago. The earliest evidence of life comes from the southwestern coast of Greenland. There are no fossils to be found there, at least not in the conventional sense. An organism can leave behind a visible part of its body—a skull, a shell, the impression of a flower petal—but it also leaves behind a special chemistry, and scientists now have the means of detecting it.

The ratio of carbon 13 to carbon 12 is lower in organic carbon, such as wood or hair, compared to inorganic carbon that escapes out of a volcano as carbon dioxide. This makes it possible to tell if the carbon in a rock had ever been inside a living thing. Consider, for instance, a leaf growing on an elm tree. It builds up a low ratio of C-13 to C-12. A caterpillar that nibbles that leaf will incorporate the carbon in its prey into its own tissue, and it will take on a low C-13 ratio as well, as will the bird that eats the caterpillar. Birds, caterpillars, and leaves all die sooner or later, and when they do, they all become part of soil, which eventually washes out to the ocean and becomes sedimentary rock. And even those rocks, made partly from carbon that has cycled through life's metabolism, will bear life's low C-13 ratio. Any sedimentary rocks that formed before life appeared on Earth would have the high C-13 ratio of a volcanic origin.

In 1996 a team of American and Australian scientists traveled to the twisted fjords and bare islands of southwestern Greenland, where the oldest sedimentary rocks on Earth can be found. A layer of volcanic rock cuts through them, and the scientists used the uranium-lead clock inside its zircons to date it to 3.85 billion years. They then sifted through the surrounding rock. Over its lifetime, it has been cooked, compressed, and otherwise ravaged almost beyond recognition. But the researchers found microscopic bits of carbon in a mineral known as apatite in the sedimentary rocks. They brought these samples back to their labs and blasted off bits of the apatite with a beam of ions and counted up the carbon isotopes it contained. They found that carbon in the apatite had the same low C-13 ratio as biological carbon today—a ratio that could only have come from life.

Scientists cannot say just how long life existed on Earth before it left this mark in the rocks of Greenland, because no rocks older than 4 billion years have survived. But it's safe to say that life must have had a hellish birth. Giant asteroids and miniature planets pummeled Earth for its first 600 million years. Some of them were big enough to boil off the top few meters of the oceans and kill any life it held. Perhaps life survived these cataclysms hidden around the thermal springs at the ocean floor, where bacteria can be found today. When rains filled the seas again, the microbes were able to emerge from their refuges.

However life got started, it had to have been in full swing by the time it left its mark in the Greenland rocks. At the time, the oceans were teeming with bacteria generating their own food as they do today, either from sunlight or from the energy contained in the chemistry of hot springs. These self-sustaining microbes were probably food for predatory bacteria, as well as hosts for viruses.

The oldest actual fossils of bacteria date back 3.5 billion years, about 350 million years after the earliest chemical signs of life. These fossils, discovered in the 1970s in western Australia, consist of delicate chains of microbes that look exactly like living blue-green algae (otherwise known as cyanobacteria). For billions of years, these bacteria formed vast slimy carpets in shallow coastal waters; by 2.6 billion years ago they had also formed a thin crust on land.

Of course, life is not limited to bacteria. We humans belong to an enormous group of organisms called eukaryotes, which include animals, plants, fungi, and protozoans. The evidence for the oldest eukaryotes doesn't come from traditional fossils, which date back only about 1.2 billion years. It comes, once again, from molecular fossils. Among the many things that distinguish eukaryotes from bacteria and other life-forms is how their cell membranes are constructed. Eukaryotes stiffen them with a family of fatty acids known as sterols. (Cholesterol belongs to the sterol family; while it may be dangerous when too much of it gets into the bloodstream, you couldn't live without it. Your cells would simply disintegrate.)

In the mid-1990s, a group of geologists led by Jochen Brocks of Australian National University drilled 700 meters down into the ancient shales of northwest Australia to formations that have been dated with uranium and lead to 2.7 billion years ago. Inside the shale, the geologists found microscopic traces of oil that contained sterols. Because eukaryotes are the only organisms on Earth that can make these molecules, Brocks's team concluded that eukaryotes—probably simple, amoeba-like creatures—must have evolved by 2.7 billion years ago.

WHEN LIFE BECAME LARGE

For another billion years or so, eukaryotes remained, like bacteria, microscopic. But around 1.8 billion years ago the first multicellular fossils appear, in the form of mysterious coils measuring about 2 centimeters long. The oldest recognizable multicellular organisms are red algae dating back 1.2 billion years.

The earliest known communities of animals lived 575 million years ago. Known as Ediacarans, they were a strange collection of mostly immobile creatures. Some of them may have been relatives of the major groups of animals alive today; others are extinct mysteries.

Our own multicellular lineage, the animals, don't leave fossils until about 575 million years ago. "Animal" is a generous description for the creatures that left these marks. There are disks with three-pronged ridges on them, like coins from a vanished empire. There are fronds with rows of slits that must have looked like underwater Venetian blinds. There are ribbed impressions on the old seafloors shaped like gigantic thumbprints.

These creatures are known as the Ediacaran fauna (named after the Edi-acara Hills in Australia where they have been found in abundance). In the past paleontologists have tried to classify them as plants, lichens, or even some failed experiment in multicellular life. These days, many experts think that at least some of them are early relatives of the major groups of living animals (known as phyla). Some of the fossils might be related to jellyfish. The giant thumb-print may actually be kin to the annelids, a group that includes earthworms and leeches. The fronds may be sea pens, which live today on coral reefs. But there are still many Ediacarans that remain up for grabs. No one has taken on the coins yet.

Tucked away among these fossils of Ediacarans are signs of the future of the animal kingdom. In rocks as old as 550 million years there are traces of burrows and tracks that could only have been made by muscular animals. They point to the existence of complex animals that could dig and crawl, as opposed to the rooted Ediacarans and the drifting jellyfish. These phantom animals may have already evolved many of the hallmarks of complex animals, such as a muscle wall and a gut. Primitive animals such as jellyfish lack these structures, while animals such as insects, flatworms, starfish, and people have them. The difference comes down to how their embryos form. Jellyfish are composed of two body layers (biologists call them diploblasts). Other animals have three layers: an ectoderm, which ultimately produces skin and nerves;

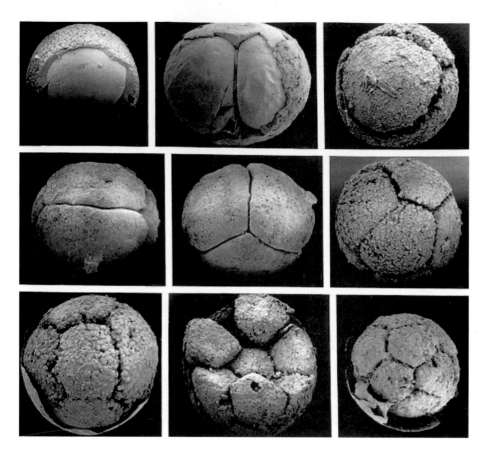

In 1998 paleontologists discovered microscopic fossils of embryos dating back 570 million years. The arrangements of their cells are identical to those of living animals.

mesoderm, which forms muscle, bone, and many internal organs; and endoderm, which builds the gut. These three-layered animals are triploblasts. Triploblasts presumably made the burrows and tracks 550 million years ago, but paleontologists have yet to find their fossils.

In 1998 paleontologists in search of these early triploblasts made a promising discovery: Precambrian embryos. A team of American and Chinese researchers found a cache of microscopic fossils dating back 570 million years. Some are single-celled fertilized eggs. Some are at the next stage of division, a two-cell ball. Some have four cells, eight cells, sixteen, and so on. The paleontologists have no idea what exactly these embryos would have grown up to be, but judging from their size and the pattern of their division, the best candidates would be triploblasts.

By 530 million years ago, in the early Cambrian period, the Ediacarans had declined and disappeared. At the same time the fossil record of triploblasts exploded. Among the fossils of this time you can find the earliest clear-cut relatives of many of the major groups of living animals. Our own phylum, the chordates, is represented by fossils of creatures that looked like lampreys and hagfish—Owen's vertebrate archetype made flesh.

Other phyla made a more extravagant entrance. A relative of today's mollusks looked like a pincushion studded with arrowheads. Living lamp shells were foreshadowed by *Halkieria,* which looked like an armored slug. *Opabinia* had five eyes sprouting like mushrooms from its head, and it stirred up the seafloor with a clawed nozzle that it could also use to grab prey and stuff into its mouth. *Opabinia* now appears to be an early relative of living arthropods. Other phyla that today live in humble obscurity—velvet worms or peanut worms, for instance—gloried during the Cambrian explosion in a diversity that they would never again enjoy.

Darwin's worries over the Cambrian turn out to be unfounded. Now that scientists can read isotopic clocks and recognize molecular fossils, they have shown that the world did indeed swarm with life for billions of years before the Cambrian, as Darwin proposed. The Precambrian, far from some mysterious prologue to evolution, actually takes up 85 percent of the history of life. And paleontologists now have a marvelous collection of Precambrian fossils, including bacteria, protozoa, algae, Ediacarans, burrow makers, and animal embryos. But even with a much smoother fossil record, the Cambrian period is clearly the most remarkable episode in animal evolution. No matter how long animals were already lurking in the oceans, their diversification accelerated 535 million years ago in a tremendous explosion. Thanks to precise uranium-lead dating, scientists have determined that the Cambrian explosion took only 10 million years.

The Cambrian explosion took place completely underwater. As these new animals came into existence, the continents were bare, except for bacterial crusts. But it was not long, geologically speaking, before multicellular life spread onshore. First came plants. Around 500 million years ago green algae gradually evolved waterproof coats that allowed them to survive exposed to air for longer and longer periods of time. The first land plants probably looked like today's moss and liverworts, forming a low, soggy carpet along the banks of rivers and coastlines. By 450 million years ago centipedes and other invertebrates were beginning to explore this new ecosystem. New species of plants evolved that could hold themselves upright, and by 360 million years ago trees were growing 60 feet high. Out of the coastal swamps would sometimes slither our ancestors—the first vertebrates that could walk on land.

By 535 million years ago, during the Cambrian period, a huge array of new animals began to appear. Most of the major groups of living animals emerged during this time.

Plants first colonized dry land 500 million years ago; by 360 million years ago, the first forests had evolved. They in turn supported a diversity of land animals, including insects and vertebrates.

In the history of life, the move to land is a brief coda. Nine-tenths of our evolution took place completely underwater. But from our point of view, the last few hundred million years on land are the most interesting period of all. Fossils of the earliest land vertebrates show that they had branched into two lineages by 320 million years ago. One branch, the amphibians, produced some lumbering giants early on, but today is represented only by frogs, salamanders, and other small creatures. They generally need to stay moist and lay soft eggs that can dry out easily. The other branch, the amniotes, evolved a strong water-tight eggshell. From among

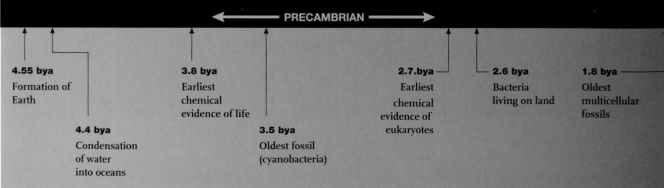

BILLIONS / MILLIONS / THOUSANDS OF YEARS AGO (bya, mya, kya)

← PRECAMBRIAN →

4.55 bya
Formation of Earth

4.4 bya
Condensation of water into oceans

3.8 bya
Earliest chemical evidence of life

3.5 bya
Oldest fossil (cyanobacteria)

2.7.bya
Earliest chemical evidence of eukaryotes

2.6 bya
Bacteria living on land

1.8 bya
Oldest multicellular fossils

these amniotes the dinosaurs emerged around 250 million years ago; they evolved into the dominant land animals and stayed that way until 65 million years ago, when most dinosaur branches became extinct (the only survivors were birds, which are just feathered flying dinosaurs). Although the first mammals appeared alongside the first dinosaurs, they didn't dominate the land until their reptile counterparts disappeared. Our own primate lineage probably emerged around then, but it wasn't until 600,000 years ago that the oldest fossils of *Homo sapiens* were buried in the earth. All humans alive today can trace their heritage back to a common ancestor that lived only around 150,000 years ago.

While these few pages can't do full justice to the majestic depths of life's history, one thing is clear: our own time in this universe is almost inconceivably brief. No longer can human history match the scale of natural history. If the 4 billion years that life has been on Earth were a summer day, the past 200,000 years—which saw the rise of anatomically modern humans, the origin of complex language, of art, religion, and trade, the dawn of agriculture, of cities, and all of written history—would fit into the flash of a firefly just before sundown.

In the end, Darwin got the luxury of time he craved. But the fossil record, though it documents the pattern of life's evolution, does not reveal the details of just how it evolved. And on this point, Darwin also never managed to close his case during his lifetime, because he never understood how heredity works.

While geologists and paleontologists were charting the history of life, other scientists in the twentieth century solved heredity's mystery and linked it to natural selection. The connection lies among molecules and atoms, as did the proof of life's antiquity. But the atoms of heredity are not embedded in rocks for billions of years. They are lodged in the core of our own cells.

A timeline of life's 4-billion-year history.

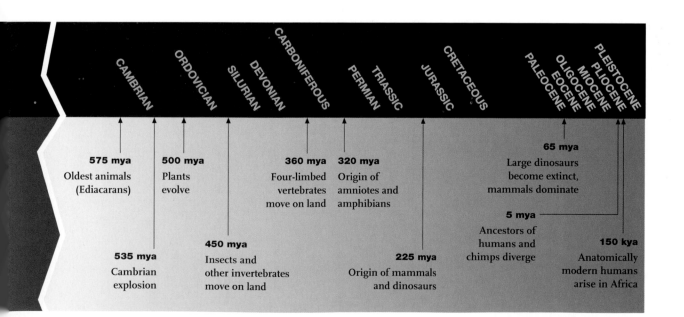

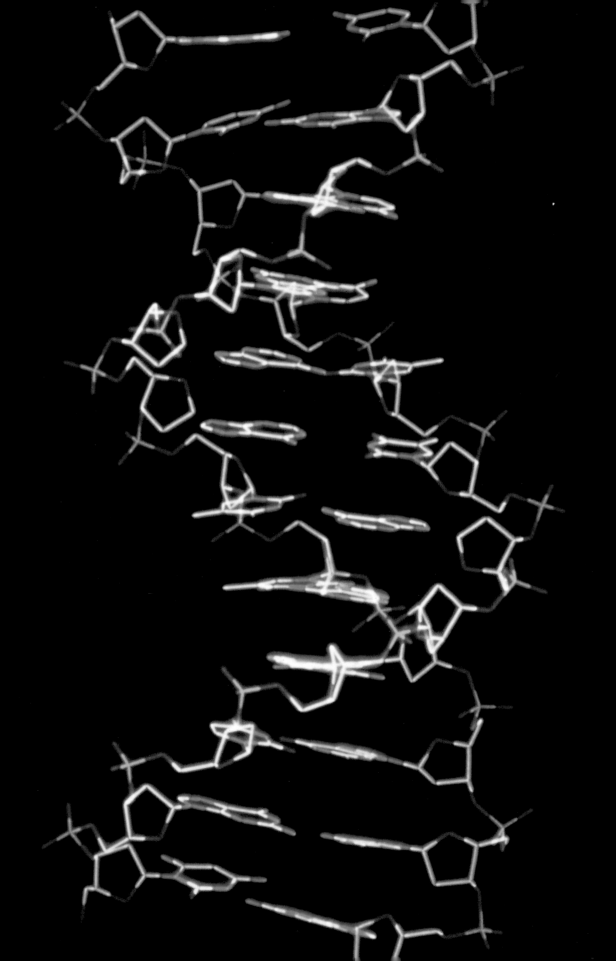

Witnessing Change

GENES, NATURAL
SELECTION, AND
EVOLUTION IN ACTION

THE PUZZLE OF heredity—how two people can create a child with qualities of both parents—inspired a lot of wild ideas in the 1800s. One of the wildest, at least from our perspective today, was known as pangenesis. It held that heredity is carried by tiny particles that bud from cells throughout a person's body. These particles (called gemmules) supposedly stream like trillions of migrating salmon to the sex organs, where they concentrate inside sperm or eggs. And when a sperm fertilizes an egg, the gemmules of both parents blend together. Since each particle comes from a cell from a particular part of a parent's body, they combine together into a new person with traits of both parents.

Pangenesis turned out to be a failure, but the scientist who proposed it was not relegated to history's ranks of scientific crackpots. His reputation survived, thanks to a few other ideas

When the structure of DNA
changes, life evolves.

that have withstood the test of time. Pangenesis was the work of Charles Darwin.

Along with Earth's age, heredity was one of Darwin's great frustrations. *Origin of Species* persuaded most scientists by the end of the nineteenth century that evolution was a reality, but many of them were skeptical of Darwin's own mechanism of change, namely, natural selection. Many of them resurrected Lamarck's old ideas instead. Perhaps there was a built-in direction to evolution, they claimed, or perhaps there were ways that adults could acquire traits in their lifetimes that they could pass on to their children. If Darwin could have shown that heredity forbids these ideas but allows for natural selection, he could have refuted his critics. But that knowledge was beyond Darwin and all the other scientists of his day.

In the years after Darwin's death, biologists finally began to learn how heredity works. Only then could they recognize that the neo-Lamarckians were wrong. Only then could they recognize how heredity makes natural selection not just possible but inevitable and how it allows new species to form. It took the work not just of geneticists to make this discovery, but of zoologists and paleontologists as well. By the middle of this century, they had combined their research into a collective understanding of evolution that became known as the "modern synthesis." Younger scientists have used the modern synthesis as the foundation for their research. They're beginning to understand how evolution happens on a molecular level, and as a result, natural selection is no longer the elusive, imperceptible force Darwin imagined it to be. In fact, scientists can witness natural selection happening in the wild today, as well as the branching of old species into new ones. Scientists don't even have to watch animals or plants or microbes to see natural selection play out: they can watch it take place within our bodies, or even among artificial lifeforms within a computer.

HEREDITY'S MONK

If history had played out differently, scientists might have cracked the secrets of heredity in Darwin's own lifetime. Even as he was writing *Origin of Species,* a Moravian monk was already discovering the fundamental laws of genetics in his garden.

Gregor Mendel was born to a poor peasant farmer in 1822 in what is now the Czech Republic and grew up in a two-room house. When his teachers recognized Mendel's quick intelligence, they arranged for him to become a novitiate at the monastery in Brno, then in Moravia. The monastery was filled with monks as dedicated to science as to prayer, who read deeply in geology, meteorology, and physics. From the monks, Mendel learned about the latest advances in botany, such as new techniques for artificially fertilizing plants in order to breed better and better strains. Eventually they sent Mendel to the University of Vienna, where he continued to study biology. But it was the physics and mathematics that he learned there that shaped him as a scientist. The physicists in Vienna showed Mendel how to test a hypothesis with experiments—something that few biologists at the time were doing. The mathematicians meanwhile taught Mendel how to use statistics to find order hidden in seemingly random collections of data.

In 1853 Mendel came back to Brno. He was by now a man in his thirties, broad-shouldered and a little corpulent, with a high forehead and twinkling blue eyes behind gold-rimmed glasses. He worked as a schoolteacher, teaching natural history and physics to second and third graders. Although he had 100 students and six days of classes a week, he still managed to live the life of a scientist, taking regular readings of the weather and keeping up with scientific journals. And during that time he decided to set up an experiment to learn about the heredity of plants.

In Vienna, some of Mendel's professors struggled to understand what made species distinct, about how one generation produced a new generation that looked like itself. These questions fused together in the mystery of hybrids. Breeders knew how to develop distinct varieties of flowers, fruits, and other plants, and they could cross varieties to produce hybrids. Many were sterile, and among the hybrids that could produce offspring, the new generations often reverted back to the forms of their ancestors. But if plants could somehow form stable hybrids, it was possible that species were not eternal or unchanging. In the 1700s, the Swedish biologist Carl Linnaeus had speculated that species of plants belonging to the same genus had developed from a common ancestor by hybridization.

For most of the nineteenth century, scientists generally thought that heredity worked by blending the qualities of parents in their offspring. But Mendel came up with a radically different idea: parents could pass down traits to an offspring, but their traits did not blend. To test his idea, Mendel planned out an experiment to cross varieties of plants and keep track of the color, size, and shape of the new generations they produced. He chose peas, and for two years he collected varieties and tested them to see if they would breed true. Mendel settled on 22 different varieties and chose seven different traits to track. His peas were either round or wrinkled; yellow or green. Their pods were yellow or green as well, and they were also smooth or ridged. The plants themselves might be tall or short, and their flowers might blossom at their tips or along their stems. Mendel would record the appearance of the traits in each generation.

The Moravian monk Gregor Mendel (1822–1884) uncovered clues during the 1850s as to how heredity works. But his discoveries were neglected until 1900, 16 years after his death.

Delicately placing the pollen of one plant on another, Mendel created thousands of round-wrinkled pea hybrid seeds. He then waited for the plants to bloom in the monastery garden. When he shucked the pods a few months later, he saw that the hybrid peas were all round. The wrinkled trait had utterly disappeared from sight. Mendel then bred these round hybrids together and grew a second generation. Some of their offspring were wrinkled (and just as deeply wrinkled as their wrinkled grandparents). The wrinkled trait had not been destroyed in the round generation; it had gone into hiding in the hybrids and then reappeared.

The number of peas that ended up wrinkled would vary on each plant, but as Mendel counted up more and more of them, he ended up with a ratio of one wrinkled seed for every three round ones. He crossed varieties to follow the fate of other traits, and the same pattern emerged: one green plant for every three yellow ones, one white seed coat for every three gray ones, one white flower for every three violet ones.

Mendel realized that he had found an underlying regularity to the confusion of heredity, but contemporary botanists pretty much ignored his work. He died at his monastery in 1884 with a reputation as little more than a charming putterer. But he was actually a pioneer in genetics, a field that didn't even come into existence until 16 years after his death. After a hundred years of research, it is now clear why Mendel's peas grew the way they did.

Peas, like every other organism on Earth, carry in each of their cells a molecular cookbook for creating their bodies. This information-bearing molecule is called deoxyribonucleic acid, better known as DNA. It is shaped like a twisted ladder; its information is inscribed on its rungs, made from a pair of chemical compounds known as bases. Bases serve as the letters in life's recipes, but unlike the 26 letters of the English alphabet, DNA is written with only 4—adenine, cysteine, guanine, and thymine.

A gene—a stretch of DNA usually spanning a few thousand base pairs—is a recipe for a protein. To create a protein, our cells construct a single-stranded version of a gene (called RNA), which gets shuttled to a protein-building compartment. Proteins are long molecular chains, as are DNA and RNA, but they are not made out of bases. Instead, proteins are formed from another group of compounds called amino acids. Our cells use the information encoded in the bases of RNA to grab the appropriate amino acids and assemble them into a chain; once a piece of RNA has been read to its end, the new protein is complete. The attraction of the atoms in the new protein to one another make it collapse, like spontaneously folding origami. Because they take on thousands of different structures, proteins can play thousands of roles, from serving as pores in cell membranes to stiffening fingernails to carrying oxygen from the lungs through the bloodstream.

It is the way the DNA cookbook is passed down from one generation to the next that creates the 3-to-1 ratio Mendel discovered in his peas. In plants

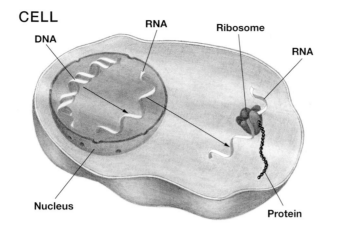

CELL

DNA

RNA

Ribosome

RNA

Nucleus

Protein

Cells build proteins in a three-step process. First, the cell copies its double-stranded DNA into a single-strand version called RNA. Then it ships the RNA to a protein-building factory called a ribosome. There, in the final step, the RNA acts as a template for the ribosome as it builds a string of amino acids, creating a protein.

and animals, the recipes of the genes are organized into volumes, each of which is called a chromosome. We humans, for example, have 30,000 genes arranged on 23 pairs of chromosomes. A pair of chromosomes may carry identical versions of a given gene, or they may bear different ones. When a normal cell divides in two, both new cells get a complete collection of genes. But when cells give rise to sperm or egg, each sex cell only receives a single chromosome from each pair. Which of the two it receives is a matter of chance, and when a sperm fertilizes an egg, the two sets of chromosomes unite into a new pair, creating the genetic code for a new organism.

The colors of Mendel's pea plants, their textures, and all the other traits he recorded were controlled by different versions of pea genes. One of the genes that his peas inherited comes in two different versions, one making them smooth, the other wrinkled. A purebred smooth pea carries two copies of the smooth gene; two copies of the wrinkle-causing gene can be found in purebred wrinkled peas. When Mendel crossed these breeds of peas together, he got hybrids, each carrying one smooth gene and one wrinkled gene, but bearing a smooth coat. For reasons that geneticists still don't fully understand, genes like the one that makes peas smooth can dominate their partners.

But while the wrinkled gene was silenced in the hybrids, it did not disappear. Each of the hybrid's eggs and pollen grains received only one form of the gene, so each offspring had a 50-50 chance of inheriting a particular gene from each of its parents. Thanks to these odds, a quarter of the new peas received two wrinkled genes, a quarter received two smooth ones, and half received one of each. Because the new hybrids were also smooth, in the second generation the wrinkled peas were outnumbered 3 to 1.

The way most traits are inherited is a lot more complicated than what Mendel recorded when he grew his peas. Very often, a species carries many more than two different versions of a gene. And it is rare that a single gene is alone responsible for a trait. In most cases, many different genes are involved.

The human race isn't divided into those who carry a "tall gene" that makes them 6 feet tall and those with a "short gene" that makes them only 5 feet. Many genes help determine a person's height, so changing just one of them will make only a slight difference. If our DNA is a cookbook, our bodies are a smorgasbord. Using salt instead of yeast in the bread will make a big difference to the meal, but confusing thyme and oregano in the chili probably won't raise an eyebrow.

REWRITING LIFE'S COOKBOOK

The variations that Darwin saw among his pigeons and his barnacles—and which he was at a loss to explain—come into existence as DNA's sequence changes. Cells can duplicate their DNA almost flawlessly, but every now and then a mistake creeps in. Proofreading proteins can find most of these mistakes and fix them, but a few slip through. Some of these rare changes—known as mutations—may alter a single letter in the recipe of DNA, but others can be far more drastic. Many pieces of DNA can spontaneously cut themselves out of one position and splice themselves back in elsewhere, altering the gene where they make their new home. Sometimes when DNA is being copied as a cell divides, an entire gene can get duplicated, or even an entire set of genes.

As early as the 1920s, scientists began to realize that mutations had huge ramifications for evolution. These researchers—foremost among them the British mathematician Ronald Fisher and the American biologist Sewall Wright—synthesized natural selection and genetics, putting Darwin's theory on a far more solid foundation.

When DNA mutates, a cell may simply malfunction and die, or it may multiply madly into a tumor. In either case, the mutation disappears when the organism that carries it dies. But if the mutation happens to alter the DNA in an egg or a sperm, it gets a chance at immortality. It may get carried into the genes of an offspring, and that offspring's offspring. The effects that a mutation has—favorable, unfavorable, or neutral—will influence how common it becomes as generations pass. Many mutations have harmful effects, often killing their owner before it is born or interfering with its ability to reproduce. If a mutation cuts down reproductive success, it will gradually disappear.

But sometimes instead of harm, a mutation does some good. It may change the structure of proteins, making them more efficient at digesting food or breaking down poisons. If a mutation's effects allow an organism to have more offspring, on average, than the organisms that lack it, it will gradually become more common in a population. (Biologists would say that this mutant has a higher fitness than the others.) As the mutant's offspring thrive, the mutation they carry becomes more common, and it may do so well that it drives the older version of the gene to extinction. Natural selection, Fisher and Wright showed, was largely a matter of the changing fortunes of different forms of genes.

Fisher made a particularly important breakthrough when he demonstrated that natural selection progresses by the accumulation of many small mutations rather than by a few giant ones. Fisher used some esoteric math to prove this point, but a simple hypothetical example can make it clear. Consider a dragonfly's wings. If a dragonfly has particularly short wings, it may not be able to generate enough lift to stay off the ground, but if it has wings that are too long, they may be too heavy to flap. Somewhere between short wings and long ones is the length that brings the greatest fitness. If you chart out this relationship between length and fitness on a graph, you draw a hill, with its peak at the optimal length of the wings. If you were to actually measure the wings of dragonflies and plot them on this graph, they might cluster as points near the hill's peak.

Now imagine that a mutation cropped up that changed the length of the dragonfly's wings. If the insect's fitness gets lowered as a result, insects with a better wing design may outcompete its offspring. But if the mutation pushes the dragonfly closer to the peak of fitness, natural selection will favor it. In other words, natural selection tends to push life up the hillside of fitness.

On such a landscape a giant mutational leap might seem like the best strategy for evolving quickly. Instead of natural selection's gradual creep, it could catapult a dragonfly's fitness to the top of its hill. But mutations are catapults with no sense of aim. They occur randomly, hurling the dragonflies in random directions across the evolutionary landscape. Rather than landing squarely on the hilltop, they might end up somewhere far away, their wings much too long or much too short. Mutations with small effects, on the other hand, can nudge the dragonflies uphill much more reliably. Even a slight advantage, translating into just a few extra offspring, may allow a mutation to spread through a population after a few dozen generations.

Of course this uphill journey is only a metaphor, and a simplistic one at that. For one thing, the terrain of evolution is not fixed. As the environment changes—as temperatures rise and fall, as competing species invade or retreat, as other genes evolve—hills can become valleys and valleys hills. The landscape is more like the surface of a slowly surging ocean.

Nor does evolution always produce the best possible combination of genes. Genes can, for example, sometimes spread without any help from natural selection whatsoever. Heredity is like a ball on a roulette wheel. If the ball is thrown on the wheel often enough, it will land half the time on red and half the time on black. But if you play only a few spins, your ball might end up landing on red every time. The same thing happens with genes. Say two smooth hybrid peas produce four new plants. Each new plant has a 25 percent chance of inheriting two smooth genes, a 25 percent chance of inheriting two wrinkled ones, and a 50 percent chance of being hybrids. But that doesn't mean they will turn out to be one smooth pea plant, one purebred wrinkled one, and two hybrids. They might all end up smooth, or even all wrinkled. Every pea plant is a genetic roll of the dice.

These statistical flukes don't happen in big populations, but small ones can defy Mendel's probabilities. If a few dozen frogs living on a mountaintop breed only among themselves, a mutant gene may emerge among them and spread without any help of natural selection—thanks only to an odd spin of the evolutionary roulette wheel. And once the mutant gene has spread through the entire population, the gene it replaced is gone for good.

THE MODERN SYNTHESIS

Fisher, Wright, and the other scientists who were the first to show how genetics fuel evolution were not field biologists. They were mainly lab-bound experimentalists and mathematically inclined theoreticians. But by the 1930s, other researchers began to apply their ideas to the real world: to the patterns of diversity found among living species and to the evidence in the fossil record. Just as Fisher and Wright had melded genetics and evolutionary theory, this next generation of scientists added new ingredients from ecology, zoology, and paleontology. By the 1940s, non-Darwinian explanations for evolution— ideas about inner forces directing some sort of Lamarckian transformation, or giant mutations creating new species in a single generation—had become hopelessly old-fashioned.

One major step toward the modern synthesis came in 1937 with the publication of a book called *Genetics and the Origin of Species,* by a Soviet scientist named Theodosius Dobzhansky. Dobzhansky had come to the United States nine years earlier to work in the laboratory of Thomas Hunt Morgan at Columbia University, where biologists studying the fruit fly *Drosophila melanogaster* were discovering the true nature of mutations. Dobzhansky was an oddball in the lab; to the other members of the Fly Room, fruit flies lived only in milk bottles in cluttered laboratories. But Dobzhansky had studied insects in the wild since he was a child in Kiev. When he was a teenager, his goal in life was to collect every species of ladybird beetle in the region. "Seeing a lady beetle still produces in me a flow of a love hormone," Dobzhansky would say many years later. "The first love is not easily forgotten."

Dobzhansky developed a sharp eye for the natural variations among different populations of ladybird beetles, and when he read about Morgan's work on mutations, he wondered if it might reveal their underpinnings. But the genetics of the ladybird beetle were too complicated for Dobzhansky to work out, so he switched his research to *Drosophila melanogaster,* Morgan's well-studied fly.

Dobzhansky quickly earned a reputation as a brilliant student of genetics, and at age 27 he was invited to come to New York to learn the newest methods of the Fly Room. When Dobzhansky and his wife arrived at Columbia, they found the Morgan lab a dreary, roach-infested place. But in 1932 things improved: Morgan packed up shop and moved out to the California Institute of Technology. Dobzhansky followed him and settled happily among the orange groves.

In California, Dobzhansky finally began to answer the questions he had asked as a teenager: What were the genetics that determined the differences between populations of a species? Most biologists at the time assumed that in any given species, the animals all had practically identical genes. After all, it had taken Morgan years to find a naturally occurring mutation arising in his fruit flies. But these were assumptions bred in the lab.

Dobzhansky began studying the genes of wild fruit flies, traveling from Canada to Mexico to catch members of the species *Drosophila pseudoobscura.* Today biologists can sequence every letter in a species' entire genetic code, but in Dobzhansky's day, the technology was far cruder. He could only judge differences between chromosomes by looking at them through a microscope. Even with such simple methods, he found that different populations of *D. pseudoobscura* did not have identical sets of genes. Each population of fruit flies he studied bore distinctive markers in their chromosomes that set them off from other populations.

Decades later, when geneticists had invented more precise ways to compare DNA, they would find that the variability Dobzhansky had found among his fruit flies was the rule, not an exception. Among humans, for example, many biologists once thought that races carried dramatically distinct sets of genes. Some even went so far as to claim that they were separate species. But research on human genetics now shows that these old ideas were wrong. "The biological notion that we used to have of races is not compatible with the reality of the genetics that we're finding today," says Marcus Feldman, a geneticist at Stanford University.

Out of 30,000 or so genes in the human genome, an estimated 6,000 genes exist as different versions (known as alleles). The distinctions that we conventionally use to divide the species into races—skin color, hair, and the shape of faces—are controlled by only a few genes. The vast majority of variable genes do not respect so-called racial boundaries. There is far more variability within any given population of humans than between populations. If all the humans on Earth were wiped out except a single tribe in a remote New

Guinea valley, the survivors would still preserve 85 percent of the genetic variability of our entire species.

Dobzhansky's discovery of so much genetic variability in a single species raised a deep question: If there was no standard set of genes that distinguished a species, what kept species distinct from one another? The answer, Dobzhansky correctly realized, was sex. A species is simply a group of animals or plants whose members reproduce primarily among themselves. Two animals belonging to different species are unlikely to mate, and even if they do, they will rarely produce viable hybrids. Biologists already knew that hybrids often died before they hatched or they grew into sterile adults. Dobzhansky ran experiments on fruit flies that demonstrated that this incompatibility is caused by specific genes carried by one species that clash with the genes from another species.

In *Genetics and the Origin of Species* Dobzhansky sketched out an explanation for how species actually came into existence. Mutations crop up naturally all the time. Some mutations are harmful in certain circumstances, but a surprising number have no effect one way or the other. These neutral changes pop up in different populations and linger, creating a variability that is far greater than anyone had previously imagined. And all this variability can be a good thing, evolutionarily speaking, because if conditions change, a mutation's neutral effects can become useful and be favored by natural selection.

Variability is also the raw material for making new species. If a population of flies starts breeding only among themselves, their genetic profile will grow distinct from the rest of its species. New mutations would crop up in the isolated population, and natural selection might help them to spread until all the flies carried them. But because these isolated flies were breeding only within their own population, the mutations could not spread to the rest of the species. The isolated population of flies would become more and more genetically distinct. Some of their new genes would turn out to be incompatible with the genes of flies from outside their own population.

If this isolation lasted long enough, Dobzhansky argued, the flies might lose the ability to interbreed completely. They might simply lose the ability— or interest—to mate with the other flies. Even if they did produce offspring, the hybrids might become sterile. If the flies were now to come out of their isolation, they could live alongside the other insects but still continue mating only among themselves. A new species would be born.

Dobzhansky's 1937 book captivated biologists far beyond the confines of genetics. In the mountains of New Guinea, an ornithologist named Ernst Mayr found *Genetics and the Origin of Species* to be an enormous inspiration. Mayr specialized in discovering new species of birds and mapping out their ranges. The work was hard, and not just because of the malaria and the headhunters. Like other ornithologists, Mayr had a difficult time determining exactly when a group of birds deserved the title of species. A bird of paradise species might be recognizable by the color of its feathers, but from place to place it might have a huge amount of variation in other traits—on one mountain it might have an extravagantly long tail and on another its tail would be cut square.

Biologists typically tried to bring order to this chaos by recognizing subspecies—local populations of a species that were distinct enough to warrant a special label of their own. But Mayr saw that the subspecies label was far from a perfect solution. In some cases, subspecies weren't actually distinct, but graded into each other like colors in a rainbow. In other cases, what looked like a subspecies might turn out to be a separate species of its own.

When Mayr read *Genetics and the Origin of Species,* he realized that these puzzles of species and subspecies shouldn't be considered a headache: they were actually a living testimony to the evolutionary process Dobzhansky wrote about. Variations emerge in different parts of a species' range, creating differences between populations. In one part of a bird's range they may create a long tail, in others a tail that is cut square. But because the birds mate with their neighbors, they do not become isolated into a species of their own.

One of the starkest examples of what this flow of genes can do is a phenomenon known as the "ring species." In the North Sea, for instance, there

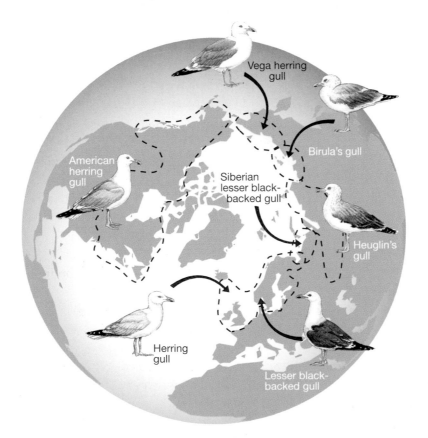

The range of gulls forms a ring around the North Pole. Within the ring, neighboring birds can mate with each other, even though they look slightly different. But the birds at the two ends of the ring—the Herring gull and the lesser black-backed gull—are so distinct that they can't mate with each other although they live side by side. Ring species are some of the best evidence for how new species evolve.

is a species of bird known as the herring gull. It has a gray mantle and pink legs. If you move west through its range, you come across more herring gulls in Canada, which look essentially the same as the ones in the North Sea, except for a few minor differences in their coloring. But by the time you reach Canada, the differences become stark, and in Siberia, the gulls have a dark gray mantle and legs that are less pink than yellow. Yet despite these differences, they are still scientifically classified as the herring gull (although their common name is the vega gull). Keep moving through Asia and into Europe, and the gulls continue to get darker and more yellow-legged. You will find dark, yellow-legged gulls extending even farther west, all the way to the North Sea where your trip began. Here these gulls, known as the lesser black-backed gulls, live alongside the gray-mantled, pink-legged herring gulls.

Because the two groups of birds look so different and do not mate, they are treated as two separate species. Yet lesser black-backed gulls and herring gulls live at two ends of a continuous ring, inside of which all the birds can mate with their immediate neighbors. A ring species is exactly what you'd expect given the way mutations arise and spread.

A population of birds can evolve into its own species if it gets cut off from its neighbors. Mayr argued that the easiest way to cut them off is by geographical isolation. A glacier may thrust across a valley, isolating the birds on the mountains on either side. A rising ocean may turn a peninsula into a chain of islands, stranding birds on each of them. This sort of isolation doesn't have to last forever; it need only form a barrier long enough to let the isolated population become genetically incompatible with the rest of its species. Once the glacier melts, or the ocean drops and turns the islands back into a peninsula, the birds will be unable to interbreed. They will live side by side but follow separate evolutionary fates.

Biologists such as Mayr and Dobzhansky helped assemble the modern synthesis by studying living animals. But if they were right, then the same processes should have been at work for billions of years, and the fossils should record their effects. Yet even in the 1930s, many paleontologists were not yet convinced that natural selection could account for what they saw among their bones. They saw long-term trends in the evolution of animals that seemed to follow a built-in direction. Horses seemed to evolve steadily from dog-sized creatures into bigger and bigger forms; at the same time, their toes steadily shrank away until their feet turned into hooves. The ancestors of elephants were originally the size of pigs, and their descendants evolved into colossal sizes over tens of millions of years; at the same time, their teeth seemed to steadily expand and become more complex. There was no sign, paleontologists claimed, of the open-ended, irregular experimentation that natural selection might produce.

Henry Fairfield Osborn, the president of the American Museum of Natural History, declared that these trends were proof that much of evolution was not governed by natural selection. Each of the mammal lineages began with the potential already within them to become horses or elephants—"something which in time *may* appear," as he put it. Only through a struggle against the elements and other animals could a species discover that potential. "Disprove Lamarck's principle and we must assume that there is some third factor in Evolution of which we are now ignorant," he declared in 1934.

But one of Osborn's students, the paleontologist George Gaylord Simpson, never thought much of this sort of reheated Lamarckism. Simpson was more impressed by Dobzhansky's ability to link genetics and natural selection. After Simpson read *Genetics and the Origin of Species,* he decided to see whether the fossil record could be accounted for in the same way.

Simpson took a closer look at the trends in the fossil record that Osborn claimed were evidence for directed evolution. Under his scrutiny, the linear trends broke up into bushy trees of lineages branching off in many directions. Horses, for example, had evolved into many different sizes and hoof anatomies over the past 50 million years; many of these branches are long extinct and had nothing to do with the origin of living horses.

If the natural selection that scientists were studying in labs was in fact responsible for the transformations of the fossil record, it would have to work at a rate that was fast enough to produce the changes paleontologists could see. The Fly Room researchers had made careful measurements of how quickly mutations emerged in fruit flies, and how quickly they could spread with the help of natural selection. Simpson invented his own ways to measure rates of evolutionary change in fossils. He looked over the enormous collections of bones that paleontologists had gathered in the previous century, measuring their dimensions, and plotted how they changed over time. Simpson found that lineages could evolve at fast or slow rates, and even within a single lineage evolution could speed up and slow down over time. And Simpson discovered that the fastest rates of change he found in fossils were outstripped by the speed of evolution documented in fruit flies. The modern synthesis, and not some mysterious Lamarckian process, was all Simpson needed to make sense of his bones.

By the 1940s, the architects of the modern synthesis had shown that genetics, zoology, and paleontology were all telling much the same story. Mutations are the foundation of evolutionary change; combined with Mendelian heredity, the flow of genes, natural selection, and geographical isolation, they could create new species and new forms of life; and over millions of years they could create the transformations recorded in fossils. The success of the modern synthesis has turned it into a driving force behind the evolutionary research of the past 50 years.

BIRD BEAKS AND GUPPY LIFETIMES

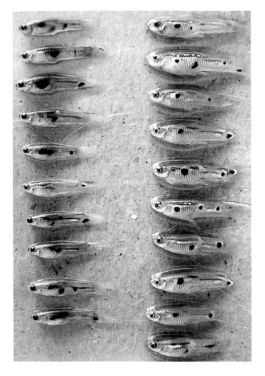

Guppies that live in Trinidad evolve into different sizes depending on their habitat. The ones that live in streams free of predators (right column) are bigger than the ones in streams that are full of them (left).

Darwin never imagined that anyone could witness natural selection. He thought that the variations in his pigeons were the closest that he could get. In the wild, Darwin assumed evolution was too slow and gentle for our short-term minds to perceive, just as we can't see rain wash away a mountain. But these days biologists building on the modern synthesis can actually witness flashes of evolutionary change taking place before their eyes.

David Reznick, a biologist at the University of California at Riverside, gets a glimpse in the forests of Trinidad, where guppies swim in the streams and pools. At the lower elevations, the guppies face the assault of predatory fishes, but the ones in higher waters live in peace, because few of the predators can move upstream past the waterfalls and craggy rocks. In the late 1980s, Reznick began to use the guppies as a natural experiment.

Like all animals, guppies have a timetable for their lives—how long they take to reach sexual maturity, how fast they grow during that time, how long they live as adults. Theoretical biologists have predicted that the life history of animals can evolve if mutations that alter it bring the animals more reproductive success. Reznick put their predictions to the test.

In ponds with a lot of predators, guppies that live fast should be more successful than slow-growing ones. With the threat of death hanging over a guppy, it will grow as quickly as possible so that it can start mating as soon as possible and have as many offspring as possible. Of course, this strategy comes with a heavy price. By growing so quickly, a guppy may shorten its own natural lifespan, and by quickly giving birth to babies, a female guppy can't take time to supply her offspring with much energy, which puts them at risk of dying young. But Reznick reasoned that the threat of an early death offset these risks.

To see whether this trade-off was real, Reznick rescued guppies that were being terrorized in the downstream pool and put them in pools with relatively few predators. Eleven years in these conditions produced guppies that were, on average, in less of a rush. They took 10 percent longer to mature than their ancestors and were over 10 percent heavier by the time they were fully grown. They were also laying smaller broods of eggs, but each of the new guppies that hatched from those eggs was bigger.

To spend 11 years watching guppies become 10 percent bigger may seem at first like a dull way to pass the time. But in the history of life, 11 years is a fraction of a flash. The rate of evolution that Reznick has witnessed is thousands of times faster than the rate that George Gaylord Simpson documented in the fossil record. When Simpson estimated the rate of evolutionary change among fossils, he could compare it only to the rate at which fruit flies evolved in laboratories. No one could say whether the flies were evolving unnaturally or not. But now scientists like Reznick have shown that even in the wild, animals can change rapidly.

Sometimes nature runs evolutionary experiments of its own, without any help from humans whatsoever. In these cases, biologists simply have to observe. After Darwin left the Galápagos, scientists came back every few decades to study his puzzling finches. In 1973 Peter and Rosemary Grant, husband and wife biologists now at Princeton University, arrived on the islands to study the effects of natural selection on the birds.

Most years on the Galápagos, the weather follows a standard pattern. For the first five months of the year it is hot and rainy, followed by a cool, dry period. But in 1977 the wet season never came. A periodic disturbance of the Pacific Ocean called La Niña altered weather patterns over the Galápagos, causing a disastrous drought.

On Daphne Island, where the Grants worked, the drought was lethal. Out of the 1,200 medium ground finches (*Geospiza fortis*) that lived on the island, more than 1,000 died. But the Grants discovered that the decimation wasn't random. *G. fortis* lives mainly on seeds, which it cracks with its strong beak. Small *G. fortis* can break only small seeds, but larger birds have beaks that are strong enough to break big ones. After the drought had lingered for a few months, the small finches ran out of small seeds and began dying off. But the big finches managed to survive, because they could eat seeds that the smaller ones couldn't get to. (In particular, they depended on a plant called caltrop, which grows spiked shells to protect its seeds.)

The survivors of the 1977 drought mated in 1978, and the Grants could see evolution's mark on their offspring. A new generation of *G. fortis* was born, and the Grants' student Peter Boag discovered that on average their beaks were 4 percent larger than those of the previous generation. The big-beaked finches, which had fared better during the drought, had passed their trait to their offspring and altered the profile of the entire population.

In the years since the drought, the finches have continued to change. In 1983, for example, a season of heavy rains and abundant seeds favored finches with smaller beaks, and the Grants found that by 1985 their average size had dropped 2.5 percent. The finches can change quickly, but it seems that they are swinging back and forth like a pendulum. After tracking 4,300 medium ground finches on Daphne Island between 1976 and 1993, the Grants have found no overall trend in their beak size. If a finch has a beak that helps it survive its first crucial year of life, it will probably go on to have lots of offspring. But big beaks are good in some years, and small ones are good in others.

Short-term climate fluctuations can make natural selection drive a population of animals in circles. But under other conditions, it can push them in one direction for a long time. Instead of a cycle of droughts and rains, for example, the climate of an island may become progressively damper for centuries. It's also possible for a group of finches to settle on an island where other finches are already specialized for eating certain kinds of seeds; in that case, evolution might favor genes in the newcomers that allowed them to eat other kinds of food. They would be able to avoid having to compete with the resident finches and risk being driven to extinction. And in either case, with enough time a new kind of finch might emerge.

HOW SPECIES ARE MADE

While the Grants don't know exactly which long-term pressures have acted on Darwin's finches since they arrived on the Galápagos, they do know that evolution has not always gone in circles. It has turned a single common ancestor into 14 species, each with its own peculiar adaptations. The evidence of this evolution is inscribed in the genes of the birds.

As populations of finches experience natural selection and become isolated from one another, their DNA becomes more and more distinct. The Grants enlisted the help of German geneticists to look for the genetic differences between all 14 species of Darwin's finches. They also compared their DNA to an Ecuadorian bird called the grassquit, which ornithologists have suggested might be the closest living relative of the Darwin finches on the mainland. The researchers then compared the sequences and drew a family tree. Whenever they found two species with genes that were more similar to each other than to any other species, they joined their branches together, the node representing their common ancestor. They then joined them to more distant relatives, until all of the birds were united on a single tree.

Their results, which they published in 1999, show that all of the finches do indeed descend from a common ancestor. All 14 species are more closely related to one another than any of them is to the grassquit. An ancestral population of grassquit-like birds arrived at the Galápagos Islands a few million years ago and gave rise to four different lineages of finches. The first to branch off on its own were the warbler finches, a group of species that use a slender beak to catch insects. The next branch to split off were the vegetarian finches, which have a stubby beak that they use to eat flower blossoms, buds, and fruit. Finally, two more lineages evolved: the tree finches adapted to catching insects in trees (the woodpecker finch, for example, ferrets out insects

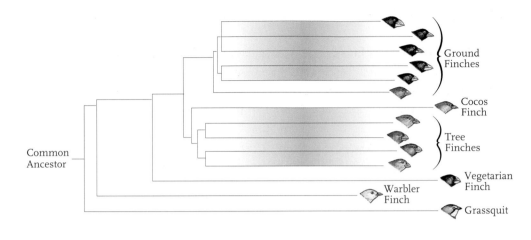

from cracks with cactus spines, which it holds in a chisel-shaped beak), while the ground finches emerged, including the seed-eating *G. fortis*.

Ornithologists have classified the ground finches into six different species, but according to the tree built by the Grants and their German colleagues, they are barely formed. Their genes are clearly distinct from the other finches on the Galápagos, but it is next to impossible to tell them apart from one another. Although the ground finches look and behave differently, they are still mating with one another and successfully producing hybrids. They are, in other words, six species in the process of being born.

While the finches of the Galápagos may be diverging quickly, the biggest explosion of speciation occurring on Earth has been taking place in Lake Victoria and the other Great Lakes of East Africa. Lake Victoria spreads over 27,000 square miles of East Africa, its floor practically as flat as a pool table. It is home to a group of fishes known as cichlids. There are 500 species of these small, brightly colored animals in Lake Victoria, none of which live anywhere else on Earth. Each species has some feature that makes it unique from the other residents of the lake. Some of them scrape algae off rocks with their teeth. Some of them crush shellfish, some pluck the eyes of other species of cichlids. Some species perform courtship rituals in which males build underwater sand castles for females to inspect. Some of them carry their young in their mouths.

A group of geologists came to Lake Victoria in 1995 hoping to look back at the past few hundred thousand years in the lake's mud. The rivers that flow into it carry with them pollen, dust, and dirt, which gets buried year after year on the lake bottom. The geologists thought that if they drilled into the lake bed, they would draw up a core of mud recording hundreds of thousands of years of river flow, which would tell the story of the surrounding woodlands and savannas during that time. But they had drilled down only about 9

The cichlids of the Great Lakes of East Africa have exploded into an unparalleled diversity of species. In Lake Victoria alone, more than 500 species have evolved from a single ancestor in less than 14,000 years.

meters—in other words, into mud that had formed about 14,500 years ago—when all traces of a lake disappeared.

Their drills showed that 14,500 years ago, the deepest parts of Lake Victoria were covered in grass. It seems that during the Ice Age a cool, arid climate dried up the rivers that fed the lake, and its vast supply of water simply evaporated. Over the past few million years ice ages have come and gone, and Lake Victoria has emptied out and filled up again along the way. The last time the glaciers melted, the lake swelled to its current size in a few centuries.

A dry lake is no home for a fish. The ancestors of Victoria's cichlids must have lurked in nearby streams, and when the waters came back to the lake, a single species of cichlid slipped in as well. The cichlids that live in Lake Victoria today are all close relatives of one another, and only distantly related to cichlids in other lakes and rivers. Just as brothers and sisters have similar genes, so do these fish. Their genes show that a single lineage of mouth-brooders came to the lake after it refilled, and then, in the time that it took for humans to build civilization, 500 species were born. Look into the waters of Lake Victoria with an understanding of evolution, and you see a biological explosion.

This evolutionary boom seems to be a case of the right animal showing up at the right place at the right time. Cichlids are the perfect fish for quick specializations. For one thing, they have an extra set of jaws at the back of their mouth that can be used to break up food, leaving their front jaws free to evolve into new kinds of grabbing tools. Meanwhile, their teeth have shown a surprising evolutionary flexibility, having turned into pegs, spikes, and spatulas. As a result, their bodies can be reshaped by evolution into a staggering number of forms.

Cichlids may also have been primed for an evolutionary explosion by their elaborate sex lives. Male cichlids go to great lengths to attract females, dancing complicated jigs or building bowers out of sand and gravel. If a female likes what she sees, she releases her eggs, which the male fertilizes. The choices that females make for their mates are influenced by their genes; as a result, some females may have a preference for a particular shade of red, or a particularly steep angle of a bower wall, or a particular jig in the mating dance. These preferences may spread through the females until they become indifferent to any other male. With enough time, these tastes may isolate a population of fish and turn it into a new species.

When cichlids entered Lake Victoria 14,000 years ago, they were released from evolutionary constraints that their ancestors had faced in their rivers. Rivers are rapidly changing places, subject to sudden floods, droughts, and shifting courses. Under such conditions, evolution doesn't favor fish that are too specialized for one particular part of a river; ones that are adapted to surviving all sorts of unexpected conditions will thrive. But the cichlids that colonized Lake Victoria entered a much more stable place, where they could adapt to special habitats such as rocky shores or deeper waters with sandy bottoms. They could rapidly evolve specialized ways of living and not get punished for their change.

Biologists are now studying the genetic differences of cichlids to figure out exactly how different species formed in Lake Victoria, but they are running out of time. In the 1950s and 1960s, a new fish was introduced to Lake Victoria. The Nile perch, which lives in some other lakes of East Africa, can grow up to 2 meters long on a diet of fish such as cichlids. It was brought to

Victoria as a new source of food for the people who live around the lake. The Nile perch thrived and the catch of fishermen on the lake multiplied tenfold. But it thrived by devouring cichlids.

Meanwhile, farming and logging have caused massive soil erosion, sending topsoil into Lake Victoria, turning what were once clear waters dark. The cichlids, so attuned to the appearance of their potential mates, can't distinguish their markings anymore and end up mating with other closely related species. The reproductive isolation that was pushing these fishes into hundreds of new forms is breaking down.

The silty murk and the Nile perch together have eliminated over half of the cichlid species in Lake Victoria in only 30 years. Humans may have ended this burst of speciation just as they've gotten acquainted with it.

FIGHTING COLDS WITH NATURAL SELECTION

As a concept, natural selection has come a long way this century. In 1900 many scientists wondered if it was significant, or even real. By 2000, it was possible to witness natural selection reshaping life and helping to carve out new species. The resurgence of natural selection this century has even led scientists to discover it at work in some unexpected places. Anything that fulfills Darwin's three basic requirements—replication, variation, and rewards through competition—may feel natural selection's power.

Our bodies, for example, fight off diseases with an immunological version of natural selection. When a virus or some other parasite invades, your immune system tries to mount an attack. But in order to fight off the invader, the immune system has to be able to recognize its opponent. Otherwise it will attack anything it encounters, including the body itself. The immune system harnesses the power of evolution to fine-tune its assault.

When foreign substances enter the body, they encounter a class of immune cells called B cells. B cells have receptors that can snag foreign substances—a toxin made by bacteria, for instance, or a fragment of a virus's protein coat. When a B cell snags these substances (known as antigens), a signal travels into its interior, causing it to multiply into millions of new cells.

The new cells start spewing out antibodies, which are free-floating versions of the receptor that snagged the antigen in the first place. The antibodies course through the body, and when they encounter their antigen, they can lock on to it. While one end of an antibody grabs an antigen, the other end gets rid of it. It may neutralize a toxin, drill holes in the walls of bacteria, or catch the attention of the immune system's assassin cells, which can swallow a parasite.

B cells can produce antibodies that precisely match any one of billions of antigens, produced by parasites ranging from viruses to fungi to hookworms.

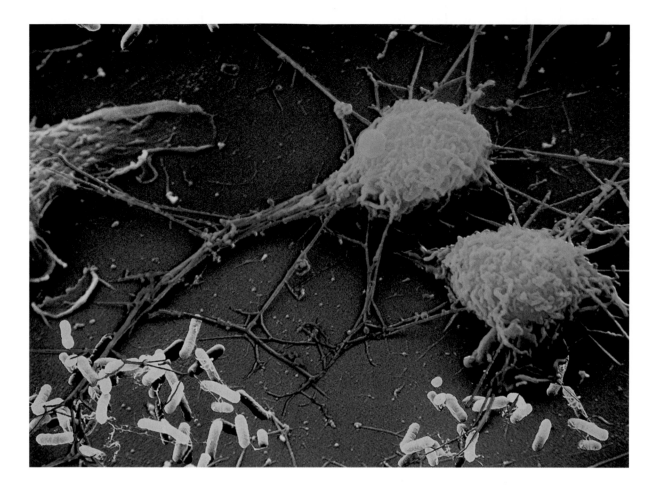

Their precision allows the immune system to recognize and destroy only these invaders, without also latching on to the cells of our own bodies and destroying them as well. Yet our DNA does not contain instructions for building an antibody for every antigen our B cells might encounter. There are billions of antigens, but only about 30,000 genes in human DNA. Our immune system uses a different, far more efficient way to create antibodies: our B cells evolve.

This evolution begins when B cells first form deep inside our bone marrow. As they divide, the genes that build their antigen receptors mutate rapidly, randomly creating billions of receptors with different shapes. They thus take the first step of any evolutionary process: the generation of variation.

The young B cells creep from the bone marrow to the lymph nodes, where many antigens are drifting by. Most of the B cells will be unable to latch on to the antigens, but on rare occasions, a B cell will have the right receptor to grab one. The fit doesn't have to be perfect; if a B cell manages to grab anything at all, it is stimulated to start multiplying madly. You can tell when a B cell has gotten lucky this way: as it proliferates, the lymph node swells into a lump.

White blood cells (shown here killing bacteria) harness the creative powers of natural selection to destroy invading pathogens.

Some of the successful B cell's descendants immediately start releasing antibodies with the same structure as their antigen-grabbing receptors. But others continue dividing without making antibodies. These B cells begin to multiply, mutating more than a million times faster than normal human cells. The mutations alter only the genes they use to build their antigen receptors and their antibodies. In order to survive, these hypermutating B cells must grab an antigen. If a cell fails, it dies. After successive rounds of mutations and competition, evolution produces B cells that can lock on to antigens more and more precisely. Less-adapted cells fail to grab antigens and die. Within a matter of days, this evolutionary process can improve the antigen-snagging ability of B cells by 10 to 50 times.

Imagine if Paley had been introduced to antibodies, so well designed for fighting particular diseases. He might have said that antibodies must be the work of a designer, that nothing so well crafted, so well suited to its own antigen, could have been formed on its own. And yet every time we get sick, he is proved wrong.

EVOLUTION *IN SILICO*

The powers of natural selection can be seen not only in our own bodies but in computers. Life as we commonly know it is written in only one language: the bases of DNA and RNA. But some scientists have been creating what they claim are artificial life-forms in computers that have no need for biochemistry. And like DNA-based life, they can evolve. While critics may question just how alive these creations really are, they nevertheless show how mutations combined with natural selection can turn randomness into complexity. They are even showing how natural selection can create new kinds of technology.

One of the most complex forms of artificial life can be found dwelling in the computers at the California Institute of Technology. There, Christoph Adami, Charles Ofria, and other scientists have created a wildlife refuge they call Avida (*A* stands for artificial, and *vida* is the Spanish word for life). The organisms that live in Avida each consist of a program made up of a series of commands. During the organism's life, an "instruction pointer" moves line by line through the program, carrying out each command that it reads, until it gets to the end, whereupon it automatically returns to the beginning and the program repeats itself.

A digital organism's program can create a copy of itself that becomes a self-sustaining organism of its own, multiplying until it takes up all the computer space that Avida can spare. And by allowing the programs of these digital organisms to mutate as they reproduce, Adami can make them evolve. On rare occasions, a line in a digital organism's program may spontaneously switch to another command; sometimes when an organism tries to copy itself, it may accidentally misread a command and put in the wrong one; it may insert an extra random command or erase one. Just as mutations are usually harmful to biological creatures, most changes to the programs in Avida are harmful bugs that slow down an organism or kill it. But sometimes a mutation can make a digital organism reproduce faster.

Adami can set up experiments with Avida that mimic the evolution of biological organisms. In one early experiment, Adami created a digital organism that could replicate, but which also carried several useless (but harmless) commands in its program. This progenitor gave rise to millions of descendants, which diverged into mutant strains. Within a few thousand generations some of the strains became far more common and successful than the others. What the successful digital organisms shared in common was a short program. In each case, mutations had stripped them down to the simplest program that could still replicate—about 11 steps long.

Evolution drives the digital organisms toward simple genomes in this experiment because they live in a simple environment. In more recent experiments, Adami has made Avida more like the real world, by requiring his digital organisms to eat. In Avida, numbers are food—an infinite string of 1s and 0s that digital organisms digest and turn into new forms. Just as bacteria can eat sugar and transform it into the proteins they need to survive, a digital organism with the proper commands in its program can read the numbers Adami supplies it with and transform them into various forms.

In the natural world, evolution favors organisms that can turn their food into proteins that help them reproduce more successfully. Adami created a similar system of rewards for his digital organisms in Avida. He set up a list of tasks for the organisms to perform, such as reading a number and transforming it into its opposite, so that 10101 becomes 01010. If an organism evolves the ability to do this, he rewards it by speeding up the rate at which its program runs. With a faster-running program, a digital organism can replicate faster. And the rewards for carrying out more complex operations are bigger than for carrying out simpler ones. This reward system radically changes the direction of evolution in Avida. Instead of becoming stripped-down virus-like organisms, the organisms evolve into sophisticated data processors.

Avida is essentially evolving new pieces of software, although they are unlike any program written by a human being. The alien structure of Avida's programs has attracted the attention of Microsoft, which has funded some of Adami's research. They recognize that our DNA is in some ways like an extraordinary computer program, but it can keep a trillion-cell human body going for 70 years without crashing. There seems to be something about the way that evolution produces information processing that makes it more robust than human creations. Microsoft would like to know whether they might someday be able to evolve software rather than write it. The programs that evolve inside Avida today are as simple compared to a spreadsheet as bacteria are to a blue whale. Yet evolution has produced blue whales, and it's conceivable that in an artificial world like Avida, it might be able to produce spreadsheets as well. The challenge will be to landscape the evolutionary hills and valleys of artificial life in the right way, so that a spreadsheet design represents the highest peak of fitness.

Avida is part of an infant science, known as evolutionary computing. Its disciples are discovering that natural selection can shape not only software but hardware as well. A computer can be challenged to come up with thousands of different designs for a device, which it can then test in a simulation. The ones that run best are saved, and then randomly altered in small ways to create a new generation of designs. Without any more guidance than this, computers can evolve some extraordinary inventions.

In 1995, for instance, the engineer John Koza used evolutionary computing to design a low-pass filter, a device that can cut off sounds above a certain frequency. Koza chose 2,000 cycles a second as his cutoff. After 10 generations, his computer produced a circuit that muffled frequencies above about 500 cycles and only completely extinguished them above about 10,000. After 49 generations, it had created a circuit that produced a sharp drop-off at 2,000 cycles. Natural selection had created a design for a seven-rung ladder made out of inductors and capacitors. The same design had been invented in 1917 by George Campbell of AT&T. The computer, without any direction from Koza, had infringed on a patent.

Since then, Koza and others have evolved thermometers, amplifiers complete with woofers and tweeters, circuits that control robots, and dozens of other devices, many of which replicate the work of great inventors. It won't be long, they predict, before evolutionary computing will create devices that warrant patents of their own.

For the moment, this kind of evolution remains trapped inside computers, dependent on human programmers and engineers for its existence. But within a few decades, independent robots may be able to evolve on their own, transforming themselves into new forms that humans could never imagine. In a sign of things to come, Hod Lipson and Jordan Pollack, two engineers at Brandeis University in Massachusetts, announced in August 2000 that they had programmed a computer to use evolution to design a walking robot.

Lipson and Pollack's computer evolved 200 robot designs, each starting completely from scratch. Using a simulation program, Lipson and Pollack scored the robots by how fast they could move across the floor, replaced low-fitness robot designs with ones of higher fitness, and mutated all the remaining robots again. After several hundred generations, the computer then built some of the most successful robots out of molded plastic. These evolved robots walk like inchworms, crabs, and other real animals, yet they look unlike any real animal (or, for that matter, the animal-like robots that humans have built).

The dawn of artificial evolution is a triumph Darwin could not have imagined. Four billion years ago, a new form of matter emerged on this planet: a substance that could store information and replicate itself, that could survive as that information gradually changed. We humans are made of that mutable stuff, but we may now be carrying its laws into new forms, into silicon and plastic, into binary streams of energy.

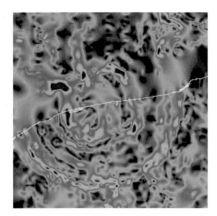

PART TWO

Creation and Destruction

Rooting the Tree of Life

FROM LIFE'S DAWN TO
THE AGE OF MICROBES

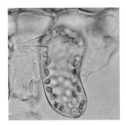

NATURAL SELECTION is not unique to guppies in Trinidad or finches on the Galápagos Islands. It is at work on all species everywhere on the planet, and it has been at work ever since life first emerged. Scientists have traced life back at least 3.85 billion years, and the fossil record now reveals the emergence of new forms of life—eukaryotes, animals and plants, fish, reptiles and mammals—in the eons that have passed since. Generation upon generation of evolution transformed those earliest organisms into all the new forms that came afterward.

Darwin was never eager to speculate on how these great transformations took place. The natural selection taking place in his own day was enough of a puzzle for him, unable as he was to comprehend heredity. But now enough evidence is emerging—in the form of gene sequences, fossils, and ancient traces of Earth's chemistry—

An underwater plant releases bubbles of oxygen as it carries out photosynthesis. Thanks to their bacterial partners, plants provide us with the oxygen we need to breathe.

to let scientists begin deciphering the evolution of life. In the process, today's evolutionary biologists are pushing beyond the limits of the modern synthesis, discovering that evolution's reign has been quirkier and more remarkable than previous generations could have guessed.

THE TREE OF LIFE

The history of life has not unspooled as a simple line. As Darwin proposed, it has grown like a tree over time, as new species have branched off from old ones. Most of those branches have been pruned by extinction, but not before they gave rise to life as we see it around us on Earth today.

Scientists have been drawing and redrawing the tree of life for decades. At first they could only compare different species by looking at their anatomy, such as the sutures of skulls and the twists of wombs. But this method failed when scientists tried to step back and look at life on its broadest scale. You can compare elm leaves to the leaves of maples or pines, but there are no leaves on humans to compare them with. Fortunately, elms and humans are both based on DNA. By sequencing snippets of genetic material from hundreds of species, ranging from frogs to yeast to cyanobacteria, scientists over the past 25 years have assembled the tree of life. The latest version appears below.

This tree is not an icon, but a scientific hypothesis. It offers the simplest interpretation of the genetic sequences that scientists have studied, how genes have mutated from one form to another. As new species are discovered and new genes are sequenced, the simplest interpretation may demand that some

This evolutionary tree encompasses all living species on Earth. The common ancestor (the base of the tree) gave rise to three great branches: bacteria, microbes known as archaea, and eukaryotes (a group of species that includes us). The lengths of the branches reflect how much the DNA of each lineage has diverged from their common ancestor. They demonstrate that most of life's genetic diversity turns out to be microbial; the entire animal kingdom (shown at the upper right) are just a few twigs at one end of the tree.

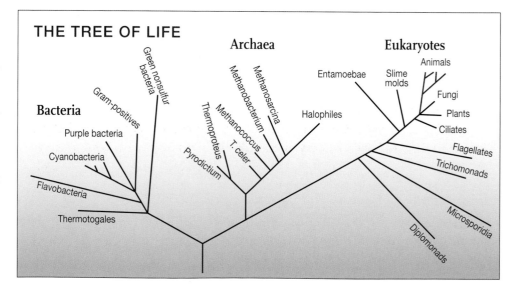

of the branches be rearranged. But despite a surge of new data, this tree has retained its basic structure, which suggests that it is fundamentally sound.

This tree is a strange thing to behold. In the late nineteenth century, evolutionary biologists drew the tree of life as if it were a mighty oak, with branches coming off a main trunk. The simplest organisms such as bacteria sprouted near its base, and humanity was placed at its very crown, the pinnacle of evolution. But instead of a single shaft of evolution ever ascending, scientists now see life splayed out into an unruly thicket.

The tree is split into three main branches. Our own, the eukaryotes, includes plants, fungi, and animals, as well as single-celled protozoa, such as amoebae that live in the forest soil and the oceans, and parasites that cause diseases like malaria, dysentery, and giardia. Eukaryotes all have a distinct sort of cell. They keep most of their DNA balled up in their nucleus, and their cells contain many other compartments where new proteins are built and energy is generated.

Biologists once thought that all of the species that were not eukaryotes fell into a single group, known as prokaryotes. After all, they all seem to look the same. Their DNA, for instance, floats loose inside their membranes, not coiled in a nucleus. But the genes tell another story. Bacteria form their own branch, while there is a third major branch on the tree of life that is more closely related to us than to bacteria. First identified in the 1970s by University of Illinois biologist Carl Woese, these organisms may look like bacteria, but they have cellular machinery that is radically different. Woese named these microbes archaea, meaning "first," for the branch on which they appear.

Another great surprise of the new tree of life is just what a small space we multicellular eukaryotes take up in the story of evolution. It's almost impossible to make out the difference between ourselves and the elm trees. And meanwhile the diversity within the bacteria, archaea, and single-celled eukaryotes turns out to be stunning. Microbiologists are continually dredging up new species, new families, even new kingdoms of microbes, which have colonized the deepest reaches of Earth's crust, the boiling water of hot springs, and the acid-drenched warmth of the human gut. Most of the diversity of life, not to mention the sheer physical mass, is microbial.

The base of the tree of life represents the last common ancestor of all life on Earth today. All living species share certain things in common. All of them, for example, carry their genetic information as DNA and use RNA to turn them into proteins. The simplest explanation for these universal properties is that all living species inherited them from a common ancestor. That common ancestor therefore must have been a relatively sophisticated creature. In turn, it must have descended from a long line of ancestors. For all we know, there were deeper branches that we can't see now because they've become extinct. And beyond these vanished ancestors lies the origin of life itself.

IN SEARCH OF LIFE'S ORIGINS

Even if the tree shown on page 102 doesn't reach back to life's beginning, it can help scientists who are trying to reconstruct that first great biological transformation: from nonlife to life. Along with the geological record, it can offer clues and constraints. Any explanation of how life began has to account for the evidence that has been left behind.

Although scientists are a long way from knowing the precise history of life's early evolution, they can study it in the same way they study later transitions. As we'll see in chapter 6, new groups of animals didn't emerge in one giant leap; rather, pieces of their new body plan were added on step by step until the forms we see in living animals took shape. Scientists have found compelling evidence that life could have evolved into a DNA-based microbe in a series of steps as well.

The first step in the rise of life was to gather its raw materials together. Many of them could have come from space. Astronomers have discovered a number of basic ingredients of life on meteorites, comets, and interplanetary dust. As these objects fell to the early Earth, they could have seeded the planet with components for crucial parts of the cell, such as the phosphate backbone of DNA, its information-bearing bases, and amino acids for making proteins.

Comets may have seeded the early Earth with many of the building blocks of life.

As these compounds reacted with one another, they may have produced more lifelike forms. Chemical reactions work best when the molecules involved are crowded together so they bump into one another more often; on the early Earth, the precursors to biological matter might have been concentrated in raindrops or the spray of ocean waves. Some scientists suspect that life began at the midocean ridges where hot magma emerges from the mantle. The branches nearest the base of the tree of life, they point out, belong to bacteria and archaea that live in extreme conditions such as boiling waters or acids. They may be relics of the earliest ecosystems on the planet.

Scientists suspect that prebiological molecules became organized into cycles of chemical reactions that could sustain themselves independently. A group of molecules would fashion more copies of itself by grabbing other molecules that surrounded it. There may have been many separate chemical cycles at work on the early Earth. If they used the same building blocks to complete their cycles, they would have competed with one another. The most efficient cycle

would have outstripped the less efficient ones. Before biological evolution, in other words, there was chemical evolution.

Ultimately, these molecules gave rise to DNA, RNA, and proteins. Scientists have debated for decades which of the three emerged first. DNA can carry information for building bodies from one generation to the next, but it is helpless without the help of RNA and proteins. It cannot, for example, join molecules together or cleave them the way enzymes do. Proteins have the opposite shortcoming: they do the work that's required to keep a cell alive, but it is very difficult for them to carry information from one generation to another. Only RNA can play both roles, carrying a genetic code and doing biochemical work. The twin abilities of RNA make it the leading candidate for life's first molecule.

When scientists first uncovered RNA's role in the cell in the 1960s, few thought it might have been the primordial stuff of life. Delivering information from the genes to the protein-building factories of the cell, it seemed like a lowly messenger. But in 1982 Thomas Cech, then at the University of Colorado, discovered that RNA is actually something of a molecular hybrid. On the one hand, it can carry information in its code. On the other hand, Cech found that it can also act as an enzyme, able to alter other molecules. One of the jobs that enzymes do, for example, is to edit out useless sequences after DNA is copied into RNA. Cech discovered that some versions of RNA can loop back on themselves and edit their own code, with no help from enzymes.

In the late 1980s biologists realized that thanks to RNA's two-faced versatility they could make it evolve in their labs. One of the most successful teams

Hot springs in Yellowstone are home to some of the most primitive microbes on Earth. Researchers suspect that life may have begun 4 billion years ago in near-boiling water.

was led by biologist Gerald Joyce, who works at the Scripps Research Institute in La Jolla, California. Joyce began with Cech's original RNA molecule and replicated it into 10 trillion variations, each with a slightly different structure. He then dumped DNA into the test tubes that held these variants and waited to see whether any of them could cut off a piece of it. Because Cech's RNA was adapted for cutting RNA, not DNA, it wasn't a big surprise that none of the RNA variants did a good job. In fact, only one in a million of them managed to grab DNA and slice it. And these few successful molecules were so bad at the job that they needed an hour to do it.

Joyce saved these clumsy RNA molecules and replicated each of them into a million new copies. Once again, the new generation was rife with mutations, and some of the new variants were able to cut DNA faster than the previous generation. Joyce saved these slightly superior RNA molecules and replicated them again. After he had carried out the entire process for 27 generations (a process that took two years), the evolved RNA could cut DNA in only five minutes. In fact, their ability to cut DNA was equal to their natural ability to cut RNA.

Joyce and other biologists can now make RNA evolve much faster than in these first experiments. Producing 27 generations of RNA takes three hours instead of two years. The biologists have found that in the right environment, evolution can make RNA do things it has never been known to do naturally. Evolved RNA can slice apart not only DNA but many other molecules. It can bind to single atoms or entire cells. It can join together two molecules to create a new one. With enough evolution it can even join together amino acids—the crucial step in creating proteins. It can join a base to its phosphate backbone. In other words, it can evolve the ability to carry many of the jobs that it would have to do if cells had only RNA, without DNA or proteins.

RNA is so evolvable that biotechnology companies are now trying to transform it into anticoagulants and other drugs. The work of Joyce and others suggests that RNA could have played the role of both DNA and protein on an early Earth. Many biologists now refer to this early stage of life as the "RNA world."

After RNA had evolved, proteins might have emerged next. At some point in the history of the RNA world, new forms of RNA might have evolved the ability to connect amino acids together. The proteins they created might have helped RNA replicate faster than it could have managed on its own. Later, the single-stranded RNA might have constructed its double-helix partner, DNA. Less likely to mutate than RNA, DNA would have proved a more reliable system for storing genetic information. Once DNA and proteins had come into existence, they would have taken over many of RNA's chores. Today RNA is still a vital molecule, but only a few vestiges of its former power have survived, such as its ability to edit itself.

At this point, life as we know it truly began. But for RNA, life was never again the same. The RNA world had met its Armageddon.

THE MANGROVE OF LIFE

Most of the scientists who crafted the modern synthesis were zoologists or botanists, with a deep knowledge of plants and animals. Plants and animals generally transmit their genes by mating and producing offspring bearing a combination of their DNA. Their evolution unfolds as mutations rise and spread through the generations. But animals and plants are relative newcomers in the history of life. Evolution has been—and continues to be—mainly a story about microbes. Bacteria and other single-celled organisms do not obey the same laws as we do when it comes to how they replicate their genes. As evolutionary biologists discover just how different microbes are, they are redrawing parts of the tree of life.

Bacteria and other microbes can reproduce as the cells of our own body do, by dividing themselves into two copies, each with its own set of DNA. If a bacterium incorrectly duplicates one of its genes, it will create a mutation in one of its offspring, and every offspring of that mutant will receive its mutation as well. But microbes can acquire new genes even after they're born.

Many species of bacteria carry genes on loops of DNA that are separate from their chromosomes. A bacterium can pass these loops—called plasmids—to another bacterium, belonging either to its own species or to a different one altogether. Viruses can also carry DNA between bacteria, as

New evidence suggests that the tree of life may be more complicated than the version shown on page 102. As indicated in the tree shown here, early life may not have existed as distinct species; instead early organisms may have traded their genes promiscuously. Life may descend from a huge primordial menagerie (the crossed blue lines at the bottom of the tree) rather than from a single common ancestor. Billions of years later, after the three great branches of bacteria, archaea, and eukaryotes split apart, distantly related species still joined together sometimes, as bacteria were swallowed up by other organisms. Two of the most important of these fusions—bacteria giving rise to mitochondria and chloroplasts—are shown here.

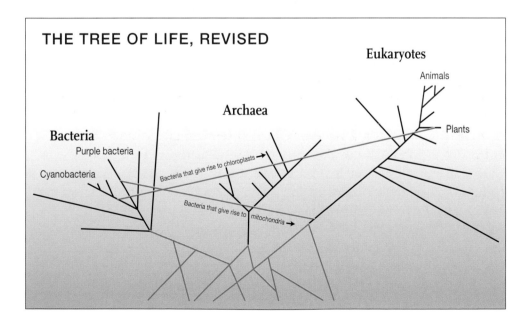

THE TREE OF LIFE, REVISED

Eukaryotes

Animals

Archaea

Plants

Bacteria

Purple bacteria

Cyanobacteria

Bacteria that give rise to chloroplasts ➔

Bacteria that give rise to mitochondria ➔

they pick up genetic material from their hosts and inject it into new ones. It's even possible for some of the genes on a bacterium's own chromosomes to slice themselves free and make their way into a different microbe. And when a bacterium dies and its DNA gushes out of its ruptured cell walls, other bacteria sometimes scoop up its genes and integrate them into their own genomes.

As early as the 1950s, microbiologists could watch bacteria trade genes, but they had little idea of what sort of effect these kinds of exchanges had had on the history of life. Perhaps microbes traded genes so rarely that their swaps left no mark. Only in the late 1990s did scientists get a chance to find out, as they began sequencing the entire genomes of microbes. The results were astounding. A sizable fraction of many bacterial genes originally belonged to distantly related species. Over the last 100 million years, for example, *Escherichia coli* has picked up new DNA from other microbes 230 times.

The evidence for this transfer of genes can be found even at the deepest branches of the tree of life. *Archeoglobus fulgidus* is an archaean that lives around seeps of oil on the seafloor. It has all the qualities you would expect in an archaean, particularly in the molecules it uses to build its cell walls and the way it copies information out of its genes and turns them into protein. But it eats oil using enzymes that are found only in bacteria, not in other archaea. Our own genes have a split heritage as well. The ones that handle information processing, such as copying DNA, are closely related to archaean genes. But many of the genes responsible for housekeeping—in other words, making the proteins that help process a cell's food and clean up its waste— are more like those of bacteria. The discovery of these alien genes now makes the evolution of early life much more complicated—and much more interesting.

These results have inspired Carl Woese, the microbiologist who first recognized the three branches of life, to offer a new vision of the common ancestor of life on Earth. As life emerged from the RNA world into the DNA world, it was still immensely sloppy in the way it replicated itself. It had none of the careful proofreading enzymes and other mechanisms that ensure that our cells make faithful copies of their DNA. Without these safeguards, mutations were rampant. The only proteins that could survive for more than a few generations without being destroyed by the mutations were simple ones; any complex proteins requiring a lot of genetic instructions were vulnerable.

With such a fragile system of replication, these primordial genes were more likely to move from one microbe to another than to be passed down from one generation to another. Because these early microbes were so simple, wandering genes could easily fit into their new home and help with the housework—with breaking down food, hauling out waste, and other chores.

Parasitic genes could have invaded the microbes as well, manipulating other genes in order to make extra copies of themselves that could then escape and infect other microbes.

On the early Earth, Woese claims, there was no genealogy. Life had not yet separated into distinct lineages, and thus no single species lies at the base of the tree of life. Our common ancestor was every microbe that lived on the early Earth: a fluid matrix of genes that covered the planet.

But there came a time when wandering genes found it more difficult to find a home in a new host. More complicated systems of genes began to evolve, able to do a better job than the simple collections had in the past. It would be as if a migrant farmhand, able to pick fruit or bale hay or shovel manure, showed up on a farm where the workers had learned how to control their equipment with computers. He wouldn't be able to fit in. As these systems of genes became more specialized, they did a better job of accurately replicating DNA. Genes could be passed down through the generations, forming clear lines of descent. Out of the blurry pond of early evolution, three main branches of life emerged: eukaryotes, archaea, and bacteria. But although they became distinct branches, each of them carried a jumble of genes as a reminder of our promiscuous past.

If Woese turns out to be right, the tree of life will have to be redrawn yet again. Instead of a bush, it will have to look more like a mangrove, with a tangle of roots at its base representing the mingling of early genes. Gradually three trunks emerge, but their branches entwine with one another many times.

EVOLUTION ON DOUBLE TIME

It may not have taken very long for life to evolve from the earliest organisms, with just a handful of genes, to full-blown microbes such as cyanobacteria, which contain more than 3,000 genes. Scientists still have few clues to this chronology, but what they do know suggests that early evolution ran at a brisk pace. Fossils from Australia, for example, show that there were definitely cyanobacteria-like microbes on Earth 3.5 billion years ago. Molecular fossils from Greenland tell us that *some* kind of life was on Earth 350 million years earlier, by 3.85 billion years ago. Scientists can't say exactly what sort of life was leaving its mark in Greenland, but it was already altering the chemistry of the oceans and atmosphere on a global scale. Perhaps it was cyanobacteria-like microbes, or perhaps only RNA-based organisms, or perhaps something in between.

Now compare what we know about life's history to what we know about Earth's history. Earth is 4.55 billion years old, and for the first few hundred million years, enormous impacts regularly melted the planet all the way through. Any life that had formed during this violent time might have been eradicated. Even after Earth had reached its current size and oceans had begun to form, million-ton rocks continued to fall out of the sky every few million years. If life existed on Earth when these impacts occurred, it might have survived in hidden refuges, perhaps in the chambers of undersea volcanoes. But it might also have become extinct. The last storm of titanic impacts occurred 3.9 billion years ago; 50 million years later, life was well established, and 350 million years later sophisticated microbes were definitely alive.

How could such a complex genetic system have evolved so quickly? The biologists who forged the modern synthesis of evolution mainly studied how minor genetic changes—a switch of A to G at one position in a gene, for instance—could add up to big evolutionary results. But it turns out that there's another important ingredient to evolution: the accidental duplication of entire genes.

Gene duplications occur at about the same rate as single-base mutations. Once a new copy of a gene appears, it may end up with one of several fates. It may make more of the protein the original gene made, which may raise the organism's fitness. The protein may be essential for processing food, for example, and making more copies of the protein lets an organism eat more

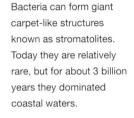

Bacteria can form giant carpet-like structures known as stromatolites. Today they are relatively rare, but for about 3 billion years they dominated coastal waters.

efficiently. In that case, natural selection will hold on to the duplicated gene in a form much like the original.

But the extra gene may instead be superfluous. In these cases, a mutation that strikes the new copy won't affect the fitness of the organism that carries it, because the original gene is still doing its job. Most of the time a mutation to a duplicated gene will simply render it useless. Our DNA brims with these genetic ghosts, known as pseudogenes. But sometimes mutations can transform a duplicated gene so that it makes new proteins that can do new jobs.

The genomes of bacteria, archaea, and eukaryotes all contain hundreds of duplicated genes, which can be grouped into families in much the way species are grouped together. And in both cases, the grouping reflects a common descent. Gene families are the work of many rounds of gene duplication, reaching back to the earliest stages of life. Genes did not simply mutate on the early Earth: they multiplied.

FUSION EVOLUTION

Even after the tree of life had split into its three main trunks, evolution was still able to fuse together its distant branches. We should be grateful that it could, because we are a product of one of these unions; other fusions gave rise to plants and algae. If life hadn't combined in these ways, there'd be little oxygen on Earth for us to breathe, nor would we be able to breathe what little oxygen there was.

Our respiration depends on sausage-shaped blobs in our cells called mitochondria. Almost all eukaryotes have mitochondria, which use oxygen and other chemicals to create fuel for our cells. When mitochondria were first discovered at the end of the 1800s, many scientists were struck by how much they looked like bacteria. Some even went so far as to say that they *were* bacteria— that somehow every cell in our body was invaded by oxygen-breathing microbes, providing them shelter in exchange for fuel.

Scientists already knew that other bacteria could live inside animals or plants and not cause disease. In many cases they actually had a mutually beneficial existence, known as symbiosis. Some bacteria live inside cows, for example, and digest the tough tissues of the grass their hosts eat; the cows then eat some of the bacteria. Still, it was one thing to say that bacteria lived inside our bodies, and another to say that they lived inside our cells. Many scientists remained skeptical.

But meanwhile more bacteria-like things were turning up inside cells. Plants, for example, have a second set of blobs in their cells that they use to carry out photosynthesis. Known as chloroplasts, they capture incoming sunlight and use its energy to combine water and carbon dioxide into organic matter. And like mitochondria, chloroplasts bear a striking resemblance to bacteria. Some scientists became convinced that chloroplasts were, like mitochondria, a form of symbiotic bacteria—specifically, that they descended from cyanobacteria, the light-harnessing microbes that live in oceans and freshwater.

Until the early 1960s, the symbiotic theory sputtered in and out of scientific fashion like a weak flame. Most scientists were so focused on discovering how DNA in the nucleus of our cells stored genetic information that the symbiotic theory, claiming that our cells were made up of more than one organism, sounded absurd. But in the 1960s, scientists discovered that mitochondria and chloroplasts have genes of their own. They use their DNA to make their own proteins, and when they duplicate themselves, they make extra copies of their DNA, just as bacteria do.

Yet in the 1960s scientists still lacked the tools for finding out exactly what sort of DNA mitochondria and chloroplasts carried. Perhaps, some skeptics argued, their genes had originated inside the nucleus, and at some point evolution had moved it into outlying shelters. Then in the mid-1970s two teams of microbiologists, one headed by Carl Woese and the other by W. Ford Doolittle at Dalhousie University in Nova Scotia, showed that this was not so. They studied the genes inside the chloroplasts of some species of algae, and

Plants convert sunlight and carbon dioxide in structures called chloroplasts, which descend from light-harvesting bacteria.

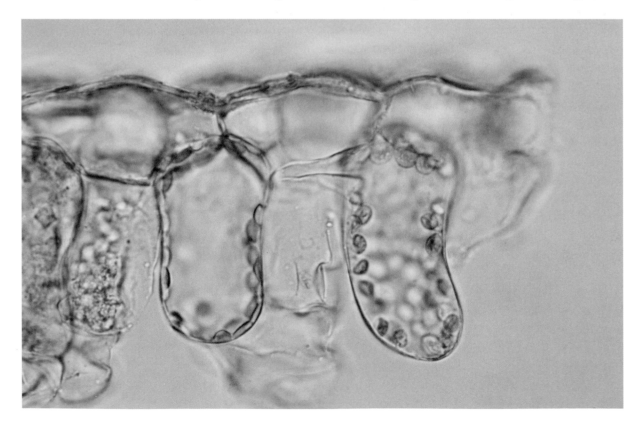

they found that they bore little resemblance to the genes in the algae's nucleus. Chloroplast DNA, it turns out, is cyanobacteria DNA.

The genes of mitochondria had an even more astonishing story to tell. In the late 1970s Doolittle's team showed that they were also bacterial genes, and in the years that followed, other scientists zeroed in on exactly which kind of bacteria they belonged to. In 1998 Siv Andersson of Upsalla University in Sweden and her colleagues discovered the closest relative of mitochondria yet known: *Rickettsia prowazekii*, a vicious bacteria that causes typhus.

Rickettsia is carried by lice and normally makes rats its home, but the parasite can also live in humans. And when humans live in filthy, cramped conditions where lice and rats can thrive—such as slums and army camps—a typhus epidemic can break out. Once the bacteria enter a person's body through the bite of a louse, they push their way into the cells of their host, feeding on them and making more copies of themselves. A raging fever and unbearable aches follow, and sometimes even death.

Typhus is a disease so lethal that it can steer the course of history. When Napoleon set out to defeat Russia, he marshaled an army of half a million soldiers. In 1812 they marched east across Poland. The Russian army retreated from them, never engaging in battle, and when Napoleon arrived in Lithuania, he took the capital, Vilnius, without firing a shot. Yet by the time his army entered the city, 60,000 French soldiers were already dead of typhus.

The Russian army fell farther back, burning the croplands as they retreated. Without food, the French forces grew weak, and the typhus epidemic gained more force. Napoleon left behind his dying soldiers in makeshift hospitals and marched on. When he finally reached Moscow, the city had been emptied out and the Russians then proceeded to burn down two-thirds of it. Napoleon realized he had to get out of Russia before winter came or his entire army would perish.

The army retreated along the same route that they had taken into Russia, surviving on horse meat and melted snow. At the hospitals they had set up on their way east, dead bodies were strewn in the corridors. The army had no choice but to leave behind more of its sick, and soon they died as well. Moving west across Poland and Prussia, the army fell apart into small bands of soldiers that straggled back on their own. They became human lice, triggering new epidemics of typhus in the villages they passed through. "Wherever we went," one French soldier wrote, "the inhabitants were filled with terror and refused to quarter the soldiers." Only 30,000 soldiers returned home. Nineteen out of every twenty men who had left for Russia had died. Napoleon never recovered from the losses that *Rickettsia* inflicted on him, and his empire soon crumbled.

It now turns out that Napoleon's dying soldiers carried within their cells the closest known relatives to the bacteria that killed them.

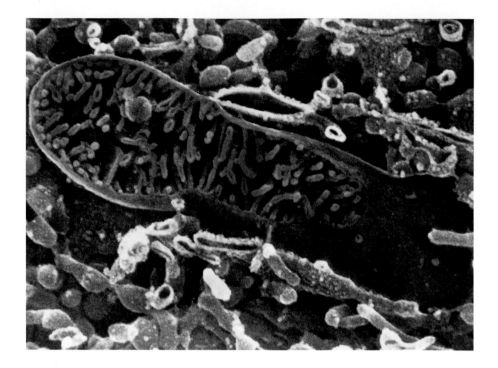

Mitochondria (shown here in red) use oxygen to create fuel supplies for eukaryote cells. Their DNA reveals that they descend from free-living, oxygen-breathing bacteria.

At some point in the distant past, a long-lost oxygen-breathing bacterium gave rise to the ancestors of both *Rickettsia* and mitochondria. Both lineages were originally free-living microbes, feeding on the nutrients that surrounded them. Eventually each lineage began to live inside other organisms. *Rickettsia* evolved into a ruthless parasite that could plunge into its hosts and ravage them. But the bacteria that invaded our ancestors ended up in a kinder relationship. Miklos Müller of Rockefeller University has suggested that protomitochondria may have hung around early eukaryotes to feed on their wastes, and the eukaryotes—which could not use oxygen for their metabolism—came to rely in turn on the wastes of the oxygen-breathing protomitochondria. Eventually the two species merged and the exchanges between them began to take place with a single cell.

Evolution by fusion was not part of the modern synthesis. It is a way to change a species without gradually accumulating mutations to its DNA: just meld two species into one, and suddenly create a new genome. But symbiotic evolution, as strange as it may seem, still follows Darwin's basic rules. Once bacteria get settled in their new host, natural selection continues to shape their genes. Mitochondrial DNA can mutate, and if a mutation interferes with the job of creating energy for a cell, it can be weeded out by natural selection. On the other hand, if a mutation helps mitochondria do their job better and raises an organism's fitness, natural selection can spread that gene instead. Mitochondria have also lost many of the genes that their free-living ancestors used to survive in the outside world. These genes proved to be an unnecessary burden once mitochondria could rely on their hosts to handle those chores. Eventually evolution stripped the superfluous genes away.

For the first 3 billion years or so of life's history, microbes were the sole inhabitants of Earth. These were not eons of evolutionary ennui, no matter what we may anthropocentrically assume. Biochemically speaking, the age of microbes was a time of remarkable change—a global flux of genes that invented countless ways of turning energy into life. It was only after microbes had gone through all of this evolution that our own multicellular ancestors—the first animals—could appear.

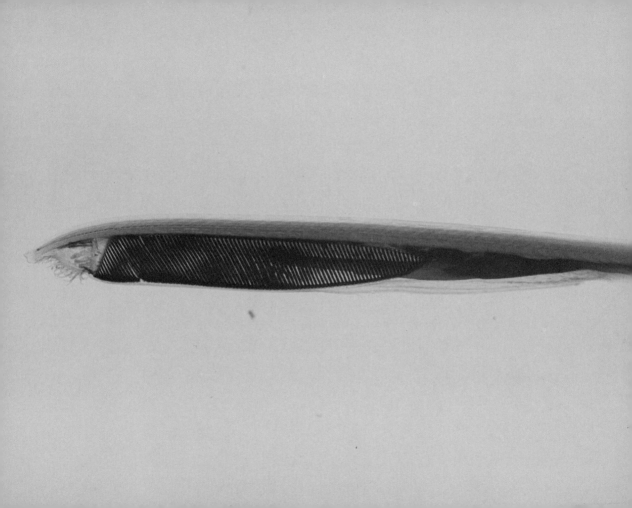

The Accidental Tool Kit

CHANCE AND
CONSTRAINTS IN
ANIMAL EVOLUTION

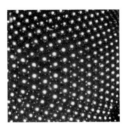

OF ALL THE DIFFERENCES between ourselves and our amoeba-like ancestors a billion years ago, one is paramount: we have bodies. Instead of a single cell, we are made of trillions. This vast collective isn't just a bag of identical copies, but a menagerie of dozens of types of cells, organized into hundreds of body parts from spleens to eyelashes to skeletons to brains. And most remarkable of all, each of our bodies is built out of a single original cell. As that cell multiplies into an embryo, genes begin producing proteins that control its development. Some of the proteins switch on other genes or shut them down; some leave the cell where they were made and spread away, to act as signals to neighboring cells, which respond by taking on new identities, or by crawling through the embryo to find a new home. Some divide madly, while others commit suicide. And when this dance is over, our bodies have taken shape.

Lancelets are the closest living relatives to vertebrates. Although they have a nerve cord with a brain-like tip, they lack a head or limbs.

There are millions of other kinds of bodies on Earth, from tentacled squids to quilled porcupines to mouthless tapeworms. They are marvels to behold; their origins are a grand challenge to comprehend. All animals descend from a common single-celled ancestor, but scientists are still learning how it was that they diversified into so many different bodies. The answer lies both within animals and outside them, in their genetic history and the ecology in which they lived.

Scientists have only just begun exposing the ways that genes build animals, but their results have already proved to be revolutionary. Most animals, including us, use a standard tool kit of body-building genes. It contains tools for marking off the coordinates of an animal's body—front and back, left and right, head and tail. It also contains a set of genes that control the development of entire organs such as eyes and limbs. The tool kit is remarkably unchanged from one species to another—a gene that controls the growth of eyes in a mouse can be donated to a fly and build its eye instead.

Judging from the fossil record, this tool kit must have gradually evolved in the millions of years that preceded the Cambrian explosion. It gave animals an extraordinary flexibility for evolving new forms. Simply by making a few changes—altering the timing of a gene's activity, for example, or the places where it became active—the tool kit could produce dramatically new body plans. On the other hand, as diverse as animals have become, they have obeyed certain rules. There are no six-eyed fish or seven-legged horses. It appears that the tool kit shuts down certain paths of evolution.

The diversification of animals has also been controlled by the environment in which animals evolve. Any new kind of animal has to find a place in its ecosystem where it can survive. Otherwise it simply disappears. The fate of any new kind of animal is far from predictable, often depending on random strokes of luck and accidental fortune. Consider vertebrates that live on land. They all have four limbs with digits (or, in the case of snakes, descend from ones that did). But that doesn't mean that this design evolved simply because it was the best possible way to walk on land. In fact, legs and toes evolved on fish millions of years before they left the water. Only later did they turn out to allow vertebrates to move on dry land. The great transformations that animals have gone through all contain the same lesson: evolution can only tinker with what the history of life has already created.

EVOLUTION'S MONSTERS

To learn how animals evolve, biologists make monsters. They create flies with legs sprouting from their heads. They cover their bodies with eyes. They put extra toes on a mouse or a spinal cord in a frog's belly.

Biologists need no surgery to construct these monsters. In each case, all they have to do is alter a single gene, either by shutting it down or by changing the time or place where it makes its protein. These genes, biologists have discovered, control the development of animal bodies.

The quest to make these monsters reaches back more than a century. In the 1890s an English biologist named William Bateson catalogued every sort of hereditary variation known to science. Bateson was particularly struck by animals that were born with a body part in the wrong place. A spiny lobster had an antenna where its eye should have been. A moth grew wings instead of legs. Sawflies had legs for antennae. Among these monsters, there were even some humans. On rare occasion, people are born with little ribs sprouting from the neck, or an extra pair of nipples.

Somehow, these mutations were able to construct an entire body part in a place where it didn't belong. Bateson called the process that created these freakish variations "homeosis." The first clue to how homeosis works came in 1915, when Calvin Bridges of Columbia University traced a case to a particular mutation. He discovered mutant fruit flies that grew an extra pair of wings. These double-winged flies passed down their mutant gene to their double-winged offspring, Bridges discovered; ever since, geneticists have kept their descendants alive.

Normally, fruit flies are born with a pair of wings and a pair of club-shaped structures called halteres that help them stay balanced in flight (left). But a mutation to one of the genes that controls its development turns the halteres into wings (top right). A mutation to another gene can make legs grow out of a fly's head (lower right).

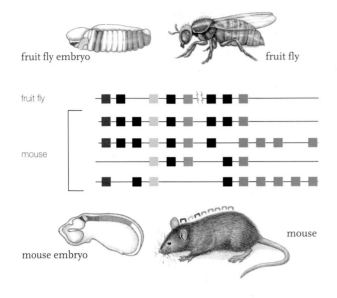

fruit fly embryo

fruit fly

fruit fly

mouse

mouse embryo

mouse

Hox genes determine the head-to-tail anatomy of many animals. Here Hox genes are represented by a series of colored squares, and the parts of the embryos whose development each gene controls are coded in the same color. Insects and other invertebrates have a single set of Hox genes, while vertebrates have four sets of them. But their Hox genes are almost identical to the corresponding genes in insects.

Yet it wasn't until the 1980s that biologists finally figured out how to isolate the gene that was responsible for Bridges's mutants. They discovered that it was only one of an entire family of related genes, which are now called Hox genes. Biologists found that by altering other Hox genes, they could create equally grotesque flies, with legs sprouting out of their heads or antennae where their legs had been.

By studying these sorts of mutants, biologists were able to figure out how normal Hox genes work. Hox genes become active early in a fly embryo's development, when it still assumes a nondescript football shape. The embryo begins to divide into segments, and although the segments all look identical, each is already fated to become a particular part of the fly's body. It's the job of Hox genes to tell the cells in each segment what they are going to become, whether they will become part of the abdomen or a leg, a wing, or an antenna.

Hox genes exert their power by acting like master control switches for other genes. A single Hox gene can trigger a chain reaction of many other genes, which together form a particular part of the body. If a Hox gene gets mutated, it can no longer command those genes properly. The error may end up making the segment grow a different body part. That was the secret of Calvin Bridges's double-winged flies.

Hox genes are surprisingly elegant. Biologists can tell which cells in a fruit fly larva have active Hox genes inside them by making them glow. They inject special light-producing proteins that bind with the proteins produced by Hox genes. The glow of each Hox gene marks a distinct band of segments. Some Hox genes are active in the segments near the head of a fly, while others switch on in segments closer to the tail. Remarkably, the Hox genes themselves reflect this head-to-tail order: they are lined up on their chromosome in the same order as they are expressed in a fruit fly larva, with the head genes in front and the tail genes at the end.

When biologists first discovered Hox genes in fruit flies in the 1980s, they knew almost nothing about how genes control the development of embryos. They were overjoyed to be able to study the process even in a single species. But they assumed that the genes that built fruit flies would be peculiar to insects and other arthropods. Other animals don't have the segmented exoskeleton of arthropods, so biologists assumed that their very different bodies must be built by very different genes.

Joy turned to shock when biologists began to find Hox genes in other animals—in frogs, mice, and humans; in velvet worms, barnacles, and starfish. In every case, parts of their Hox genes were almost identical, regardless of the animal that carried them. And the genes were even lined up in the chromosomes of these animals in the same head-to-tail order as they are in a fly.

Biologists discovered that the Hox genes did the same job in all of these animals: specifying different sections of their head-to-tail axis, just as they do in insects. Hox genes in these different animals are so similar that scientists can replace a defective Hox gene in a fruit fly with the corresponding Hox gene from a mouse, and the fly will still grow its proper body parts. Even though mice and fruit flies diverged from a common ancestor more than 600 million years ago, the gene can still exert its power.

THE MASTER-CONTROL GENES

In the 1980s and 1990s, scientists discovered many other master-control genes at work in animal larvae, each just as powerful as Hox genes. While Hox genes work from head to tail, other genes mark out the left and right sides of the body, and still others establish top and bottom. The three dimensions of fruit fly legs are mapped out by master-control genes as well. Master-control genes help build organs. Without the Pax-6 gene, a fly is born without eyes. Without the tinman gene, a fruit fly has no heart.

And just as with Hox genes, each of these master-control genes also exists in our own DNA, often doing the same jobs they do in flies. A mouse version of Pax-6, for example, can cover a fly's body with extra eyes. As biologists explore the genes of other animals—whether they are acorn worms or sea urchins, squid or spiders—they are finding that they share these master-control genes as well.

Master-control genes are able to use the same body-building instructions to build very different kinds of animal bodies. A crab's legs are hollow cylinders with muscles running along their interior. Our own legs have a beam of bone at their core, and muscles running along their exterior. But crabs and humans still share many master-control genes for building limbs. The same goes for eyes, even though a human eye is a single ball of translucent jelly with an adjustable pupil, while a fly has hundreds of compound eyes that together form an image. A human heart is a set of chambers that sends blood coursing into the lungs and back into the body, while a fly's heart is a tubular two-way pump. In all these cases, the master genes that help build them are the same.

This common genetic tool kit is so intricate that it could not have evolved independently in every lineage that uses it. It must have evolved in the common ancestor of these animals. Only after that common ancestor gave rise to the different animal lineages did the master-control genes begin to control different kinds of body parts. Yet as different as these animals became, their tool kit barely changed over hundreds of million years. That is why the master-control genes from a mouse can build a fruit fly's eye.

THE GENES BEHIND THE CAMBRIAN EXPLOSION

Once biologists discovered the genetic tool kit, they realized that it might have made the Cambrian explosion possible 535 million years ago. The first animals to appear in the fossil records include primitive animals such as jellyfish and sponges—diploblasts whose embryos form from only two layers. Biologists have looked for master-control genes in these animals as well but have mostly been disappointed. Diploblasts have only a handful of these genes and don't seem to use them in the same tightly organized way that triploblasts do.

That's not surprising when you consider the simplicity of the jellyfish's body. It does not have a body axis with left and right sides. Instead, its body is radially symmetrical, like a bell or a sphere. Its mouth is also its anus. Its nervous system is a decentralized web, rather than branches running off a central cord. It does not have the complex organization of a lobster or a swordfish.

Only after the primitive diploblasts branched off on their own did the genetic tool kit emerge in the common ancestor of all other animals. It made more complex bodies possible in these new animals: they could set up a grid of coordinates in a developing embryo, dividing the body into more parts, more sensory organs, more cells for digesting food or making hormones, more muscles for moving through the ocean.

Exactly what kind of body that common ancestor had is difficult to say. But paleontologists shouldn't be surprised to unearth some inch-long creature that lived not long before the Cambrian explosion with a wormlike body; a mouth, a gut, and an anus; muscles and a heart; a nervous system organized around a nerve cord and a light-sensing organ; and, finally, some kind of outgrowths on its body—if not actual legs or antennae, then perhaps appendages around its mouth to help it eat. It might be the creature that left those anonymous trails among the Ediacarans.

Paleontologists now believe that only after the genetic tool kit was complete could the Cambrian explosion take place. Only then was it possible for dozens of new animal body plans to emerge. Evolution did not build a new network of body-building genes from scratch in the process; it simply tinkered with the original genetic tool kit to build different kinds of legs, eyes, hearts, and other body parts. These animals took on dramatically different appearances, but they still held on to an underlying program for building bodies.

One of the most dramatic examples of this flexibility lies in the origin of our own nervous system. All vertebrates have a nerve cord running down their backs (the dorsal side, as biologists call it), while the heart and digestive tract are on the front (or ventral) side. Insects and other arthropods have an opposite arrangement: the nerve cord is on the ventral side, and the heart and gut are on the dorsal.

These mirror-image body plans inspired a fierce debate between Georges Cuvier and Geoffroy Saint-Hilaire in the 1830s. Cuvier found their anatomy so fundamentally different that he decided vertebrates and arthropods belonged in two completely distinct groups. But Geoffroy claimed that if you transformed the arthropod body plan drastically enough, you would end up with the vertebrate body plan. It turns out that Geoffroy was right, but in a way he couldn't have imagined. The nervous systems of vertebrates and arthropods are indeed starkly different. But the genes that control their development are the same.

When a vertebrate embryo begins to form, the cells on both the dorsal and the ventral sides have the potential to become neurons. Yet we do not have spinal cords running down our bellies, because the cells on the ventral side of vertebrate embryos release a protein called Bmp-4, which prevents cells from becoming neurons. Gradually Bmp-4 spreads from the ventral cells toward the dorsal side of the embryo, blocking the formation of neurons as it goes.

If Bmp-4 spreads all the way to the other side, no neurons could form at all in a vertebrate embryo. But as the embryo develops, its dorsal cells release

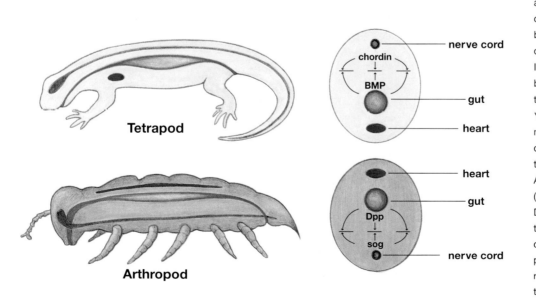

All vertebrates (side view at top left) have spinal cords running along their backs, and hearts and digestive systems in front. Insects and related invertebrates (bottom left) have the opposite arrangement. Yet the same kind of mechanism controls their development, as shown in the cross sections at right. A nerve-blocking protein (Bmp in vertebrates and Dpp in insects) spreads through the body until it is countered by a second protein (chordin or sog, respectively), which allows the spinal cord to develop.

a protein that blocks Bmp-4. Known as chordin, it protects the dorsal side of the embryo from Bmp-4, leaving the cells there free to turn into neurons. Eventually they give rise to the spinal cord that runs along a vertebrate's back.

Compare that sequence of events to what happens in a fruit fly. When a fruit fly embryo first forms, it can also form nerves on both its dorsal and ventral sides. But then a nerve-repressing protein called Dpp is made on its dorsal side, instead of the ventral side where Bmp-4 first appears in vertebrates. As Dpp spreads toward the ventral side of the fly, it is blocked by the protein sog. Protected from Dpp, a fly's ventral side can form a nerve cord.

These sets of genes not only perform similar jobs in insects and vertebrates, but their sequences are nearly identical. The nerve-blocking gene Dpp and the nerve-blocking gene Bmp-4 are matches, as are their antagonists, sog and chordin. They are so similar, in fact, that if a sog gene from a fly is inserted into a frog embryo, a second spinal cord will start taking form in the frog's belly. The same genes are building the same structures in insects and frogs, but they're flipped.

Such similar genes doing such similar jobs must have a common ancestry. John Gerhart at the University of California at Berkeley has proposed how this transformation took place. The first animals with the genetic tool kit grew several small nerve cords running along the sides of their bodies rather than a single big one. These ancestral animals carried a gene that was the ancestor of both chordin and sog, and it promoted the growth of neurons at all the places where a nerve cord was to form in their embryos.

This common ancestor gave rise to all the lineages that appeared during the Cambrian explosion. In the lineage that led to arthropods, the nerve cords all coalesced into a single one running on their ventral side. In vertebrates, the cords all migrated to the back. But the original genes for building nerve cords didn't disappear; the place where they became active changed. And so, over time, they became the mirror images that so impressed Geoffroy.

GENE DUPLICATION AND THE DAWN OF VERTEBRATES

Vertebrates acquired more than just spinal cords running down their back during the Cambrian explosion. With some tinkering to their genetic tool kit, they evolved eyes, complex brains, and skeletons. In the process, vertebrates became powerful swimmers and excellent hunters and have remained the dominant predators of the ocean and land ever since.

The oldest known vertebrate fossils—lamprey-like creatures found in China—date back to the midst of the Cambrian explosion, 530 million years ago. In order to understand how those first vertebrates emerged from their ancestors, biologists have studied our closest living invertebrate relative. Known as the lancelet, it's not a very impressive cousin. It actually looks like a headless sardine pulled from a can. The lancelet starts out life as a tiny larva, floating in shallow coastal waters and swallowing bits of food that drift past it. When it grows to be half an inch long, the adult lancelet burrows into the sand, sticks its head up into the water, and continues to filter-feed.

But as unassuming as the lancelet may look, it shares some key traits with vertebrates. It has slits near the front of its body that correspond to the gills of fish. It has a nerve cord running along its back, which is stiffened by a rod called a notochord. Vertebrates have notochords as well, but only while they are embryos. Over time, the notochord withers away as the spinal column grows larger.

In other words, certain pieces of the vertebrate body plan had already evolved in the common ancestor of lancelets and vertebrates. Yet lancelets also lack much of the anatomy that make vertebrates so distinct. They have no eyes, for example, and their nerve cord ends in a tiny bump, not in a proper mass of neurons one would, at first glance, call a brain.

But it's possible to see precursors of brains and eyes in lancelets. The lancelet can detect light with a pit lined with photosensitive cells, and these cells are wired up like a retina in a vertebrate eye and connected to the front of the nerve cord in much the same way as our eyes are to our brains. That tiny bump at the front of the lancelet's nerve cord may consist of only a few hundred neurons (our brains have 100 billion), but it is divided into simplified versions of parts of our own vertebrate brains.

The similarity between the lancelet's nerve cord and vertebrate brains extends to the genes that build them. Hox genes and other master-control genes that map out the vertebrate brain and spinal cord do the same job in the lancelet embryo, in almost precisely the same head-to-tail order. In the cells of the developing lancelet eyespot, the genes are the same as those that build a vertebrate eye. It's a safe bet that the common ancestor of lancelets and vertebrates had the same genes for constructing the same basic brain.

Once the ancestors of vertebrates and lancelets branched apart, our ancestors went through an extraordinary evolutionary experience. Lancelets have 13 Hox genes, but vertebrates have *four* sets of those genes, each arranged in the same head-to-tail order. Mutations must have caused the original set of Hox genes to be duplicated. After they evolved into four sets, the new genes met various fates. Some of them went on carrying out the jobs of the original Hox gene. But other copies of Hox genes evolved until they were able to help shape the vertebrate embryo in new ways.

Thanks to this explosion of gene duplication, our ancestors began to evolve more complex body plans. Vertebrates were able to grow noses, eyes, skeletons, and powerful swallowing muscles. At some point early in vertebrate evolution, the Hox genes that controlled their head-to-tail development were borrowed to help build fins. Fins helped vertebrates control their swimming and maneuver more effectively than their lancelet-shaped ancestors.

Instead of passively filtering food out of seawater, early vertebrates could now start hunting. Able to chase down big animals, they could evolve to be bigger themselves. Thanks to their genetic revolution, the early vertebrates eventually gave rise to sharks, anacondas, humans, and whales. Without those new Cambrian genes, we might still be like the lancelets, our tiny brainless heads still swaying in the tides.

LIGHTING THE CAMBRIAN FUSE

The evolution of the genetic tool kit was a key ingredient in the Cambrian explosion. But once it had evolved, the explosion didn't happen right away. The earliest animals that possessed the genetic tool kit probably lived tens of millions of years before the first record of the Cambrian explosion 535 million years ago. If animals already had so much evolutionary potential within them, something must have prevented them from taking off.

The tool kit of these early animals may have been like a fuse waiting for a match. Before the Cambrian, the oceans weren't a propitious place for animal evolution. Big, active animals like the ones that appeared in the Cambrian explosion need a lot of energy, and to generate it they need to absorb oxygen. But the chemistry of rocks formed on the seafloor during the Precambrian suggests that there wasn't much oxygen for them to take up. Photosynthetic algae and bacteria at the ocean's surface were releasing oxygen in abundance, but little of it was getting very far below. Oxygen-breathing scavenger bacteria on the surface fed on these photosynthesizers after they had died, leaving the rest of the ocean oxygen-starved.

Around 700 million years ago oxygen levels began to rise, eventually reaching perhaps half the concentration found today. The rise of oxygen has been linked to the breakup of a supercontinent at the time. As it disintegrated, more carbon may have been carried down into new ocean basins, leaving behind more free oxygen in the atmosphere. Some of that extra oxygen then managed to build up in the ocean as well.

After the oxygen levels rose in the oceans, the planet as a whole appears to have gone through some violent times. Ice ages overwhelmed the earth, according to Harvard geologist Paul Hoffman, bringing glaciers close to the equator. They melted away only after volcanoes released enough carbon dioxide to warm the atmosphere. As life became isolated in refuges during the global ice ages, evolution may have taken place at a faster pace, creating new species with new adaptations. Because animals already had their complex genetic circuits in place, they could respond to this evolutionary pressure by flowering into all the forms of the Cambrian explosion.

Genes and physical conditions may have started the Cambrian explosion, but ecology may have determined how big it eventually became. Among the new kinds of animals that emerged in the early Cambrian were creatures that could, for the first time in the history of Earth, eat algae. These invertebrates used feathery appendages to catch their food and became enormously successful. (Today, vast armies of fairy shrimp, water fleas, and other algae eaters continue their success.) Once algae eaters began to thrive, they spurred the appearance of large, fast-swimming predators, which in turn could have been devoured by larger predators still. The ocean's food web quickly wove itself into a complex tangle.

The new pressures of grazing and hunting may have spurred even more diversification, not just among animals but in algae as well. The most common group of algae found in the early fossil record is known as the acritarchs. Before the Cambrian, acritarchs were small and nondescript, but during the Cambrian explosion they suddenly evolved spikes and other ornaments and took on much bigger forms. They were probably evolving defenses against the grazers, making themselves harder to swallow. The grazers evolved ways to get around these defenses and evolved defenses of their own—spikes, shells, and plates of armor—against their predators, which in turn had to find new ways to attack them, evolving claws and crushing teeth and drills and sharper senses. The Cambrian explosion became a fire that could feed itself.

THE PARTY'S OVER

Yet within a few million years the fire had gone out. Paleontologists recognize only one new phylum that evolved after the Cambrian explosion—bryozoans, colonial animals that form mats on the ocean floor. That's not to say that animals have not changed since then. While the first vertebrates were all lamprey-like creatures, they've since evolved into a staggering variety, from the snowy egret to the tree kangaroo, the hammerhead shark, the vampire bat, and the sea snake. But all of these animals have two eyes, a brain housed in a skull, and muscles surrounding their skeletons. Evolution may be a creative force, but it does not have infinite scope. In fact, it works under tight constraints and can fall into many different traps.

When life goes through a burst of evolutionary transformations, the new species start seeking out ecological niches. In Lake Victoria, the cichlids evolved to scrape algae off rocks, eat insects, and take advantage of other food in the lake. The first algae scrapers may not have done a very good job, but without any other algae scrapers around, not very good was good enough. As these cichlids evolved, they created new ecological niches for even more species of cichlids: the predators that swallow them whole, or the scale rakers, or the egg stealers. Life creates new niches, but it probably can't create an infinite number of them. Sooner or later, species will begin competing with one another for them instead. Some will win and others will lose. In the older lakes of Malawi and Tanganyika, the cichlids have had millions of years more time to evolve, but they have not invented any ecological niches beyond those found in young Lake Victoria.

The Cambrian explosion may have come to an end when its ecosystem filled up in much the same way, only on a far grander scale. In the Cambrian explosion, big mobile predators appeared on Earth for the first time, as did burrowers and algae grazers. It's possible that these animals filled all the possible ecological niches and evolved to be so proficient that they shut out newcomers. Without an opportunity to explore new designs, new kinds of animals can't establish themselves.

Sometimes an explosive burst of evolution may stop because the genetic complexity it creates blocks its own path. The earliest animals were extremely simple, with only a handful of different kinds of cells, assembled by relatively few developmental genes. By the end of the Cambrian explosion, their descendants had evolved many different types of cells and used a complicated network of interacting genes to build their bodies. Often a gene that has evolved to help build one structure was borrowed to build several others. Hox genes, for example, build not only the brains and spines of vertebrates but also their fins and legs. When a gene is responsible for several different jobs, it becomes harder to change. Even if a mutation improves one structure the gene helps to build, it may completely destroy the other ones. The difference between the way evolution operated at the beginning of the Cambrian and at the end may be akin to the difference between trying to remodel a one-story house and remodeling a skyscraper.

Because evolution can only tinker, it cannot produce the best of all possible designs. Although it has come up with many structures that make engineers coo in admiration, it is often stuck making the best of a bad situation. Our eyes, for example, are certainly impressive video cameras, and yet in some ways they are fundamentally flawed.

When light enters a vertebrate eye, it travels through the jelly and strikes the photoreceptors of the retina. But the neurons in the retina are actually pointed backward. It's as if we were gazing at our own brain. Light has to make its way through several layers of neurons and a web of capillaries before it finally gets to the nerve endings that can detect it.

Once light strikes the backward-pointing photoreceptors of the retina, the photoreceptors then have to send their signals back up through the layers of

The eyes of (top to bottom) humans, octopi, frogs, insects, and crustaceans are very different, yet many of the genes that build them are identical. This eye-building genetic network may have originated 600 million years ago and has been conserved ever since.

the retina toward the front of the eye. As they travel, the neurons process the signals, sharpening the image. The uppermost layer of retinal neurons connects to the optic nerve, which sits on the top of the retina. The nerve burrows back down through all the neurons and capillaries in order to leave the eye and travel to the back of the brain.

This architecture is, as the evolutionary biologist George Williams has bluntly put it, "stupidly designed." The layers of neurons and capillaries act like a mask, degrading the light that finally gets through to the photoreceptors. In order to compensate, our eyes are continually making tiny movements so that the shading shifts around the image we see. Our brains can then combine these degraded pictures, subtract away the shading, and create a clear image.

Another flaw arises out of the way the retinal neurons attach to the optic nerve on top of the retina. The optic nerve blocks even more incoming light, creating a blind spot in each eye. The only reason blind spots don't cause us much trouble is that our brains can combine the images from both eyes, canceling out each blind spot, and create a full picture.

Yet another clumsy element of eye design is the way that the retina is anchored. Because the photoreceptors have delicate, hairy nerve endings, they can't be cemented firmly in place. Instead, they are loosely joined to a layer of cells lining the wall of the eye called the retinal pigment epithelium. The retinal pigment epithelium is essential for the eye. It absorbs extra photons so that they don't bounce back at the photoreceptors and blur the image they receive. It also houses blood vessels that supply the retina with nutrients, and as the retina sloughs off old photoreceptors, it can carry away the waste. But the connection between the epithelium and the retina is so fragile that our eyes can't withstand much abuse. A swift punch to the head can detach the retina, leaving it free to float around the inside of the eye.

Eyes can work perfectly well without taking this shape. Just compare the vertebrate eye to the eye of a squid. Squid eyes are powerful enough to let them track their prey in near darkness. They are spherical and have lenses, just like vertebrate eyes, but when incoming light strikes the inner wall of a squid's eye, it does not have to struggle through a tangle of backward neurons. Instead, it immediately strikes a vast number of light-sensitive endings of the squid's optic nerve. The signals

run directly from the nerve endings into the squid's brain, without having to travel backward over to any intervening layers of neurons.

To understand the flaws of the vertebrate eye (as well as its strengths), evolutionary biologists look back at its origins. The best clues to the early evolution of vertebrate eyes come from lancelets, the closest living relatives to vertebrates. The nerve cord of a lancelet is actually a tube, and the neurons that line it have hairlike projections called cilia that project into its hollow center. Some of the neurons at the very front of the tube serve as a light-sensing eyespot. Like other neurons in the lancelet, these light sensors are pointed inward, which means that they can only detect light that strikes the opposite side of the lancelet's transparent body and pass into the hollow tube.

Just in front of these light-sensing neurons, the neural tube ends. The cells that line the front of the tube have dark pigment inside them, which scientists suspect acts as a shield, blocking out light striking the front of the lancelet. Since light can't hit its eyespot from all directions, the lancelet can then use it to orient itself in the water.

Thurston Lacalli, a biologist at the University of Saskatchewan, has discovered some remarkable similarities between the structure of the lancelet eyespot and a vertebrate embryo's eye. A vertebrate brain forms initially as a hollow tube, just like the one in lancelets, with inward-pointing nerve cells. Eyes develop at the front end of this tube, as its walls project outward into a pair of horns. At the tip of each horn, a cup forms. On this cup's inner surface, the retinal neurons establish themselves, their nerve endings still pointing inward. On the outer surface, the pigmented cells take hold.

If you slice this eyecup and look at the arrangement of cells, you'll find it has the same topography as the lancelet eyespot. The retinal neurons still point inward, toward the center of the neural tube, as the light-sensing neurons do in lancelets. The rods and cones of the retina are highly evolved versions of the lancelet's cilia. As the tube changes shape in vertebrates, they end up pointing toward the wall of the eye. Moreover, the retinal neurons in a

A cross-sectional view of the front tip of a lancelet (left) shows striking similarities to a cross section of a developing vertebrate embryo's eye (right). A lancelet uses photosensitive neurons (red and blue) to detect light. The neurons that form the retina of a vertebrate eye follow much the same arrangement—and are built with nearly identical genes. This homology suggests that the vertebrate eye evolved from a lancelet-like eyespot.

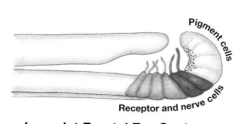

Lancelet Frontal Eye Spot

Pigment cells

Receptor and nerve cells

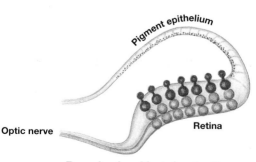

Pigment epithelium

Optic nerve

Retina

Developing Vertebrate Eye

vertebrate embryo are still positioned between the pigmented cells and the optic nerve in the same way as they are in the lancelet head.

These connections are clearest when a vertebrate is still a new embryo. The more it develops, the harder it becomes to see the similarities. The walls of the cup grow so thin that eventually the cells on the inner edge and the outer edge press against each other. In the process, they create the peculiar, delicate connection between the retina and the pigmented epithelium.

These similarities between vertebrate embryo eyes and lancelet eyespots offer clues to how our eyes acquired their strange shape. The eyespot of a lancelet-like ancestor evolved into a pair of cup-shaped light detectors branching off the nerve tube. Their cup shape allowed them to capture more light than a flat eye-spot. It gradually curved into a spherical eye that could begin to form images on its retina. But because the vertebrate eye was built on a basic lancelet design, it was stuck with retinal neurons pointing away from the incoming light.

The structure of eyespots in the ancestors of vertebrates constrained the forms that they could later take. Evolution could only make the best of a lancelet-like anatomy and the rules of development that built it. In order to turn an eyespot into a true eye, we have to put up with blind spots, detached retinas, and degraded light. Yet the advantages of having *any* ability to form images outweighed the unavoidable shortcomings of the eye's design.

Once the vertebrate eye had evolved, complete with lens, jelly, and back-ward retina, many lineages evolved new versions that work better in their own environment. For example, three different lineages of fishes have each evolved double eyes. Their eyes have two pairs of lenses rather than one; when these fishes float at the water's surface, one pair of eyes gazes up into the air, while the other looks down into the water. The upward-pointing eyes are shaped to focus light as it passes through air, while the other pair is designed to handle the optics of water.

Meanwhile a few vertebrates on land—most notably birds and primates like ourselves—have evolved extraordinarily powerful vision. They developed a dense patch of photoreceptors in a small region of the retina called the fovea; the neurons that would normally block the path of incoming light to the fovea are pushed to the sides. And yet, despite all these innovations, the backward retina endures. Thanks to 530 million years of evolutionary con-straints, our children will never be able to see like squid.

FISH FINGERS AND LIFE ON LAND

When a roulette ball lands on its wheel, its fate is not absolutely random. It does not bounce off the wheel and stick to the ceiling. It does not end up perched on the border between two numbers. The force of gravity, the energy of the throw, and the instability of the wheel's borders push the ball onto one of the numbers. Its fate is constrained, although it remains unpredictable.

The same holds for evolution. It is channeled within certain constraints, but that doesn't mean that its transformations unfold with steady, predictable progress. The internal forces of evolution—the way genes interact to build an organism—meet up with the external forces of climate, geography, and ecology, like advancing weather fronts. When they collide, they produce evolutionary tornadoes and hurricanes. As a result, scientists have to be on their guard when they try to reconstruct how evolutionary transformations took place, because it is easy to impose a simple story on a counterintuitive reality.

If 530 million years ago marks one major milestone in our evolution—the dawn of vertebrates during the Cambrian explosion—the next must be 360 million years ago, when vertebrates came on land. During those intervening 180 million years, vertebrates evolved into a vast diversity of fishes—including the ancestors of today's lampreys, sharks, sturgeon, and lungfish, as well as extinct forms, such as the jawless, armor-plated galeaspids and placoderms. But during that entire time there was not a single back-boned creature walking on dry land. Only 360 million years ago did vertebrates finally emerge from the ocean. From them, all terrestrial vertebrates (known as tetrapods) are descended—everything from camels to iguanas to toucans to ourselves.

Early descriptions of this transition were infused with a heroic tone, as if it were part of some foreordained step toward the rise of humanity. From the squirming fish of the sea, the story went, pioneering species emerged onto dry land, struggling with their fins and evolving lungs and legs to let them conquer dry land, rising high and standing tall. In 1916, the Yale paleontologist Richard Lull wrote, "The emergence from the limiting waters to the limitless air was absolutely essential to further development."

In fact, the origin of tetrapods was a far different story, one that paleontologists themselves did not even begin to understand properly until the 1980s. Before then, evidence about what the earliest tetrapods were like was hard to come by. Researchers knew that of all fishes, the ones most closely related to tetrapods were an ancient lineage known as lobe-fins. The living lobe-fins include lungfishes, which live in Brazil, Africa, and Australia. These freshwater fishes can breathe air if their ponds dry up or if the oxygen levels of their water drop dangerously low. The other lobe-fin is the coelacanth, a hulking, wide-mouthed creature that lives hundreds of feet below the ocean's surface off the coasts of southern Africa and Indonesia.

The skeletons of lobe-fins bear some special similarities to those of tetrapods. Their stout, muscular fins, for example, have the same basic arrangement as legs and arms: a single bone closest to their body, which connects to a pair of long bones, which in turn connect to a group of smaller bones.

Although lungfish and coelacanths are the only lobe-fins alive today, 370 million years ago lobe-fins were among the most diverse groups of fish. And paleontologists discovered that some of those extinct lobe-fins were even more like tetrapods than living lobe-fins are.

As for the oldest tetrapods, paleontologists knew of only one species: a 360-million-year-old creature called *Ichthyostega*. Discovered in the mountains of Greenland in the 1920s, this 3-foot-long, four-legged creature was clearly a tetrapod, but it had a flat-topped skull that looked more like lobe-finned fish than it did later tetrapods.

Paleontologists concluded that *Ichthyostega* was the product of a long struggle to adapt to dry land. The American paleontologist Alfred Romer sketched out the most thorough scenario for this origin for tetrapods. Their lobe-fin ancestors lived in freshwater rivers and ponds, but a change in the climate brought on seasonal droughts that made their homes evaporate every year. The fish that could drag themselves to the next pond survived, while the ones that were stranded died. The more mobile lobe-fins were more likely to survive, so over time their fins evolved into legs. Eventually these fish became so good at moving on land that they could hunt the insects and other invertebrates crawling around on the ground, and they gave up life in the water altogether.

Romer's scenario seemed logical enough, at least until a second early tetrapod was discovered in Greenland. In 1984 Jennifer Clack, a paleontologist at the University of Cambridge, was perusing the notes from an expedition of Cambridge geologists in the 1970s. They had discovered *Ichthyostega*-like fossils and had simply stored them away right under Clack's nose. Clack

Ichthyostega, one of the earliest vertebrates with legs and toes, lived 360 million years ago. Despite its anatomy, it probably spent little if any time out of water.

returned to their site in 1987 and found the complete skeleton of another 360-million-year-old tetrapod, named *Acanthostega*.

Acanthostega had all the required hallmarks of a tetrapod, such as legs and toes, but it was an animal that could only have lived underwater. For one thing, Clack and her colleagues discovered bones in its neck that supported gills. For another, its legs, shoulders, and hips were all far too weak to hold up its weight on dry land.

Acanthostega made no sense in Romer's scenario, but paleontologists were realizing that some of his assumptions were wrong. *Acanthostega* and other early tetrapods did not in fact live in harsh, drought-plagued habitats. Instead they lived in lush coastal wetlands, a habitat that was coming into existence for the first time on Earth as large trees began to grow along coasts and rivers. There were no droughts to drive fish toward a tetrapod's body.

Clack and other paleontologists now argue that fish evolved legs and toes not to walk on land but in order to move underwater. With a few minor changes to the ways in which master-control genes built fins, evolution rearranged the bones into toes. These tetrapod-like fish could then have used their fins to clamber through reedy marshes, over fallen logs and other debris. They could grip on to rocks to lie still as they waited to ambush passing prey. While this sort of locomotion may seem strange to us, some living fish do much the same thing. Frogfish have finger-like projections on their fins that they use to walk slowly over coral reefs.

Whatever they were originally used for, feet and toes didn't evolve in response to the demand for walking on land, their current use. As of 2000, paleontologists had discovered a dozen or so species of early tetrapods, and they all appear to have been aquatic. (Clack and her colleagues have taken a fresh look at *Ichthyostega* and have concluded that it may have been able to drag itself around on dry land like a seal.) Ted Daeschler of the Academy of Natural Sciences in Philadelphia has even found the fossil of a separate lineage of lobe-fins that evolved finger-like bones independently of our own ancestors. Between about 370 and 360 million years ago, it seems, walking lobe-fins went through an underwater evolutionary explosion—cichlids with legs, as it were. Only later did one branch of tetrapods move on land, their legs now taking on a new function.

Evolution often borrows things adapted for one function to perform a new one (a process known as "preadaptation" or "exaptation"). As is so often the case with evolution, this tendency was first noticed by Darwin. "When this or that part has been spoken of as adapted for some special purpose, it must not be supposed that it was originally always formed for this sole purpose," he wrote in 1862. "The regular course of events seems to be, that a part which originally served for one purpose, becomes adapted by slow changes for widely different purposes."

FORWARD INTO THE PAST: THE ORIGIN OF WHALES

The notion that evolution moved in a steady forward progression agreed well with the Victorian conception of history. People's lives in Europe were better at the end of the nineteenth century than they were at the beginning, thanks to science and industry, and they would keep getting better. That steady improvement seemed to be reflected in the history of life itself.

But Victorian biologists knew that if there was some sort of imperative for progress, many animals were ignoring it. Barnacles, for example, descended from free-swimming crustaceans, but they had given up that independent life for a lazy existence clamped to a piling or a ship hull. If evolution was a steady march, it could start moving backward at any moment. Victorian biologists didn't give up their notion of progress; they simply turned it into a two-way street, with progress in one direction and degeneration in the other. Ray Lankester, a British biologist, worried that degeneration could strike human society if people weren't careful. "Perhaps we are all drifting towards the condition of intellectual barnacles," he wrote.

But just as evolution is not a steady march of progress, it cannot run backward either. Evolution is change, nothing more or less. Tetrapods took their heroic crawl out of the water 360 million years ago, and their descendants have gone back in more than a dozen times. When they entered the water, they did not degenerate into lancelets, let alone lobe-fins. Instead, they became things altogether new, such as whales.

Whales have been trouble for scientists ever since Linnaeus put together the first modern taxonomy in 1735. "Amidst the greatest apparent confusion, the greatest order is visible," Linnaeus wrote about classification, and yet when he tried to classify whales, he seemed only to add more confusion. Were they fish or mammals? "These are necessarily arranged with the Mammalia," he demanded, "though their habits and manners are like those of fish." Whales, he pointed out, have hearts with ventricles and auricles like mammals, they are warm-blooded, have lungs, nurse their young—just like mammals on land. They even have eyelids that move.

Linnaeus's classification was hard for the public to accept. In 1806 the naturalist John Bigland complained that it "will never prevent the whale from being considered as a fish, rather than as a beast, by the generality of mankind." And in *Moby Dick,* Ishmael probably spoke for most people in the nineteenth century when he said, "I take the good old-fashioned ground that the whale is a fish, and call upon holy Jonah to back me."

Darwin saw a way out of this confusion. Linnaeus was not playing some sort of meaningless game by grouping whales with mammals. The similarities that Linnaeus found were signs that whales (including porpoises and dolphins) descended from mammals that lived on land. Evolution had produced a meta-morphosis that Ovid would have loved: it had transformed away their legs, given them flukes on their tail, had put their noses on top of their heads, and had made them so big that the heaviest whales can balance the scales with a town of 2,000 citizens. It had created a fishlike mammal, but it had not in the process gotten rid of the evidence of its ancestry.

How exactly evolution had accomplished all this, Darwin could not say. He could think of no living intermediates between whales and land mammals. But not knowing how did not bother him much, because he could imagine ways. He pointed out that bears sometimes swim for hours with their mouths open, catching insects. "Even in so extreme a case as this," he wrote in *Origin of Species*, "if the supply of insects were constant, and if better adapted competitors did not already exist in the country, I can see no difficulty in a race of bears being rendered, by natural selection, more and more aquatic in their structure and habits, with larger and larger mouths, till a creature was produced as monstrous as a whale."

The idea didn't go over well. One newspaper complained that "Mr. Darwin has, in his most recent and scientific book on the subject, adopted such nonsensical 'theories'—as that of a bear swimming about a certain time till it grew into a whale, or to that effect." Darwin dropped the example from later editions of his book.

In the 120 years that followed, paleontologists found a steady supply of fossils of whales, but even the oldest specimens, dating back more than 40 million years, were fundamentally like whales today. They had long backbones, hands in the shape of flippers, and no back legs. Their teeth were another matter, though. Living whales have either no teeth or simple pegs. The oldest whales had teeth that had the cusps and bumps of mammals on land. They looked particularly like the teeth of an extinct line of mammals called mesony-chids. These animals were hoofed mammals—relatives, in other words, of cows and horses—but they had powerful teeth and strong necks adapted for a life of eating meat, which they got either by scavenging or hunting.

Finally, in 1979, Philip Gingerich, a paleontologist from the University of Michigan, discovered a whale that lived on land.

Gingerich and his team were searching for 50-million-year-old mammal fossils in Pakistan. Today Pakistan is nestled within Asia, but when the fossil mammals were alive, it was little more than a collection of islands and coastlines. India at the time was a gigantic island drifting north toward the southern edge of Asia. Gingerich's team found many bits and pieces of mammals, most of which they could immediately identify. But a few were harder to pin down. One particularly baffling fossil was the back part of a 50-million-year-old skull. It was about the size of a coyote's and had a high ridge running like a mohawk over the top of its head, where muscles could attach and give the mammal a powerful bite. When Gingerich looked underneath the skull, he saw ear bones. They were two shells shaped like a pair of grapes and were anchored to the skull by bones in the shape of an S.

For a paleontologist like Gingerich, these ear bones were a shock. Only the ear bones of whales have such a structure; no other vertebrate possesses them. Gingerich named his creature *Pakicetus,* meaning "whale of Pakistan," and in the years since, he has found its teeth and bits of its jaw. They are intermediate between mesonychids and later whales, confirming that *Pakicetus* was in fact a 50-million-year-old whale—the oldest whale known at the time. Yet the rocks where *Pakicetus* fossils have been found showed that this coyote-like creature had lived and died on land, among low, shrubby plants and shallow streams only a few inches deep. *Pakicetus* was a terrestrial whale.

Fifteen years later, in 1994, a former student of Gingerich's named Hans Thewissen discovered another primitive whale. Thewissen found not just bits and pieces of his creature but what would ultimately turn out to be its entire skeleton. This 45-million-year-old whale had giant feet that looked as if they could fit in clown shoes and a bulky skull in the shape of an alligator's head. He named it *Ambulocetus,* meaning "walking whale." By the close of the twentieth century, Thewissen, Gingerich, and other paleontologists had found several other species of whales with legs in Pakistan, India, and the United States. What once seemed an impossibility is now commonplace.

Paleontologist Phil Gingerich (left) has discovered ancient whales with legs in Egypt (shown here) as well as Pakistan.

To understand how these early whales evolved into their fishlike life, paleontologists have compared the fossil whales to living species and mesonychids to create an evolutionary tree, shown on the following page. It tells a provisional story of how whales came to be. Darwin shouldn't have been thinking of bears, it turns out; he should have been thinking of cows and hippos. These hoofed mammals are among the closest living relatives of whales. The closest extinct relatives were the mesonychids. Mesonychids took many forms, from creatures the size of squirrels to a terrifying monster called *Andrewsarchus* that measured 12 feet long, the biggest known carnivorous mammal of all time. Among their ranks, the first whales wouldn't have drawn much notice.

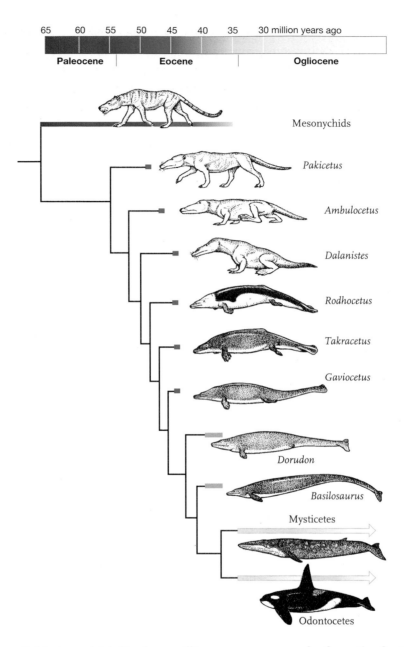

65   60   55   50   45   40   35   30 million years ago

Paleocene   **Eocene**   **Ogliocene**

Mesonychids

Pakicetus

Ambulocetus

Dalanistes

Rodhocetus

Takracetus

Gaviocetus

Dorudon

Basilosaurus

Mysticetes

Odontocetes

The evolutionary tree of whales offers clues to how coyote-like mammals gradually adapted to life in the ocean.

Pakicetus, which lived 50 million years ago, was far from the first whale. Mesonychids and whales all share a common ancestor, and that common ancestor must have lived before both the oldest known whale and the oldest known mesonychid. While the oldest known whales are about 50 million years old, there are fossils of mesonychids dating back 64 million years. Thus, whales must have diverged from mesonychids before 64 million years ago, more than 14 million years before *Pakicetus* lived.

Early whales still had legs attached to their shoulders and hips, which in turn were firmly attached to their spines. Their ears still resembled the ears of land mammals, which can pick up sounds in the air. Their teeth still resembled those of mesonychids in their general outline, but they were already changing. Maureen O'Leary, a paleontologist at the State University of New York at Stony Brook, has looked closely at the teeth of early whales and has discovered long gouges running along the outward sides of the lower molars. These gouges formed as the whales scraped their molars with their upper teeth. The whales had to have been making only vertical bites, not side-to-side chewing, to form them. There's fossil evidence that later whales, which also had these gouges, fed on fish. That has led O'Leary to suggest that *Pakicetus* and its contemporaries had already started eating fish or other aquatic animals. Even without a modern whale's body, an early whale could have swum, if only in a dog paddle.

But soon after *Pakicetus* emerged, evolution began changing other parts of whale anatomy, producing animals better adapted to swimming. *Ambulocetus,* Hans Thewissen's walking whale, has short legs, a long snout, big feet, and a powerful tail. That sort of anatomy would have suited it for swimming like an otter—pushing back with its hind feet and adding more thrust by moving its tail up and down. Yet *Ambulocetus,* like otters, had hips that were still articulated to its spine. In other words, it could still walk on land. It probably hauled itself out of the water to bask, sleep, mate, and bear young.

Ambulocetus, a 45-million-year-old whale, still retained the legs of its ancestors. It may have lived like an alligator, basking on land and ambushing prey in the water.

In the border zone between land and sea, many species of walking whales arose. Some were adapted for wading, others for diving. For the most part, these lineages became extinct, for reasons that may never be known. But one lineage of whales adapted to life farther out to sea. It produced species such as *Rodhocetus,* a whale Gingerich found in Pakistan that had stubby legs and hips that were barely connected to its spine. In the water it could have raised and lowered its tail and trunk together, swimming like whales do today. Modern whales vastly improve their swimming performance with the flukes at the end of their tails, made of connective tissue. Because that sort of flesh rarely fossilizes, no one knows if *Rodhocetus* already had flukes to help it along.

By 40 million years ago, whale evolution had produced fully marine whales. *Basilosaurus* measured 50 feet long, with a slender serpent body, a long snout, and arms that had become stout flippers. It lived far from shore, and dry land would have been a death sentence for it. When paleontologists look at where the stomach once was on *Basilosaurus* fossils, they sometimes find a preserved meal of shark bones. These animals were much closer to our picture of a whale. But *Basilosaurus* lived for millions of years alongside semiaquatic and seal-like whales that still retained some parts of their terrestrial past.

And *Basilosaurus* still had a few vestiges of its own. Its nostrils had moved only halfway back its snout toward the position where blow holes are on living whales. In 1989 Gingerich found a fossil in Egypt that tied *Basilosaurus* even more tightly to the past. Along its giant snaky body he found hips, and attached to the hips were hind legs. They were only a few inches long, but they even had five delicate toes.

Like all evolutionary trees, this one is a hypothesis. And like all hypotheses, it may have to be refined as more evidence comes in. It's possible, for example, that *Basilosaurus* is not the closest relative to living whales—that honor may go to a species called *Dorudon.* Meanwhile, some scientists have been uncovering fascinating information in the genes of whales. Whale DNA shows that they are hoofed mammals, just as the paleontologists had concluded years earlier. But studies of their genes suggest that whales are most closely related to one particular hoofed mammal: the hippopotamus. Paleontologists had traditionally thought that mesonychids and hippos were distant cousins. The easiest way to resolve this puzzle would be to look at the DNA of mesonychids, but given that they've been extinct for 36 million years, that's not a test anyone should hold their breath for. In the meantime, O'Leary is working on ways to combine the evidence from whale bones and whale genes in order to find the best hypothesis to explain all the evidence.

Although these open questions are important, they don't alter the basic lessons of this tree. A whale is no more a fish than a bat is a bird. Early whales evolved into remarkably fishlike forms through a gradual series of steps. But inside every whale's finlike flipper there still remains a hand, complete with fingers and a wrist. And while a tuna swims by moving its tail from side to side, whales swim by moving their tails up and down. That's because whales descend from mammals that galloped on land. Early whales adapted that galloping into an otter-like swimming style, arching their back in order to push back their feet. Eventually new whales emerged in which evolution had adapted that back-arching movement to raise and lower a tail.

Although whales went through an extraordinary burst of evolution, their history put limits on what they could become. And that was not the only constraint on how they could evolve. Whales and other mammals did not undergo their dramatic radiations until the dominant vertebrates of the day—dinosaurs and giant marine reptiles—were gone. The first mammals evolved more than 225 million years ago, and for the next 150 million years they remained squirrel-sized, barely distinguishable from one another. It was not until after the Cretaceous period ended, 65 million years ago, that the oldest fossils of most of the living orders of mammals appear. It was only then that the first primates leaped through the trees, only then that whales split off from other hoofed mammals and began their return to the ocean. In a few million years mammals became flying bats; they became gigantic relatives of rhinos and elephants; they became powerful, lion-sized predators. Mammals experienced their own explosion, one somewhere on the scale between the Cambrian explosion and the burst of cichlids in Lake Victoria. Ever since, mammals have dominated the land and thrived in the oceans. But mammals only underwent this explosion because millions of species—including marine reptiles and dinosaurs—suddenly disappeared. It was not some steady, gradual perfection that was responsible for the rise of mammals. It was, instead, an asteroid from space, sweeping away the old to make way for the new.

Extinction

HOW LIFE ENDS AND
BEGINS AGAIN

DARWIN DIDN'T think all that much of extinctions. He certainly knew about the work of naturalists like Cuvier, who argued that catastrophes had punctuated life's history, each one clearing the world for a new set of creatures to take its place. But Darwin, so impressed by Lyell's vision of gradual change, thought that the idea was hopelessly old-fashioned. "The old notion that all the inhabitants of the Earth having been swept away by catastrophes at successive periods is generally given up," he claimed in *Origin of Species*.

For Darwin, extinction was simply the exit that losers took out of the evolutionary arena. They did not leave in great stampedes; they only trickled away, as individual species were gradually outcompeted to oblivion. The fossil record might suggest that many species had gone extinct at once, but with countless fossils still undiscovered, it could well be misleading. In the future, as paleontologists

Hulky reptilian creatures called *synapsids*, which included the ancestors of living mammals, dominated the land more than 250 million years ago.

dug up more fossils, Darwin was sure that these seeming catastrophes would dissolve into a smooth, gentle continuum of extinctions.

Paleontologists since Darwin's time have indeed found many more fossils, as he would have hoped, and they've even been able to pin down their ages with great precision. Yet all this new information has shown that Darwin was wrong about extinctions. Catastrophic waves of extinctions are a reality. They have ripped through the fabric of life, destroying as many as 90 percent of all species on Earth in a geological instant. The suspects behind these mass extinctions are many, including volcanoes, asteroids, and sudden changes to the oceans and the atmosphere. All of these culprits seem to have put life under worldwide stress; once that stress passes a certain threshold, entire ecosystems collapse like a house of cards. And once mass extinctions strike, it takes millions of years for life to recover its former diversity. In the wake of mass extinctions life can change for good. They can wipe out old dominant forms and let new ones take their place. In fact, we may owe our own success to such shifts of fortune.

It also looks as if we are now entering another period of mass extinctions. But for the first time in the history of the planet, a single species—ourselves—is an agent of destruction. The overture of this extinction began thousands of years ago as humans arrived in Australia and other continents for the first time and hunted down the biggest native animals. But in the past few centuries the tempo of extinctions has been accelerating as humans have come to dominate the planet, destroying tropical forests and introducing alien invaders that are outcompeting native life. In the coming century, humans may even raise the temperature of the planet, putting more stress on species that are already on the brink of extinction. According to some estimates, more than half of the world's species will disappear in the next 100 years.

Mass extinctions remain one of evolution's great mysteries, and from the hills of northern Italy to the deserts of South Africa paleontologists are struggling to understand their role in the history of life. Their work is not merely academic. It may be able to show the direction in which humanity is steering the course of evolution. And that knowledge makes this period of mass extinctions different from past ones in another way: not only is this extinction pulse being caused by a single species, but by a species that is capable of understanding and controlling its own fate.

THE GREAT CURVE

The rough outlines of the history of extinctions became clear by the 1840s. As geologists surveyed formations, they often found that a given fossil species was limited to a particular layer of rock. Hundreds of miles away, the same fossils appeared in another span of rock layers. The geologists began to tie the rocks of the world into a single stratigraphy: a unified grand history of life. In the 1840s an English naturalist named John Phillips recognized that the fossils documented three great eras: the Paleozoic, the Mesozoic, and the Cenozoic ("ancient life," "middle life," and "early life," respectively). According to Phillips, these three eras were divided by great extinctions. He drew his argument on a piece of paper. The diversity of life rose from nothing at the beginning of the Paleozoic, dipping and rising from time to time before plunging at the end of the era. As the Mesozoic era began, life took another steep climb and then plummeted again at the boundary with the Cenozoic. The dominant life-forms of each era dwindled during the extinctions, a new menagerie appearing in their wake.

The curve that Phillips drew was correct but crude, like the outline of a distant mountain range shrouded in fog. Now, more than 150 years later, much of the fog has lifted. Geologists have unified the world's exposed rocks into a single record. They have found cliffs and outcrops where they can touch the place where the rocks of one era give way to the rocks of the next. They have looked at the atomic clocks ticking away inside those cliffs and outcrops and can put precise dates on their ages. They've loaded computers with vast databases of the planet's fossils. And, remarkably, Phillips's curve has survived relatively intact.

The latest version of the curve, shown below, is the work of a number of paleontologists, but most importantly of the late John Sepkoski of the University of Chicago. Sepkoski spent decades tallying the duration of ocean-dwelling species, which leave the best fossil record. The curve starts about 600 million

A BRIEF HISTORY OF MASS EXTINCTIONS

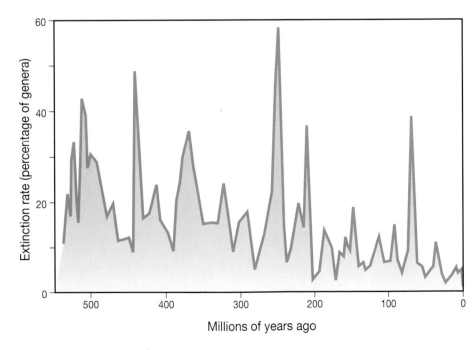

This graph charts the changing extinction rate of marine invertebrates over the past 600 million years, revealing five pulses of mass extinction.

years ago at the beginning of the Cambrian, when for the first time there are enough fossils to get a reliable picture of extinctions, and runs up to today. Along the vertical axis is the number of genera of marine animals that existed at any point.

Over much of the past 600 million years, life experienced a steady, low level of extinctions. These "background" extinctions correspond to the gradual disappearance of species that Darwin proposed. Most species live between 1 and 10 million years, and new species come into existence at roughly the same rate as older species disappear. In normal times the diversity of life is like a field full of fireflies, each flash representing a species. At any moment some fireflies are starting their flashes, while others are winking out. Yet the overall number of flashes holds roughly steady.

Now imagine that half of the fireflies in a field wink off at once. The field darkens, and the missing fireflies don't start glowing again for an hour. Something like this has happened a few times since the Cambrian period. Background extinctions have suddenly turned into mass extinctions. They have struck the ocean and land every few dozen million years, with five catastrophes standing out in particular. In each of these five mass extinctions, more than half of all species disappeared.

The wrath of these extinctions is beyond anything Darwin imagined. Just as new forms of life may come into being in ways Darwin never could have conceived, the ways they have vanished would have surprised him too. Extinction is the dark counterpart to evolution's creativity. Conditions on this planet can change so suddenly that natural selection cannot help a species adapt and survive. A pulse of extinctions can even reroute the history of life onto a new set of tracks.

THE PERMIAN-TRIASSIC EXTINCTION: BRUSHING WITH ANNIHILATION

Deaths can be hard to decipher, whether they are the deaths of individuals or of an entire species. In the Karroo Desert in South Africa there is a graveyard on an abandoned farm. Locals know that a family named the Fouches owned the farm in the late 1800s: a mother, father, and two sons. They all died in the 1890s, but no one knows how. It has been barely more than a century, and the fates of the Fouches have been lost.

Since 1991 the University of Washington paleontologist Peter Ward has been coming to the mountains near the Fouche farm. The rocks that make up the mountains contain clues to the biggest mass extinction ever recorded. John Phillips recognized this extinction, which marked the end of the Paleozoic era and the beginning of the Mesozoic era 250 million years ago. (Today, paleontologists refer to it as the boundary between two smaller units of time, the Permian period and the Triassic period.) Over 90 percent of all species on Earth disappeared, but their deaths lie in an obscurity far deeper than the Fouches', with hardly any clues as to how they vanished. Ward comes back to the Karroo every year to look for those clues. "We need to understand what happened," he explains. "It is important because the rules of evolution that held for that extinction could apply to our current event. We are undergoing a mass extinction event today, and understanding what may have happened in the past can help us predict how life on Earth will be affected."

Today the Karroo is a stark, mountainous region, but 250 million years ago, just before the mass extinctions, it would have been a very different place. "It would have been a big, wide river valley," says Ward. "The rivers would have been gigantic, Mississippi-sized. The forests lining the side were made of plants unlike those today. There were no flowers. There would have been no birds flying overhead. It would not have looked anything like our Earth as we know it today. A totally foreign, alien world."

The rocks of the Karroo Desert in South Africa hold fossils from the greatest mass extinction of all time, 250 million years ago, when 90 percent of all species disappeared.

The dominant animals of the ancient Karroo were reptilian creatures called synapsids, from which all mammals would later descend. By 250 million years ago, some synapsids had evolved into stocky hippo-like herbivores and bizarre carnivores, some of which looked like lizards with the head of a saber-toothed turtle. By 250 million years ago, these synapsids had already evolved a few of the key traits of mammals. They had evolved jaws and teeth that could chew food rather than slashing and gulping it whole, helping to create a more efficient digestive system that increased their stamina; legs that no longer sprawled to their sides but fit instead underneath their bodies, giving them a more powerful run; and a metabolism that was more warm-blooded than cold.

Synapsids were not alone in the ancient Karroo—amphibians, turtles, crocodiles, and even the forerunners of dinosaurs lived alongside them—but synapsids ruled. The fossils of the Karroo hint that synapsids swarmed over its conifer forests and fern savannas in numbers that rivaled the wildebeest and antelope that roam East Africa today. Life seemed like it could only get better for synapsids. But all of that would change in a geological instant.

"The Karroo is an amazing place," says Ward. "In paleontology, it's really one of the sacred places. There is no other place on Earth where the fossils of mammal-like reptiles are either as abundant, as easily recovered, or, to this point, as well studied. This is really the center of the Earth for understanding what happens on land during the Permian-Triassic extinction."

In a gully in the Lootsberg Pass, Ward and his colleagues can see the final years of the Permian period. The green and olive strata turn to red and purple, indications that the Karroo was turning hot and dry. The fossils of tetrapods, so easy to find in older rocks, become rarer and rarer. Eventually only three species of synapsid fossils survive—one synapsid that endured from earlier times, along with two newcomers: a predator called *Moschorinus* and an ugly, hippo-like herbivore called *Lystrosaurus*. And the rocks from the very end of the Permian, a green layer of rock, lack any sign of life whatsoever.

"The mass extinction happened in these beds," says Ward. "We have no fossils whatsoever. All the Permian creatures that we saw right down there have disappeared entirely. A few of them, we know, survived, because one or two species will be found a little higher up. But in these beds we find nothing. Not only are there no fossils, there aren't any of the burrows or the tunnels or the traces of animal activity. We see, instead, layers of rock that could only have formed in the absence of animal life. So catastrophic was that mass extinction that even the small creatures have died out. Not just the mighty, but the meek as well. This place is dead."

These rocks paint a picture of pure desolation. Ward and his colleagues can trace impressions on them that show how the disappearance of trees released the rivers of the Karroo from their narrow channels. They wandered across the basin as braided streams, clogged with eroding soil. Only in the very highest rocks of the gully do fossils of *Lystrosaurus* appear again, a rugged survivor, along with synapsids more closely related to mammals, as well as the forerunners of dinosaurs. After millions of years the trees anchor the landscape again.

Life on land suffered not just in the Karroo but around the world. Almost all species of trees on Earth became extinct, and many of the smaller plants disappeared as well. Even insects—which have not succumbed to any other mass extinctions in their 500-million-year history—vanished in great numbers. In the oceans, the destruction was even more devastating. Entire reefs died away. Trilobites, ribbed arthropods that had been among the most common marine animals for 300 million years, were claimed by the Permian-Triassic extinction. Giant sea scorpions known as eurypterids—some of which grew up to 10 feet long—emerged 500 million years ago and thrived for 250 million years. At the end of the Permian they too became extinct. All told, an estimated 90 percent of all species on Earth disappeared.

In the oceans, at least, they disappeared in a hurry. Near the village of Meishan, in southern China, there are abandoned limestone quarries whose walls record the extinction of marine animals at the end of the Permian. The carbon atoms that make up the limestone speak of a global disaster as well. Limestone is made out of the skeletons of microscopic creatures. They build skeletons by combining calcium and carbon dioxide in the seawater to form calcium carbonate. The carbon that they use may come from some living sources—a rotting leaf, a dead bacterium—or it may have an inorganic origin in a volcano. Because photosynthesis filters out a lot of carbon 13, organic carbon has a different ratio than inorganic carbon. By measuring the ratio of carbon isotopes in limestone, scientists can figure out how much organic carbon was being produced at the time the skeleton-building creatures were alive.

During the Permian-Triassic extinction, the isotopes in the limestone at Meishan went through a wild swing. Their gyrations hint that the oceans' ecosystems completely collapsed, and dead organic matter flooded the seas. Geologists have found the same isotopic lurch in rocks from Nepal, from Armenia, from Austria, from Greenland. What makes the Meishan quarries special is that their limestone is interspersed with layers of volcanic ash from eruptions that took place just before and after the extinctions. And in those ash layers are zircons, the time-telling crystals. Meishan can put a limit on just how long the collapse lasted.

In 1998 Samuel Bowring of the Massachusetts Institute of Technology and his colleagues measured the uranium and lead in zircons just above and below the extinctions and the swing of carbon isotopes. They concluded that the interval lasted less than 165,000 years, and perhaps much less. Volcanic ash layers from elsewhere in China gave the same result. In the scope of geological time, the Permian-Triassic extinction happened in a flash.

Any explanation for the Permian-Triassic extinction must be able to slip into these narrow constraints. One popular hypothesis was based on the slow but immense drop in sea levels around the extinction at the end of the Permian. About 40 percent of the continents and their surrounding shelves was covered in water 280 million years ago. By 250 million years ago that figure had dropped to 10 percent.

But if this retreat of the oceans was the cause of the mass extinctions, they should have taken millions of years to unfold, not the brief pulse that Bowring and others have documented. The world's ecosystems collapsed like a house of cards, not like a slowly eroding hillside. Scientists are looking for other culprits instead.

Volcanic eruptions might have been able to do their damage in such a short period of time. No earlier than a few hundred thousand years before the extinctions, lava began flowing out of giant vents in what is now Siberia. Over the course of about a million years, these vents belched up 11 massive eruptions, a total of 3 million cubic kilometers of lava—enough lava to cover the entire planet's surface to a depth of 20 meters. The Siberian volcanoes might have been able to cause the extinctions by wrecking the climate and chemistry that made life possible. Along with the lava, the volcanoes may have released huge clouds of sulfate (SO_4). In the atmosphere, these molecules would have acted like seeds for fine droplets, creating a haze that could have reflected sunlight away and chilled the planet. When these droplets fell out of the sky as sulfuric acid rain, they would have poisoned the ground.

In both these ways, the volcanoes could have killed off most of the trees on the planet. The insects that depended on them would have gone extinct as well, along with many vertebrates. The acid rain and cold clouds might have lasted for a few years. As they faded, the volcanoes could have wreaked havoc in another way. The Siberian volcanoes may have released trillions of tons of carbon dioxide, which gradually would have absorbed heat and created global warming. The global climate appears to have heated up quickly, perhaps in only a few decades. The heat wave would have put enormous stress on a biosphere that was already crippled.

According to Andrew Knoll of Harvard University and his colleagues, volcanic eruptions may have destroyed life in the ocean by upsetting its delicate chemical balance. There is evidence that 250 million years ago, the deep ocean had accumulated poisonously high levels of carbon dioxide. The organic carbon that fell to the seafloor produced CO_2 gas, and thanks to a sluggish circulation of water in the ocean, the gas remained trapped in their depths. The volcanic eruptions liberated the CO_2 by altering the climate, thereby stirring up the oceans. When the CO_2 reached the shallower waters, it acidified the blood of animals there, driving most of them to extinction.

Scientists have not narrowed down their search to a single culprit, and it may well be that many factors created the Permian-Triassic extinction. "If we look at the rock record across this boundary, between the Permian and the Triassic, we have so many clues that things were really getting bad, bad in many ways," says Ward. "We have this great drought, we have an increase in temperature. The very nature of how rivers worked, the ways in which sediment was accumulated on sea bottoms, the level of the ocean was changing—globally things were rapidly changing in many different directions. And this may have been a mass extinction brought about by many things going to hell in a handbasket very quickly."

REBIRTH

It strains the human imagination to picture the world just after mass extinctions. We have nothing in our experience to compare it to. But the volcanoes that have erupted during human history can offer a glimpse of what life must have been like after the dying stopped 250 million years ago.

In the Sunda Strait, running between Java and Sumatra, there once was an island called Krakatau. Before 1883, those who sailed past it could look up at the forested flanks of a quiet volcano. The Dutch set up a naval station on Krakatau in the 1600s, mined it for sulfur, and logged its trees. Indonesians lived in a few villages on the island, growing rice and pepper, until the 1800s. By 1883 Krakatau was uninhabited.

In May of that year, the volcano began to rumble. A group of Dutch volcano watchers sailed to the island and climbed the rim of one of its craters, measuring 980 meters across. They saw steam, ash, and pumice fragments the size of baseballs shoot up into the air. Then for three months Krakatau became quiet again, preparing for a climax. On August 26 the island erupted, with explosions that could be heard hundreds of miles away. A column of ash rose 20 miles. Mud rained down from the dark sky. Clouds of vaporized rock glided over the strait at 300 miles an hour. When they hit land, they raced uphill, incinerating thousands of people. Tsunamis rolled out from Krakatau, washing away dozens of villages and then heading out across the globe. They even made the English Channel bob. For months afterward, the ash from the eruption floated in the sky, turning sunsets around the world bloodred. In November 1883 fire engines in New York and Connecticut were called out because the red glow in the west looked like entire towns ablaze.

The day after the eruptions stopped, a ship named the *Gouverneur-Général Loudon*, passing by Krakatau, reported that two-thirds of the island was gone.

The eruption of Krakatau's volcano in 1883 destroyed almost every plant and animal on the island. In the decades that followed its ecosystem revived as species gradually returned.

Where the volcano had been, there was a submerged pit reaching down hundreds of feet underwater, surrounded by a frail archipelago of bare, burned earth. Nothing that had lived on Krakatau had survived, not even a fly. Nine months later a naturalist visiting the islands wrote, "In spite of all my searching I could find no sign of plant or animal life on the land, except a solitary very small spider; this strange pioneer of the renovation was in the process of spinning its web."

Within a few years a thin coat of life was covering the islands again. Cyanobacteria formed a gelatinous film over the ash, and later ferns, mosses, and a few flowering beach plants sprouted. By the 1890s, a savanna, with fig and coconut trees scattered across it, had grown on the islands. Along with the spiders lived beetles, butterflies, and even a monitor lizard.

To cover the 27 miles from the mainland to the islands, plants and animals had to travel by sea or air. Seeds of some plants could float on the currents of the Sunda Strait. The monitor lizard could swim, and other animals could ride on top of driftwood and rafts of plants. The spiders arrived on Krakatau by spinning silk balloons that carried them over the water. Birds and bats (including the Malay flying fox, with a 5-foot wingspan) could fly to Krakatau, and bring in their stomachs the seeds of fruits they had eaten on the mainland.

Yet life on Krakatau did not come back randomly. The first species to come were weedy, pioneering organisms well adapted to catastrophes. In time, other species arrived and created a succession of ecosystems, each opening the way for the next. A grassland ecosystem assembled itself first, and any animal that arrived on the islands had to be ready to survive on the food it had to offer. Emerald doves and savanna nightjars settled successfully. So did pythons and geckoes and foot-long centipedes. Many other species did not. Others had to wait as the grasslands gradually gave way to forests.

For some trees, the timing had to be exquisite. Fig trees, which were among the most successful, depend on a single species of wasp to pollinate them; if a fig arrived at Krakatau, its only hope of colonizing the island was for its pollinating wasp to come to the island soon afterward. Apparently this improbable event did happen, because figs began to spread. Animals feasted on the figs, making the diversity of the forest swell. New shade-loving species such as orchids have now established themselves. In years to come, the forest will continue to mature. Bamboo may arrive and take root, and it will allow bamboo snakes and other animals adapted to it to settle on the island.

As the forests overtook the grasslands, many of the pioneer species that had arrived on Krakatau early on disappeared. The zebra dove vanished from the island by the 1950s. Others eked out an existence by colonizing patches of the forests where a tree crashed down and opened up the canopy. Now, after almost 120 years, the flow of immigrants has slowed down considerably. Krakatau seems headed for an equilibrium.

The theory that islands have an equilibrium of diversity was pioneered by two ecologists in the 1960s. Robert MacArthur and E. O. Wilson argued that you can predict how many species an island will have from its size. The first species that arrive on an island have lots of room to spread. As more species arrive, though, they have to compete for food or sunlight, and their numbers go down. As more predators arrive, they can drive down the numbers of their prey as well. If the population of a species on that island drops too low, a hurricane or a disease can wipe out the last few individuals. The arrival of new species, in other words, raises everyone's risk of extinction.

So there are two forces pushing and pulling on the total number of species on an island—the addition of new species (those that arrive and those that form on the island) and the extinction that their competition brings. Eventually the diversity on the island reaches a point where both forces cancel each other out.

That balancing point depends on the size of the island. On a small island, there are few habitats and little space, which means that competition will be fiercer, extinctions more intense, and species fewer. Bigger islands can accommodate more species. Before Krakatau blew, it presumably had more species on it than any of the smaller islands that were left afterward.

Before Krakatau erupted, hardly anyone bothered to make note of its wildlife. What little information there is, however, suggests that ecologically it's not the same place it was before. An early explorer noted 5 species of land mussels on its beach; today, there are 19 species, none of which match the original ones. The forests that have taken hold on the new islands are not the same; they are dominated by different kinds of trees.

When an ecosystem rebuilds itself, it may follow MacArthur and Wilson's rules of diversity, but it doesn't duplicate itself. Different species scramble to take over the niches waiting to be filled. Krakatau's fate depended in large part on what plants and animals got there first and how much time they had to establish themselves before they had to face competition.

At the beginning of the Triassic period, the world was a patchwork of Krakataus. Species that could survive in awful conditions had the world to themselves, and they spread like weeds for thousands of miles. Carpets of bacteria rolled through the shallow coastal waters, unmolested by grazing animals. A few tough species of animals and plants also flourished. A single species of bivalve named *Claraia* raced across the shallow oceans of the western United States; today you can walk for miles over pavements made only of its fossil shells. On land, lush jungles were replaced by patches of quillworts and a few other weedy species, as botanically boring as an Iowa cornfield. Quillworts belong to a primitive branch of plant evolution that, by 250 million years ago, had been pretty much outcompeted by gymnosperms (a group that includes conifer trees). But quillworts can survive in harsh conditions that kill most gymnosperms, so the extinctions of other plants revived them.

For 7 million years, life on Earth remained covered in weeds. Researchers don't know why these grim conditions lasted as long as they did; the climate and the chemistry of the oceans may have been too hostile for anything but disaster-loving species to survive. Even after the physical conditions improved, it would still have taken a long time for the world's ecosystems to recover. Forests could only grow once soil had been created by the plants that came before them.

Slowly the world's ecosystems recovered, but they were never the same again. The oceans' reefs, once composed of algae and sponges, were now made up of colony-forming animals called scleratinian corals, which still make up the majority of reefs found on Earth today. Before the extinctions, the fauna that lived on a typical reef would have been dominated by slow-moving animals or ones that were rooted to the reef itself—creatures such as sea lilies, bryozoans, and lampshells. Today only a remnant of each group survives. Since the Permian-Triassic extinction, fish, crustaceans, and sea urchins have dominated the reefs instead.

On land, quillworts and other weedy plants rebuilt the soil, and conifers and other plants emerged from their refuges. They beat back the quillworts in only half a million years, rebuilding forests and shrublands. But once life on land had recovered from the mass extinctions, it was changed for good as well. Before the extinctions, the dominant insects were dragonflies and other species that keep their wings unfolded. But after the extinctions and ever since, insects with folded wings have been most common.

Almost all the synapsids, which had been diverse and dominant vertebrates before the Permian-Triassic extinction, disappeared, and during the recovery they did not take back their dominance. Reptiles became more common, evolving into new forms, like crocodiles and turtles. And about 230 million years ago, one slim bipedal reptile gave rise to the dinosaurs. Dinosaurs soon became the dominant land vertebrates, a position they would hold for 150 million years.

The Permian-Triassic extinction shows that there is something to Cuvier's revolutions after all. Millions of species can be wiped away in a geological flash, and the sort of life that takes over afterward is often profoundly different from what came before.

Mass extinctions put the normal rules of evolution on hold. At the end of the Permian period, conditions suddenly became too harsh for almost any species to survive. As species disappeared, the ecological web they helped form collapsed, and other species went extinct. Some of the survivors might have had some intrinsic qualities that kept them from vanishing. Their ranges may have spanned an entire continent or ocean, raising the chances that a few individuals might have survived in some isolated refuge. They might have been able to tolerate low levels of oxygen in the ocean or a sudden rise of temperature on land. But most of these adaptations mattered only for the short stint during which Earth became its own hell.

Once mass extinctions end, evolution returns to its normal rules. Competition between individuals and between species begins again, and natural selection invents new kinds of specialization. But a lineage that might do well playing by these normal rules can't win if it has been wiped out by a catastrophe.

Extinctions also bring bursts of change in their wake. They can clear away dominant forms of life that under normal conditions would shut out any aspiring species that have the potential to compete with them. Without this overbearing competition, the survivors are free to explore new forms. Dinosaurs may have emerged only because the dominant synapsids were overthrown.

Yet the liberation that extinctions bring to survivors isn't infinite. Even after the Permian-Triassic extinction, when competition dropped to nearly nothing, evolution did not invent any new phylum of animals. No vertebrate lineage evolved nine legs. After the Cambrian explosion, animals may have become too complex to be radically reworked by evolution. The new evolution that took place in the wake of extinctions were only variations on these basic plans.

MAMMALS: A TINY BEGINNING

If the Permian-Triassic extinction had been even a little more lethal, mammals might never have come into existence. Only a few lineages of synapsids straggled into the Triassic period, and most of them grew rarer still as the dinosaurs became more successful. But one of these lineages continued to evolve the equipment necessary for life as a mammal.

These doglike synapsids, known as cynodonts, evolved a new sort of skeleton. They evolved rib cages that could house diaphragms, allowing them to breathe more deeply and develop more stamina. They also probably evolved hair at this point, and began to nurse their young. They probably secreted fluid from glands in their skin that their young could swallow. At first this milk may have been nothing more than a liquid antibiotic that helped their young fight infections. But over time, evolution added protein, fat, and other substances that helped mammals grow quickly.

All of these innovations helped the ancestors of mammals to do a better job keeping their metabolisms high and maintaining a constant body temperature. As a result they could occupy new ecological niches that cold-blooded vertebrates could not—for example, they could hunt at night. Their high metabolism made it possible for some lineages to evolve to smaller sizes as well. (Smaller animals have a harder time holding on to body heat, because the ratio of skin surface to body mass determines how quickly heat is lost.)

These small protomammals had sharper senses than their ancestors, and to handle this new sensory flow they evolved a new rind around their brains. Known as the neocortex, this layer was dedicated to sorting out the influx of sounds, sights, and smells, turning them into intricate memories and using them to learn about their surroundings. The warm-blooded mammals could make good use of the neocortex, because their high metabolism required a constant supply of fuel. A snake can eat a rat and relax for weeks, but mam-

Dinosaurs, such as *Apatosaurus* shown here, reigned supreme for 150 million years. While *Apatosaurus* reached 30 tons, the mammals that lived alongside dinosaurs never weighed more than 5 pounds. Only after the extinction of the dinosaurs did mammals undergo an evolutionary explosion.

mals cannot go long without food. A big brain with a neocortex let mammals build a map of places to find food and remember it.

To us humans, so proud of our brains, this achievement seems like a milestone that should have instantly altered the course of evolution. But for the great synapsid dynasty, it didn't matter much. Synapsids only barely recovered from the Permian-Triassic extinction before sliding down toward extinction again by the end of the Triassic.

"We think of mammalness as the superior way to be," says Ward. "It wasn't. Dinosaurs in head-to-head competition won out. They took over the world. We talk about an age of dinosaurs—well, they wrested it away from mammals."

For 150 million years dinosaurs were the most diverse land vertebrates. Among their ranks evolved the biggest animals ever to walk on land. In 1999 researchers found pieces of the backbone of a long-necked dinosaur in Oklahoma that they named *Sauroposeidon*. Judging from the size of the vertebrae, the paleontologists estimate that it stood six stories high. *Sauroposeidon* could have crushed the biggest mammals of the Mesozoic like pinecones. None of those early mammals managed to tip the scales at 5 pounds; paleontologists who hunt for them may sieve a ton of rock and find a single tooth the size of a pinhead.

"Mammals go way back—they're as old as the dinosaurs," says paleontologist Michael Novacek of the American Museum of Natural History. "But for about the first 150 million years, they're not very dramatic animals. They really lived in the shadow of the dinosaurs, mostly small, possibly nocturnal creatures. They're not very auspicious."

Yet as inauspicious as they might have looked, mammals continued to evolve during the age of the dinosaurs. They branched into many different lineages, some still alive, some long extinct. The platypus belongs to the oldest lineage of mammals still alive today, known as monotremes. Monotremes still retain some of the characteristics of our own ancestors 160 million years ago. They have far less control over their body temperature than more recently evolved mammals. Female monotremes do not give birth to live young; instead, they lay soft-shelled, pea-sized eggs, which they carry in a slit on their belly. When the egg hatches, the infant nurses milk that oozes out of mammary glands. (Nipples had not yet evolved when monotremes branched off on their own.)

About 140 million years ago, mammal evolution produced two branches that would turn out to be the most successful of all. One was the marsupials, which include living animals such as the kangaroo, the opossum, and the koala. The male marsupial has a forked penis, which it uses to fertilize eggs in the female's twin uteruses. A fertilized marsupial egg does not develop a shell; instead, the embryo develops for a few weeks until it is the size of a rice grain, and then it crawls out of the uterus. It makes its way into a pouch on its mother's belly, where it clamps its jaws around a nipple.

The other lineage gave rise to our own sort of mammals, known as the placentals. Unlike marsupials, placental mammals keep their babies in the

The major branches of mammals evolved radically different ways of reproducing. The platypus belongs to the oldest living lineage of mammals on Earth, known as the monotremes. It still lays eggs, as did the reptilian ancestors of all mammals. Kangaroos belong to the marsupials, which give birth to live young and carry them in a pouch.

uterus until they are much larger. They can do so because the embryo is surrounded by a placenta, a special kind of tissue that can draw food from the mother. Placental mammals are born much more developed than marsupials. In some cases, such as rabbits, they are still blind and have to stay hidden in a warren. But in other cases, such as dolphins or horses, they are ready to move on their own almost immediately.

There are precious few fossils of placental mammals older than 65 million years that belong to living orders, but the few there are suggest that they began to diverge into the living orders about 100 million years ago. The first lineage to branch off would much later produce anteaters, sloths, and armadillos. These animals lack many of the traits that other placental mammals share; they don't have a cervix, for example, and their metabolism— while higher than a platypus's—is still slower than other placentals. The fact that these mammals branched off first doesn't make them missing links in our own evolution (just as monkeys are not missing links in human evolution). It doesn't mean that we descended from armadillos, sloths, or anteaters; it doesn't even mean our ancestors had armor like armadillos, claws to hang upside down from trees like sloths, or long tongues like anteaters. Once these lineages split off, the mammals went on evolving new adaptations of their own.

Paleontologists suspect that other groups of living mammals emerged around 80 million years ago. The Insectivora would eventually give rise to moles, shrews, and hedgehogs. The Carnivora produced dogs, cats, bears, and seals. The Glires would produce rabbits and rodents. The Ungulata would give rise to horses, camels, whales, rhinos, and elephants. And the Archonta would eventually include in its ranks bats, tree shrews, and our own branch, the primates. But all of this diversification was still tens of millions of years away. The ancestors of living placental mammals were practically indistinguishable from one another. It would take another mass extinction to reveal what mammals could become.

DEATH FROM ABOVE

In northern Italy you can find a beautiful rosy limestone, called Scaglia rossa, which Italian builders like to use to construct their villas. Just north of the town of Gubbio, the 1,200-foot Bottaccione Gorge is walled with the stuff. Geologists have determined that the rock at the bottom of the gorge was laid down 100 million years ago, when placental mammals were just beginning to diverge into their living groups. The rock built up continuously for the next 50 million years. It was still forming 65 million years ago at the end of the Cretaceous period, when the dinosaurs vanished, along with 70 percent of all species. It continued forming for another 15 million years, as mammals evolved into the dominant vertebrates on land. Wedged between the rock from the Cretaceous period and the rock from the Paleocene period that followed is a strip of clay only a half-inch thick, like a smear of jelly in a sandwich. Below the strip, the rocks contain calcium carbonate skeletons of plankton; their bodies make up most of the rock. In the layer of pure clay, no plankton can be found; above it, the limestone starts again, but it lacks many of the old species of plankton. In that half-inch strip our destiny may have been determined: it marks a global catastrophe our ancestors survived but the giant dinosaurs did not.

An American geologist named Walter Alvarez hammered out chunks of this clay in the mid-1970s and brought them back home with him. Alvarez hoped to find in the Scaglia rossa the precise boundary between the Cretaceous and the Tertiary and assign a date to it. If all went well, he hoped to find a way to identify the same boundary in rocks in other parts of the world. Every few million years Earth's magnetic field flips, so a compass points south instead of north. The field lines up magnetic crystals in rocks, and geologists can measure its direction millions of years later. Alvarez wanted to find layers of rock below and above the boundary that had formed when Earth's magnetic field had flipped. It might be possible then to go to other rocks and find the same sequence of flipping and use it to put a bracket around the boundary.

Alvarez showed his rock to his father, Luis, when he came home. Luis Alvarez was not a geologist himself, but he was a hungry-minded scientist nevertheless. He had won the Nobel Prize for physics in 1968; he had helped to invent the bubble chamber that would lead to the discovery of subatomic particles, and he had x-rayed the pyramids to look for hidden tombs. The rock that his son brought home fascinated him. What had happened in the oceans at the close of the Cretaceous period to stop the limestone factory and then start it again?

His son's plan to find a paleomagnetic bracket for the Cretaceous-Tertiary boundary didn't work out. North and south traded places too slowly at the end of the Cretaceous to be much use for dating. But Luis had another idea—they could use the steady rain of interstellar dust as a clock. Meteorites and other material drifting through space have chemical compositions very different from that of rocks on Earth. They have, for instance, much more of a rare element called iridium in them. (Most of the iridium that helped form the molten Earth 4.5 billion years ago sank into its core along with other metals.) Tons of microscopic debris fall into Earth's atmosphere every year, sprinkling steadily on land and sea below. Walter and Luis set out to determine how to measure the rate at which iridium landed on Earth by measuring its levels in the Gubbio rocks.

Other scientists had tried to use the same method and failed, but fortunately the Alvarezes didn't know that. They measured the iridium in the rocks from the end of the Cretaceous and found that it was enormous, 30 times the levels that were in the limestone Walter had taken from above and below the layer of clay. The steady rain from space couldn't have left that much iridium. But their measurement was not a fluke: Danish scientists had looked at rocks near Copenhagen that dated from the end of the Cretaceous and found an even bigger spike of iridium.

The Alvarezes began to wonder if Earth had gotten a giant delivery of iridium from space at the end of the Cretaceous period. That might support a wild-eyed idea from a paleontologist named Dale Russell. The dinosaurs (at least the big ones) had disappeared at the end of the Cretaceous, in a mass extinction that claimed an estimated 70 percent of all species, including giant marine reptiles and the pterosaurs that filled the skies. Russell threw out the possibility that an exploding star in the sun's neighborhood had done them in. A supernova might have released a flood of charged particles that would have raced through space and fallen into Earth's atmosphere, causing mutations and death.

The Alvarezes knew that iridium is one of the elements that are forged during a supernova. Perhaps along with the lethal charged particles, the supernova could have sent a surge of iridium to Earth. But when the Alvarezes investigated Russell's idea, they discovered that a supernova couldn't have been the cause. In addition to iridium, exploding stars also produce plutonium 244, which ought to have left a mark in the Gubbio clay. The Alvarezes found none.

Their thoughts turned instead to the possibility of a giant comet or asteroid striking Earth. Luis remembered reading about Krakatau—the eruption had launched 18 cubic kilometers of dust into the atmosphere, and 4 cubic kilometers lofted to the upper reaches of the stratosphere. Fast-blowing winds carried it for two years around the planet, masking the sun and creating blazing sunsets. Luis suggested that the impact of a giant asteroid was like a magnified Krakatau. He speculated that when the asteroid slammed into the ground, its debris flew back up into the air, along with the terrestrial rock gouged out by its crater. Together they formed a thick dark shroud around the planet. Without sun, plants withered and photosynthetic plankton in the oceans died. With nothing to eat, herbivores starved; the carnivores disappeared soon after.

The meteorite, the Alvarezes calculated, had to have been about 10 kilometers across. It would have been as if Mount Everest had been fired into the planet like a bullet. Impacts of this scale were common during Earth's early years, but they tapered off 3.9 billion years ago. Since then, giant asteroids and comets have probably struck the planet only once every 100 million years.

An impact at the end of the Cretaceous would have been a rare event, but not an unexpected one.

The Alvarezes published their impact hypothesis in 1980, and in the decade that followed, other geologists looked for more clues as to what happened at the end of the Cretaceous period (known as the K-T boundary). They found more and more evidence that something huge hit the planet 65 million years ago. At more than 100 sites around the world geologists had found the layer of clay that marks the end of the Cretaceous, and iridium consistently appears in it. Researchers have also found bits of shocked quartz in the clay that could only have been created under intense pressures such as those created in an impact.

For more than a decade, however, the Alvarezes were dogged by the absence of the crater that such an impact would have left. It was possible that the impact had occurred in the ocean and was now covered over by seafloor sediments, or that plate tectonics had sucked it into Earth's mantle, or that a volcano had covered it over. But the Alvarezes kept looking for a smoking gun. One reason that they kept searching for it was that their critics were cooking up alternative explanations. Some researchers argued that volcanoes, which are such a strong candidate for the Permian-Triassic extinction 250 million years ago, could have been the culprit at the K-T boundary. The same sort of volcanic activity that carpeted Siberia at the end of the Permian period also poured lava across India at the end of the Cretaceous. These eruptions could have brought up iridium from deep within the Earth, and could have created the intense pressures necessary for shocking quartz as well.

Geologists continued to search for a crater, and in 1985 the first clues of one emerged. Geologists found some peculiar deposits in Texas dating from the K-T boundary that contained coarse sands and pebbles. They could only have been carried there by a giant tsunami that had originated from somewhere south of the site. Perhaps, the researchers reasoned, an impact had generated the giant waves. Meanwhile in Haiti other geologists found K-T rocks containing globules of glass, which were predicted to have been formed in the impact, as molten rock hurled into the sky quickly cooled. Unlike the dust and vapor that the impact would have kicked up, these globules were too heavy to travel very far. The crater, researchers realized, must be within a few hundred miles of Haiti. Together, the tsunami and the glass pointed to an impact site somewhere around the Gulf of Mexico.

In the 1950s, Mexican geologists had discovered the remnants of a giant circular structure dating from about the end of the Cretaceous, buried off the coast of the Yucatán Peninsula. It had been pretty much forgotten, but with the discoveries in Haiti and Texas, it took on a new importance. When geologists revisited the site (called Chicxulub after a nearby town), they brought with them equipment that could detect buried rock formations by the subtle

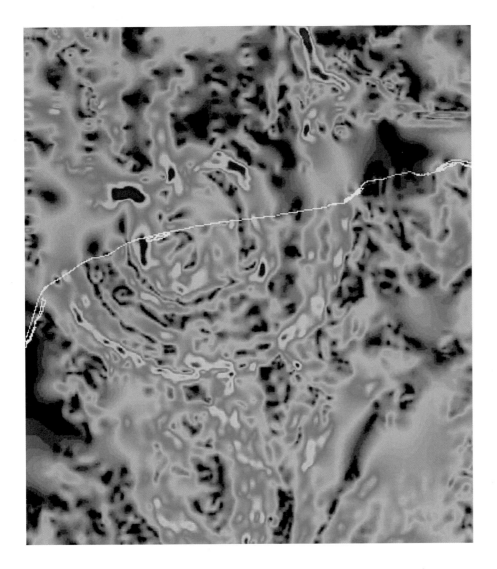

Geologists created this image from the variations in the gravitational field along the coast of eastern Mexico (shoreline marked in white). The variations reveal buried circular formations that mark a crater created 65 million years ago by the impact of an asteroid. The impact triggered one of the greatest mass extinctions of all time.

changes in the gravitational field they create. The geologists mapped out two concentric circles that looked as if they had been traced by a heavenly compass. All signs pointed to a 100-mile-wide crater buried under the sediment. Other researchers drilled into the rings and brought out rocks that they then dated. The age of the rocks—65 million years old—matched both the Alvarezes' iridium marker and the glassy globules in Haiti.

In 1998 a geologist named Frank Kyte from the University of California at Los Angeles found what may actually be a piece of the thing that hit the Yucatán. He was looking over a cylinder of rock drilled from the floor of the Pacific Ocean. The dark brown clay was loaded with iridium and shocked quartz, marking the K-T boundary. Kyte sliced the clay at the boundary, and there he discovered a lone pebble, measuring 2 millimeters across. Its chemical composition was unlike anything on Earth, but exactly like that of many

The lines in this piece of quartz were created by shock waves 65 million years ago when an asteroid hit the planet.

meteorites. Here, Kyte suggests, was a chip from a giant asteroid. It had broken free during the impact at Chicxulub and flown high over the Yucatán, arcing through the stratosphere before plopping into the Pacific.

As geologists have gotten a better sense of the asteroid that hit the planet 65 million years ago, other researchers have been looking for clues to its effect on life. At the end of the Cretaceous, they've found, the Yucatán was covered by a shallow sea, less than 100 meters deep, its bottom made of rocks rich in sulfur and carbon. Giant seagoing lizards may have swum under the shadow the asteroid cast just before it struck. The asteroid entered our atmosphere at a speed somewhere between 20 and 70 kilometers a second, creating a giant shock wave that ignited a jet of flame in its path. The fiery tail leveled trees for thousands of kilometers.

Computer models suggest that when the asteroid hit the water, it may have sent a tsunami out across the oceans, rising as high as 300 meters. The waves roared onshore, and the riptide dragged back entire forests, drowning them 500 meters underwater. An instant after the asteroid struck the water, it hit the bottom of the sea and vaporized 100 cubic kilometers of rock. The impact sprayed rock and asteroid 100 kilometers into the sky, above the stratosphere. An earthquake 1,000 times more powerful than anything in recorded history made the entire planet shiver. Geologists drilling in the Atlantic have found evidence that it triggered undersea landslides along the eastern seaboard of North America as far north as Nova Scotia, flowing 1,200 kilometers out from shore. Meanwhile, a fireball emerged from the crater and spread out hundreds of kilometers. The blackened sky was probably filled with thousands of shooting stars, molten hunks of rock that soared over the planet, igniting more fires wherever they landed.

The world burned; smoke hid the sun. Plants and phytoplankton died in the prolonged darkness, and the ecosystems that were built on them collapsed. When the smoke cleared a few months later, the world may still have been dark and cold. The impact may have vaporized the sulfate deposits in the Yucatán rocks, which combined with oxygen to form droplets of sulfur dioxide. The hazy clouds they formed may have reflected sunlight away from Earth and could have lingered for a decade. But as the haze faded, the impact ravaged the planet in yet another way, by warming it. The carbon in the limestone that was heaved into the atmosphere turned to the greenhouse gas carbon dioxide; the asteroid also sprayed the air with water vapor, an even more powerful greenhouse gas.

The heat, the cold, the fires, and the other disasters caused by the impact may have destroyed more than two-thirds of all species on Earth. The Alvarezes found a single culprit for the K-T extinction, but as in the case of the Permian-Triassic extinction, it used many weapons.

MAMMALS TAKE THE FIELD

When the skies cleared after the impact, the Cretaceous period was over. The giants were gone. The long-necked sauropods that converted whole forests into muscle and bone were extinct, along with *Tyrannosaurus rex* and the other big meat-eating dinosaurs. Giant marine reptiles and spiral-shelled ammonites disappeared from the seas. After a few thousand years the plankton in the oceans rebounded, as did the plants on land. But the ecosystems of the early Tertiary period were bottom-heavy and top-light.

Once again, a mass extinction had cleared the way for a new burst of evolution: the age of dinosaurs was followed by the age of mammals. "The death of the dinosaurs allowed mammals to evolve into many ecological niches that were not available for them," says Ward. "It was the removal of the dinosaurs through mass extinction that allowed so many lineages of mammals to come about through the evolutionary process. In that sense, it's really a good thing. There would not be humans here but for that mass extinction."

Mammals suffered along with the rest of life during the K-T extinction, as an estimated two-thirds of all mammal species disappeared. But the ones that survived inherited the earth. Within 15 million years after the K-T extinction, they had evolved into all 20 orders of the living placental mammals, along with many other orders that are now extinct. At first these new mammals remained small. Hoofed mammals as big as raccoons browsed on low leaves and were hunted by predators the size of weasels. But within a few million years they evolved out of their old niches into the ones that dinosaurs had dominated. Giant relatives of rhinoceroses and elephants browsed the leaves of shrubs and trees. Ancestors of today's cats and dogs stalked the herbivores; some mammals became scavengers, stripping corpses and crushing bones. Primates raced through the trees, using their color vision to choose the ripest fruits. Bats evolved from shrewlike tree dwellers into hundreds of flying species, some searching for fruit, others using echolocation to catch insects and frogs. Whales and the ancestors of today's dugongs and manatees colonized the oceans.

While mammals have remained the dominant land vertebrates over the past 65 million years, they've also had to weather their own evolutionary shocks. Today's climate bears little resemblance to the one in which the age began. Between 65 and 55 million years ago, volcanoes spewed carbon dioxide into the atmosphere, gradually warming the planet until palm trees could grow north of the Arctic Circle and Canada looked more like Costa Rica does today. Wyoming was home to lemur-like primates, jumping through jungle canopies.

Earth would never be so warm again. For the past 50 million years the world's average temperature has been dropping, with occasional hiccups of warmth along the way. We may have the Himalayas partly to blame. When India collided with Asia, the crash created the craggy mountain range. The rains that fell on these fresh new slopes carried dissolved carbon dioxide; the gas reacted with the rock and formed compounds that were carried away by streams and rivers to be buried in the sea. The Himalayas may have withdrawn so much carbon dioxide from the atmosphere that the climate gradually cooled. At the same time, the collision also pushed up the Tibetan plateau, just north of the Himalayas. This enormous dome began rerouting weather patterns throughout southern Asia. Air that passed over the plateau warmed and rose from the ground, pulling in moist air from the oceans to take its place. This pattern created the monsoons of India and Bangladesh and brought more rain to the Himalayas. The removal of carbon dioxide sped up even more, making the planet's greenhouse effect even weaker.

Changes were taking place in the oceans as well. During the Cretaceous, Antarctica was much farther north than it is today, and it was so warm that dinosaurs and trees thrived on its coasts. Eventually, though, the continent pulled away from Australia and moved south, becoming more and more isolated, until it reached the South Pole. Permanent ice began building up on Antarctica, reflecting sunlight back into space and cooling the atmosphere.

As winters grew colder, the tropical forests of North America disintegrated. Mammals that could not survive without them, such as primates, disappeared as well. The jungle gave way to broad-leaved trees much like those alive today, interspersed with scrublands. As carbon dioxide levels continued to fall, new kinds of plants evolved that could absorb the gas more efficiently. Among these new plants were grasses, which formed the first major grasslands about 8 million years ago.

Grass is loaded with tough cellulose and sprinkled with bits of glasslike silica, making it far harder to eat than the soft fruits and leaves that were abundant when the planet was warmer. Some mammals, such as horses, managed to survive on this diet thanks to their high-crowned teeth, which could grind down the plants. The ancestors of cows and camels were also prepared for the tough grasses because they had altered their digestive system to let bacteria help them break down the tough plants. But many lineages couldn't adapt to the cooling climate and the shifting vegetation and went extinct.

Geography also drove some mammals into oblivion. Before 7 million years ago, North and South America were separated by ocean, but continental drift gradually drew them together. At first islands formed between them, and then, 3 million years ago, the Isthmus of Panama emerged and joined the continents. The mammals from each landmass could spread into new territory to compete with species they had never encountered before.

During the 60 million years that South America had been isolated from other continents, it had developed an ecosystem unlike any other on Earth. The top predators were coyote-sized opossums and giant, fast-running flightless birds. When the continents were linked, a few species of South American mammals such as opossums, sloths, monkeys, and armadillos moved north. But the mammals that moved from north to south were much more successful. The opossum coyotes of South America were driven extinct along with all the other marsupial carnivores, their place taken by cats and dogs. The hoofed mammals of South America were replaced by horses and deer.

"The 65-million-year history of mammals, after the K-T extinction event, is marked by invasions," says Novacek. "There are lots of mammals moving from one continent to another. You can almost visualize these as armies marching across the continents. Some of these mammals that enter into new continents show a high degree of success in dominance soon after the invasion. But it's very difficult to really understand why invasions work this way. We really don't know. It may be that many of the animals that tend to invade are more mobile, or more flexible with environmental change, and that confers some kind of competitive advantage."

Around the same time as the Great American Interchange, the global climate began to swing into a new pattern. The glaciers at the poles spread toward the equator and then pulled back in a cycle of ice ages. This cycle probably is controlled by the changing orbit of Earth around the sun. Earth moves closer and

Giant ground sloths (above and below) were among the many big mammals that suddenly became extinct in the New World at the end of the Ice Age, probably due to overhunting by humans.

farther away from the sun in a 100,000-year cycle. At the same time, its tilted axis draws out a circle like a spinning top every 26,000 years. And every 41,000 years the angle of the tilt changes too, between 21 and 25 degrees (today we are at 23 degrees). These cycles combine to change the amount of sunlight that the earth receives over the course of the year.

Nicholas Shackleton, a paleooceanographer at the University of Cambridge, has studied the effects of these wobblings in ancient Antarctic ice and in seafloor mud. He has found that when the wobbles bring less sunlight to Earth, carbon dioxide levels in the atmosphere fall. Researchers don't know for sure how the former can drive the latter; it's possible that a drop in sunlight alters the way plants and plankton grow. In any case, with less carbon dioxide in the atmosphere, the planet cools. Each summer less polar ice melts, so the glaciers get larger. Eventually they march thousands of miles toward the tropics, until something triggers them to retreat—perhaps an increase in carbon dioxide in the atmosphere. As the glaciers retreat, forests can expand again. We live in one of these pleasant interglacials, which has lasted 11,000 years and may last for a few thousand more.

DIVERSITY OVER TIME

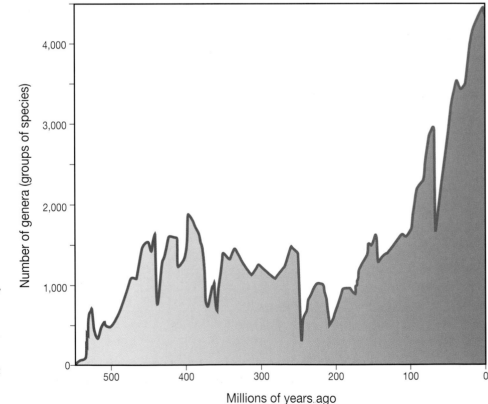

The known fossil record suggests that the diversity of life has risen over the last 250 million years. There are now as many or more species on Earth than at any other time, but they are threatened by new mass extinctions.

A snapshot of North America, taken just on the verge of the arrival of humans at the end of the last Ice Age, would have shown a landscape crowded with giant mammals. Saber-toothed cats, jaguars, cheetahs, short-faced bears, dire wolves, and other carnivores now hunted for the grazing and browsing mammals. These included mammoths grazing on the grasslands, and mastodons wandering the forests and swamps, as well as camels, horses, rhinos, and ground sloths. The North American mammals were part of a global boom of biodiversity. Studies of the fossil record suggest that the number of species has been climbing over the past 100 million years. One reason for this rise, researchers suspect, is that Pangaea has been disintegrating into smaller and smaller pieces. On a united supercontinent, there were fewer barriers to animals and plants, so versatile species could invade the territories of more specialized ones and outcompete them. As Pangaea split apart, it created more isolated habitats where more species could thrive, and more coastlines where marine life could evolve. By the time modern humans evolved around 100,000 years ago, the world may have harbored the most diversity of life in its entire history. It is a rich legacy to inherit, and a terrible one to squander.

HUMAN EXTINCTIONS: THE FIRST WAVES

The first rumblings of the latest mass extinctions began around 50,000 years ago. Up until that time, Australia was home to a collection of giants, including 1-ton wombats, kangaroos standing over 10 feet tall, marsupial "lions," and 30-foot lizards. Fossils in Australia are scarce, so it is difficult to know precisely when most species disappeared. But one species, a 200-pound flightless bird called *Genyornis,* left behind thousands of eggshell fragments. The eggshells promptly disappeared 50,000 years ago. And around that time a new species arrived on Australia's shores: humans.

This same series of events—humans arrive and big animals go extinct— was later repeated many times around the world. The oldest evidence of humans in the New World comes from a site in Chile called Monte Verde that's been dated to 14,700 years ago. Archaeologists are working on other sites that may be a few thousand years older still. These early colonizers may have traveled down North and South America by boat. Before about 12,000 years ago it simply would have been impossible for them to migrate on foot from Alaska, because the land was still cloaked in glaciers. After the glaciers had retreated, a new culture appeared in North America, bringing with it spears that could bring down a mastodon. And by around 11,000 years ago, the New World was stripped clean of its mastodons, its giant ground sloths, and just about every other mammal species over 100 pounds.

About 2,000 years ago travelers from southeast Asia landed on the shores of Madagascar. There they encountered the elephant bird, a flightless creature that weighed more than 1,000 pounds, and giant lemurs the size of gorillas. Neither lived more than a few centuries alongside humans. Until the 1300s New Zealand was home to 11 species of moa, another giant flightless bird. Although not quite as heavy as elephant birds, they would have overshadowed them at a height of 12 feet. Like the elephant birds, they lasted a few hundred years after the arrival of humans.

The sudden extinctions in North America were the first ones to be discovered, and initially many paleontologists thought they were caused by the end of the Ice Age. As the climate warmed, they argued, North American trees and grasses shifted their ranges, and the mammals that depended on them couldn't handle the sudden rearrangement. But with more research, the connection between climate change and extinction seems to be little more than coincidence. If the end of the last Ice Age was so devastating to North America, you'd expect that the ends of previous ice ages would have had a similar effect. Yet during the last million years, mammals in North America actually experienced relatively few extinctions. Despite the advance and retreat of glaciers a mile thick every 100,000 years or so, mammals have managed to endure by shifting their ranges with the shifting trees and grasses. And climatically speaking, there was nothing unusual about the end of the last Ice Age in North America compared to all the other ice ages. Casting more doubt on the climate-extinction link is the fact that while mammals were suddenly disappearing in North America 12,000 years ago, equally drastic climate changes happening in Europe, Africa, or Asia caused no significant extinctions. The one unusual thing about the end of the last ice age in North America was that it saw the arrival of humans. And as scientists have uncovered the histories of places like Australia, Madagascar, and New Zealand, they've found evidence suggesting that the big mammals and birds also went suddenly extinct soon after humans arrived—in some cases long before the end of the Ice Age and in others cases long after.

Humans may have brought about many of these extinctions with little more than spears and arrows. The ancestors of humans became hunters in Africa and spread out gradually into Europe and Asia. Over hundreds of thousands of years, the animals that they hunted had time to adapt to this new threat. But around 50,000 years ago, modern humans began moving quickly to continents and islands where humans had never been seen before. When experienced hunters arrived in Australia, North America, and elsewhere, they encountered animals unprepared for their assault. Most vulnerable would have been big, slow animals that couldn't reproduce quickly.

The extinction of big herbivores may have changed the landscapes of entire continents. According to Tim Flannery, an Australian zoologist at

Harvard University, the disappearance of Australia's grazing wombats and kangaroos allowed uneaten vegetation to pile up on the forest floors. When lightning struck, it ignited this fuel and started enormous fires where none could have burned before. The plants that had dominated Australia before humans arrived, such as southern pine and tree ferns, couldn't protect themselves well against fire; they lost ground to fire-tolerant species like eucalyptus. Now they can be found in only a few rare pockets in the outback.

The old Australian rain forests had soaked up water vapor like sponges, making the climate of Australia much moister than it is today, with flowing rivers and brimming lakes that could support pelicans, cormorants, and other birds. The eucalyptus trees that have replaced the jungles can't hold on to much water, so the lakes and rivers dried up. The leaf-browsing mammals that hadn't been wiped out in the first wave of hunting would have faced a starkly changed habitat, with nothing but low-nutrient eucalyptus trees and shrubs to feed on. The only marsupials that survived were fast, like red kangaroos, or lived in eucalyptus trees, like koalas, or could hide in burrows, like wombats. Any species that was not already preadapted to life with humans, according to Flannery, was doomed.

HISTORY IN A HOLE

Paleontologists who study extinctions millions of years ago are happy if they can prove that a die-off took less than a few thousand years to take place. But for those who study the current mass extinction, it's sometimes possible to bring that resolution down to decades, even years. All that scientists have to do is find the right place to dig.

One of those right places is a cave in Hawaii. David Burney, a paleoecologist from Fordham University in New York, has been digging a hole in it since 1997. A few years earlier Burney had been wandering with a team of scientists around the south coast of Kauai, one of the westernmost islands in the Hawaiian archipelago. They were looking for fossils and other traces of extinct life when they came across the narrow mouth of a limestone cave known locally as Mahaulepu. They squeezed through, into a corridor decorated by stalactites and flowstones. After 50 feet, they emerged into an arena of sunlight and trees. They were in what had once been a high gallery in the cave, the roof of which had collapsed thousands of years ago. Seeds had blown down over the 50-foot-high walls and sprouted, creating a sunken garden. Burney stopped, assembled a long metal tube he had been carrying, and drove it into the silty soil. When he drew it back up, he found a fossil inside: the fragile skull of a coot, a native Hawaiian bird. Here, Burney decided, he would dig a deep hole.

He mostly used his hands to dig, for fear of shattering delicate fossils with a shovel blade. Below the top few feet of silty soil, he reached black peat, and after a few more feet, the water table. From then on, sump pumps kept the hole dry for him; when he turned them off at the end of each day, the hole filled with water in a matter of minutes. Burney put the dirt in buckets, which he brought up to the surface. Volunteers washed the buckets through screens in inflatable children's swimming pools. A team of experts then sorted and bagged the fossils that were left behind and sent them to museums and labs where they could be studied closely. Burney also took samples of the soil, which he later searched for spores and pollen to determine what sort of plants grew around the cave.

The hole is now 20 feet deep and 40 feet across. The carbon isotopes in the bits of plant material at the bottom of the hole put its age at 10,000 years. For 3,000 years an underground stream slowly dropped silt on the cave floor. A rise in sea level let the ocean invade 7,000 years ago, whereupon the roof collapsed. A shallow freshwater pond formed on top of the denser saltwater that saturated the ground. Animals and plants fell over the cave walls into the pond and sank into the muck at the bottom. Caves are good for preserving bones, while lakes are good for preserving pollen from plants. A lake in a cave, like the one in Mahaulepu, is the best of all. Burney's hole provides a unified 10,000-year history of Hawaii, something no one had ever found before. He calls it "my poor-man's time machine." It is also one of the starkest examples of how humans trigger waves of extinctions.

Many of the plants and animals that Burney found at the bottom of his hole were unique to Hawaii. It is not easy for life to get to Hawaii, since it is

Paleoecologist David Burney has documented 10,000 years of Hawaiian extinctions in the mucky floor of a cave.

2,300 miles from the nearest continent. A hard-shelled seed may wheel through the ocean's gyres until it arrives on Hawaii's shores. Birds and bats blown off course sometimes settle on the islands, and migratory birds use them as a stopover on their way north or south. Sometimes on their muddy feet they carry a snail's egg or a fern spore, which can then establish themselves on the island.

The pollen and seeds that Burney has found at the bottom of his hole came from a lush coastal forest of palms, mimosa-like shrubs, and ferns. When animals arrived in the forest, they faced no mammal predators and few competitors for this lush wealth of food. Just as the finches of the Galápagos and the cichlids of Lake Victoria diversified, so did life on Hawaii. One or two species of fruit flies arrived 30 million years ago and have since evolved into an estimated 1,000 species, none of which is found anywhere else. Two dozen or so species of terrestrial snails came to Hawaii and exploded into more than 700 species. The snails probably supported the huge population of land crabs that prowled the forest floors, whose bodies Burney finds entombed in the cave.

Animals—especially birds—evolved to take advantage of every possible niche on the islands. Most of the world's owls hunt rodents and other small animals on the ground; Burney has found the skeletons of owls that had evolved to be more like hawks, grabbing other birds in flight. Three million years ago a single finch flew to Hawaii from North America and gave rise to 100 species of honeycreepers. Burney finds skulls of honeycreepers with massive nutcracker-like beaks that can open seeds too hard for any other animal on Hawaii to eat. Another bird species, the iiwi, uses a curved beak as delicate as an eyedropper to draw nectar from flowers.

The island of Kauai was created by volcanoes only 5 million years ago; in that time, some of its birds evolved into avian versions of pigs and goats. "Ducks and geese out here in the Hawaiian islands actually had the opportunity to become completely terrestrial animals," Burney explains, "to become much larger than they would ordinarily; to stop flying—they didn't need to fly; and to begin to be grazers and browsers. Among the extinct ducks and geese of these islands, birds were in a sense reinventing the goat or the pig." The ducks lost their wings and grew to the size of turkeys, with turtle-like beaks they used to crop grass. The geese lost their wings as well and grew to twice the size of today's Canada goose. Some waterfowl evolved toothlike ridges on their beaks to strip ferns clean.

At the bottom of his hole, Burney finds 45 species of birds and 14 species of land snails, in addition to other species such as bats and crabs. As Burney moves up the walls of the hole, moving up through time, he finds signs of the occasional natural disruption, such as an intrusion of the ocean marked by the bones of a mullet. But otherwise the birds, snails, crabs, palm trees, mimosas, and other species native to Kauai continue to leave behind their fossils. The land snails are particularly abundant, with more than a thousand shells in every liter of mud. For thousands of years, the fossil record is pretty much the same.

Then, about 900 years ago, a new fossil appears in the hole: a rat.

Rats arrived in Hawaii with the first Polynesian settlers who came to the islands about 1,000 years ago. The next few centuries are a blur in Burney's hole thanks to a tsunami that surged into the cave around 1500 A.D., washing away several feet of accumulated muck. But although it may have robbed Burney of a portion of history, it made up for its crime by bringing with it man-made objects. Burney finds fishhooks made of bones, and the sea urchin spines Hawaiians used to sharpen them. He has found a disk of glass-smooth basalt, which, when wet, serves as a mirror. There are tattoo needles and fragments of canoe paddles and bottle gourds with painted designs. The traces of new plants also appear for the first time in this layer, species such as bitter yam and coconut, which Polynesians brought with them to Hawaii. The bones of chickens, dogs, and pigs—all brought by the first Hawaiians as well—also turn up.

It is at this point in Burney's hole that native species start disappearing. Native land snails, once so abundant, begin to thin out. Palms and other forest trees no longer leave their pollen. The land crabs dwindle. The big grazing and browsing birds vanish. The long-legged owl disappears, replaced by the short-eared owl, which thrived on the island by eating the newly arrived rats.

In 1778 Captain Cook was the first European to visit the Hawaiian Islands, and the first place he landed was just a few miles away from Mahaulepu. As a gift to the king of the island he offered a pair of goats. Burney can find the bones of those goats' descendants in his hole, as well as of other European immigrants: animals such as horse and sheep, and plants such as Java plum and mesquite. Cattle, which grazed on the land surrounding the cave in the late 1800s, leave their bones behind as well. Giant cane toads and rosy wolf snails, brought to Hawaii this century to control pests, leave fossils in abundance.

Burney can find only a few native species in the hole after Cook's arrival. Below that mark, he has found 14 native species of land snails by the thousands; in the top few feet of the hole, he finds not one of them. Of the birds, only the migrant shorebirds and seabirds remain alive today near the cave. A few other species of birds that left fossils deep in the hole are still alive but survive only in remote mountain forests. Likewise, the plants that made up the forests are either extinct or hiding in refuges. The mimosa-like shrub Burney found, one of the dominant plants in the hole for thousands of years, can be found today only on a single rock off the island of Kahoolawe. Only two bushes remain alive.

The picture in Burney's hole is painfully clear: humans come, native wildlife goes.

THE EXTINCTIONS ACCELERATE

The extinctions at Mahaulepu appear to have a single cause, namely humans. But as with mass extinctions in the past, a single cause can have many destructive effects. In Burney's hole, he can see all of them at work. Two of the most important ways humans drove species extinct around Mahaulepu were by hunting them and by destroying their habitats.

These different kinds of destruction have emerged in a series of stages. The first animals to go extinct on Kauai were those most vulnerable to hunting. "They were slow species, they may have been particularly useful as food—a giant, flightless duck sounds great for a barbecue," says Burney. "They may have been, in particular, species that were vulnerable simply because they were on the ground all the time. They laid their eggs on the ground, they'd never seen any ground-based predators, like the rat, before. And so their eggs were just eaten up, by the rats and pigs and so forth."

The flightless ducks of Kauai probably met the same fate as the elephant birds of Madagascar and the mammoths of North America: quick destruction by humans. As the people of Kauai wiped out the easiest game, they might have turned to smaller prey such as land crabs. In Burney's hole he can watch this tragedy play out: after humans arrive, the land crab fossils get smaller and rarer, presumably as hunters were forced to take younger and younger crabs. Without enough young crabs left to reproduce, the population collapsed.

The stages of extinctions at Mahaulepu match the stages of extinction worldwide. Today, overhunting still remains a global threat to wildlife. In the deepest rain forests of central Africa, hunters are killing chimps and other primates to feed the loggers who are deforesting the region. Meanwhile, in even the remotest corner of Myanmar (formerly Burma), rare species of deer that have been discovered only in the past few years are being killed at an unsustainable rate. Hunters are killing them not to eat their meat, but in order to exchange them with Chinese traders for salt.

The second stage of extinctions at Mahaulepu—caused by the disappearance of the old habitats—came more slowly. The human population grew on Kauai, and in order to feed themselves they cleared land for crops and livestock. Without metal axes, the pioneers couldn't bring down trees very quickly; instead, they probably killed them by girdling their trunks or poisoning their roots. In place of the forests, they planted their crops of taro and sweet potato. Once Europeans arrived at Kauai, the destruction sped up greatly. By the 1840s, plantation owners had started to clear vast estates of their trees. The sandalwood was harvested for incense, and the rest was cleared so that they could graze cattle and plant pineapple and sugarcane.

The same accelerating destruction of habitats has occurred around the world. Starting about 10,000 years ago, civilizations in Mexico, China, Africa, and the Near East brought plants and animals under their control. Agriculture provides such a reliable amount of food that farmers and herders can survive at higher densities than hunter-gatherer peoples. To feed more mouths, more land had to be taken over. With cattle and sheep and goats people began grazing down meadows. To sow land for corn, rice, and wheat, they cleared away forests and grasslands. The English farmlands where Darwin had grown up had not always been farmlands; the forests that had once covered them were gradually cut, shrinking down to islands over the course of centuries. The animals that could live only in these islands were stranded inside them. The closest wild relative of cattle, the aurochs, survived in patches of forests in Poland, protected from hunting, until the 1600s, when it disappeared for good.

With a booming human population and the invention of better plows and saws in the last few centuries, wilderness has disappeared at a Malthusian rate. With more people come more farms and cities. More people need more firewood and lumber than many forests can sustainably supply. Technology has made it easier to build roads into those forests and haul the wood out. As a result, half of the world's tropical forests, where an estimated two-thirds of all species live, were logged or burned as of 2000.

As humans take over wilderness, plants and animals go extinct. In some cases—as when a dam dries up the one river where a fish species lives—the extinction is obvious. But a habitat doesn't have to disappear completely for a species to disappear. Even fragmenting a habitat is enough to cause extinctions. These fragments are like islands, and the same rules that predict how many species can live on an island like Krakatau also apply to them.

Each fragment of a forest can support only a certain number of species, proportional to its size. If it happens to contain more than its allotment of species when it becomes cut off from the rest of the forest, those extra species will vanish. If a species should be unlucky enough to disappear from all of its fragments, it is gone for good.

Species with small ranges are most vulnerable to extinction by fragmentation. Imagine that loggers cut down most of the forests along a mountain range. A salamander that lives on a single mountainside might end up in only three forest fragments, whereas a salamander that lived along the entire range might end up in 100 fragments. It's much more likely that the widespread salamander will cling to existence in at least some of its refuges than the one with a limited range. And if the widespread salamander can then travel between fragments, it may even be able to recolonize some of its former range. The salamander on the single mountainside meanwhile becomes extinct.

Most of the songbirds of the eastern United States have managed to survive forest fragmentation thanks to their big ranges. The range of most of the 200 species that lived in the forests before European contact extends beyond the eastern United States. By the twentieth century, 95 percent of the eastern U.S. forests had been cut. But they had not been logged all at once; the destruction spread out from the Northeast like a wave. By the time Ohio's forests were disappearing, New England's were beginning to recover. At any particular moment, birds could find refuges where they could survive. In this century people in the eastern United States have generally abandoned farming and the forests have recovered enough for widespread songbird species to return.

But the 28 species that lived only in the eastern United States have not been so fortunate. Their odds of survival were worse, because they had smaller ranges than other birds. As their habitat broke up into forest islands, they survived in fewer fragments, raising their odds of going extinct. Four of these birds—the passenger pigeon, Bachman's warbler, the ivory-billed woodpecker, and the Carolina parakeet—are now gone.

Many animals and plants around the world are now going the way of the Carolina parakeet. Their small ranges have been destroyed or fragmented by farming or logging. Many of those that are not yet extinct are almost certainly doomed. Research by Stuart Pimm, a biologist now at Columbia University, and his colleagues have put a time scale to this dwindling. In one study, Pimm surveyed a western region of Kenya that is known for its rich diversity of forest birds. Over the past century heavy farming and woodcutting has chopped it up into small parcels. His team studied aerial photographs dating back 50 years to figure out when these forest fragments first became isolated. They then looked through museum collections to see how many species were originally there. (Unlike insects, of which we probably know only a small fraction of the total species, birds are easy to spot. Ornithologists have identified the vast majority of the estimated 10,000 species of birds on Earth.) And Pimm visited the forest fragments to tally the species in each one.

Pimm found that the older the fragments were, the closer they were to their predicted diversity. Younger fragments still held a surplus of species, because extinctions had not yet brought them down to their equilibrium level. By comparing the older fragments to the younger ones, Pimm's team concluded that extinctions progress in a way that's similar to the decay of radioactive elements: they have a half-life, a period in which 50 percent of their species will disappear. Half of the remaining species will then disappear in the next half-life period, and so forth. The half-life of the birds Pimm studied in Kenya is roughly 50 years, a result that Pimm's colleague Thomas Brooks

The ivory-billed woodpecker is one of four species of eastern U.S. songbirds that have become extinct as their habitats have been destroyed.

has also found holds for Southeast Asian birds. In other words, the damage that has already been done to the world's forests still needs a few decades to come fully into view.

ALIEN INVASION

In addition to destroying forests and hunting animals, humans caused the extinctions at Mahaulepu in a third way: by bringing new species along with them, such as rats, chickens, dogs, and goats. Biological invaders, as these newcomers are known, are turning out to be one of the most powerful agents of global extinction at work today. Unlike hunting or deforestation, biological invasion is pretty much irreversible. If people stop cutting down forests, the trees can in time grow back. But once a biological invader has settled successfully in its new home, it is usually impossible to get it out.

Biological invasion is nothing new in the history of life. The mammals that marched from North America to South America 3 million years ago were biological invaders, suddenly confronting an isolated group of animals and driving many of them extinct. As Darwin showed, biological invasions could happen when eggs and seeds stuck to the feet of birds and traveled thousands of miles. But before humans, biological invasions were a rare event. Continents take millions of years to collide. Traveling by ocean or hitchhiking on a bird's foot is not easy: before the arrival of humans, paleontologists estimate that a new species managed to take hold in Hawaii once every 35,000 years. And those colonists were birds and bats and various small invertebrates. There was no way for a dog to ride to Hawaii on a bird's foot.

When Polynesian settlers arrived at Kauai, a host of new species arrived with them. Rats established themselves on the island and devoured bird eggs and land snails. The chickens and pigs the Polynesian settlers brought as livestock rooted up saplings and ate seeds, and may have caused even more damage to the native forests than the settlers themselves. The birds and land snails that could live only in those forests retreated from human settlements.

New species arrived at a much faster rate with Europeans. As the first truly global traders, Europeans could ferry species around the entire world in their ships. Some of the introductions were intentional—the gift of Captain Cook's goats, for example—but many were accidental. In 1826 whaling ships brought mosquitoes carrying a form of malaria lethal to Hawaiian birds. The feral pigs dug pools of standing water where the mosquitoes could breed, and the insects began biting the native birds. The malaria probably killed a large number of them; today many bird species survive only at high elevations, where the cold air kills the mosquitoes.

Biological invasions have accelerated over the past 200 years, not just at Kauai but all over the world. As sailing ships have been replaced by cargo ships and airplanes, it gets easier for plants and animals and microbes to shuttle between distant continents in staggering numbers. In one study of the ships coming into Chesapeake Bay, scientists found crabs, mullet fish, and hundreds of other species of animals in the ballast water of each ship— 2,000 animals in every cubic meter. And each year 100 million metric tons of ballast water comes to the United States. Meanwhile, insects and seeds can be carried into the country in crops and lumber. There are 50,000 alien species in the United States alone, and more coming fast. Between 1850 and 1960 San Francisco Bay received a new invader about once a year. Since then, a new species has established itself every three months. In Hawaii, a dozen new species of insects and other invertebrates establish themselves every year.

Only a few alien species become successful invaders. Weedy plants and animals can do well in human-dominated habitats because they can survive in unstable ecosystems and spread aggressively. Some alien predators also do well because their diet can include many different kinds of prey. Before World War II, for example, Guam was free of snakes. As the United States began moving military equipment onto the island, it also brought brown tree snakes, which had stowed away in the planes. When the snakes arrived in Guam they began eating any small animal they could find. Of the 13 species of native forest birds on Guam, only 3 still exist. Of the 12 native species of lizards, only 3 have survived.

Another way an invader can succeed is by escaping the restraints that kept them in check where they originally evolved. In 1935 the enormous cane toad *Bufo marinus* was brought to Australia to destroy sugarcane-eating beetles. They have since spread across northern Australia, expanding their range 30 kilometers every year. They blanket the countryside, reaching densities 10 times greater than they can manage in their native habitats. Cane toads do better in Australia than they did in Latin America, it seems, because they have escaped the forces that normally rein them in. Australian predators that try to eat the toads are killed by a venom that the toads produce in glands on their back; their New World predators have evolved an antidote to it. Meanwhile, the viruses and other pathogens that help keep the toad's numbers in check back in Latin America aren't found in Australia. The explosion of cane toads might have been tolerable if they had gotten rid of the beetles, but they showed no interest in them at all. Instead, they have eaten practically anything else they can get in their mouths, including rare marsupials and lizards.

Aliens can also succeed by changing the rules of the ecosystems where they arrive. Hawaii, for example, is an unusual place because it hasn't experienced much fire. To get fire, you need lightning, and to get lightning you need thunderstorms, which form when big landmasses heat up, churning the atmosphere above them. On most continents fire is a part of life, and plants and animals have evolved ways to defend themselves against it. But because Hawaii is a string of islands surrounded by a vast ocean, it is only rarely hit by lightning, so its plants and animals are unaccustomed to fire.

In the 1960s, two fire-adapted plants arrived in Hawaii—the bush beard-grass (*Schizachyrium condensatum*) of Central America, and molasses grass (*Melinis minutiflora*) from Africa. Their dry stems and leaves created blankets of tinder waiting for a spark. Humans obliged, with cigarettes and campfires, and fires began breaking out. The fires devastated native plants, and afterward the aliens took over the charred ground. As they spread, the fires became more intense. In some spots in Hawaii, fires now scorch 1,000 times more land every year than before the invasions. In such an inferno, native grasses have no chance of taking back their old territory.

In other ecosystems, the arrival of many aliens at once can break down a healthy ecosystem's resistance. The Great Lakes are a victim of such an invasional meltdown. Before 1900 most ships that visited the Great Lakes used rock, sand, or mud as ballast, which could carry only a few animals and seeds. In the early twentieth century, ships switched to water for ballast, and in 1959, when the Saint Lawrence Seaway was finally opened for deep-draft shipping, foreign ships began bringing a regular supply of alien species into the lakes.

Although ships come to the Great Lakes from many parts of the world, most of the successful alien species have arrived from the same region: the Black Sea and the Caspian Sea. Animals that live in those waters are adapted for handling the unexpected. The levels of the Black Sea and the Caspian Sea have risen and fallen more than 600 feet over the past few thousand years, and their waters have swung between salty and fresh. The animals that live there have evolved under these wild fluctuations, and they can now survive in a broad range of conditions. They are rugged enough to endure the long journey from Europe to North America in the ballast water, and they can multiply quickly in the freshwater of the Great Lakes.

In the mid-1980s, a small shellfish called the zebra mussel made the journey from southern Russia to Lake Saint Clair. The zebra mussel produces sticky threads that can anchor it to just about any hard surface, and it feeds by pumping water through its body to filter plankton. It can multiply until it cakes dams, water-intake pipes, and riverbeds. The sharp shells litter lake bottoms, cutting the feet of swimmers.

The zebra mussel has spread throughout the Great Lakes and into the surrounding watershed, and everywhere it goes it has turned the native ecology upside down. It covers over the shells of endangered native mussels, sealing them shut and killing them. As a rule, native mussel species disappear four to eight years after the zebra mussel arrives in a lake or a river. They are so good at filtering water that they are sucking out most of the plankton that small crustaceans depend on, starving them out of the lakes, and starving the fish that depend on them as well.

Thanks to the trail-blazing zebra mussel, other invaders from the Caspian and Black Seas are having an easier time becoming established in North America. The round goby, a major predator of the zebra mussel in Europe, was discovered in the Great Lakes in 1990. A crustacean called *Echinogammarus* that feeds on the waste of zebra mussels settled in the Great Lakes in 1995 and has increased twentyfold since then, displacing a native crustacean in the process. *Echinogammarus* is the food of choice for young round gobies back in the old country, so they have helped the alien fish increase its numbers even more. Before the zebra mussel arrived, another immigrant—a tiny colonial animal called a hydroid—had already been in the Great Lakes for decades but was still a rare species in its new home. In Europe the hydroid feeds on zebra mussel larvae, and the advent of zebra mussels in the Great Lakes triggered a burst of hydroids, which now blanket the new mussel beds. As the zebra mussels filter dirt out of the lakes, more sunlight can penetrate their waters, encouraging the growth of underwater plants. These plants offer the zebra mussels surfaces that they can attach to. In other words, by helping the plants grow, the zebra mussels are raising their own numbers. It's not just a few individual species that are alien in the Great Lakes; an entire alien ecosystem is assembling itself.

At the rate at which biological invasions are taking place today, scientists suspect they are becoming one of the most important threats to biodiversity worldwide, ranking close to habitat destruction in their deadliness. Some islands risk losing most of their native life. Native species on the island of Mauritius have declined from 765 to 685, while 730 alien species have made it their home. Half of the imperiled species in the United States are at risk thanks to biological invaders.

In the history of life, this onslaught of biological invasions is a completely new experience. Sudden catastrophes have wiped out tropical forests or coral reefs before humans evolved, but never have so many species jumped so far around the world. Biological invasions may do more than help create a mass extinction; they may leave nature altered long after we're gone.

THE FUTURE OF EXTINCTIONS

Through hunting, habitat destruction, and biological invasions, humans have driven many species to extinction and many more to its brink. But just how bad is today's extinction crisis and how bad will it get? These are supremely difficult questions. Mahaulepu is almost unique in its record of humanity's impact on a place. It is not easy, as a result, to gauge just how severe the extinctions of the past 50,000 years have been on a global scale. Making matters worse, scientists don't even know how many species exist. But the best estimates are grim. We seem to be entering a period of mass extinction on a par with the great die-offs of the past 600 million years.

Every year about 10,000 new species are reported by a small cadre of zoologists and botanists. These scientists have identified about 1.5 million species so far, and they can only guess at how many wait to be found. Based on the rate at which new species are being discovered, they estimate the total at 7 million species, although some argue the figure is as high as 14 million. That means that at least four out of every five species on Earth haven't been discovered yet, and at the rate they are currently being discovered, it will take 500 years to find them all. For many, the discovery may not come in time. If you don't even know that a species exists, you can't expect to know when its last member disappears.

Stuart Pimm has come up with a way to gauge the current extinction crisis that sidesteps our ignorance about the total number of species: he estimates the rate of extinctions, rather than their raw quantity. He has compiled data on documented extinctions in the most carefully studied groups of animals. Across all these groups—including birds, mussels, butterflies, and mammals—he has discovered the same extinction rate: somewhere in the neighborhood of 100 extinctions per million species per year. Given that all of these different groups are suffering the same rate of extinction, Pimm concludes that it's the average for all animals and plants.

It's also a rate far above the normal rate of background extinctions that can be seen in the fossil record. Except during mass extinctions, the background extinction rate has ranged between 0.1 and 1 extinction per million species per year. In other words, species are disappearing 100 to 1,000 times faster today than they did before the arrival of humans.

According to Pimm's calculations, this rate is going to accelerate in the coming years. Two-thirds of all species live in tropical forests. Half of the world's tropical forests have now disappeared, and another million square kilometers are destroyed each decade. Much of the forest that still survives has been fragmented by burning and logging. Without some dramatic conservation, tropical forests will go on disappearing until the only regions left are the ones on protected reserves—about 5 percent of their original extent. That will take only 50 years. Applying the half-life formula to this scenario, Pimm calculates that the extinction rate will rise as much as tenfold. In less than a century, Pimm estimates, half of the world's species will disappear.

Pimm's calculations, as stark as they may be, may actually underestimate the coming extinctions. He uses only deforestation in his estimates. With more planes and cargo ships shuttling between continents, for example, biological invasions are going to accelerate and cause even more extinctions. And even more species may go extinct as we alter the atmosphere.

For the past two centuries, humans have been steadily adding carbon dioxide and other greenhouse gases to the atmosphere. These gases prevent heat from escaping the planet and cause temperatures to rise. Today Earth is now on average 0.5 degrees Celsius warmer than it was in 1860. Some of the difference can be ascribed to changes in the sun and some of it to natural oscillations in the circulation of its oceans and its atmosphere. But most climatologists now agree that man-made greenhouse gases are responsible for most of the warming.

Scientists are now trying to estimate what the climate will be like in the next few decades. That answer depends in part on how much fuel humans continue burning. Will China stick with coal as its economy booms? Will electric cars become more than a publicity stunt? Adding to the uncertainty is how Earth might respond to rising temperatures. The circulation of the oceans might suddenly shift, unloading its hidden heat. The forests of the north might suck up much of the added carbon dioxide, storing away the greenhouse gases as wood. Or the Amazon might turn to a savanna. Or the melting permafrost in the Arctic might release frozen methane. The list of possibilities can fill a book. But as best as researchers can estimate, the planet will warm between 1.4 and 5.8 degrees Celsius by 2100, with the most warming happening in the far north and south.

There are already signs that life is changing as a result of global warming. The growing season in the Northern Hemisphere now starts a week earlier than it did in 1981. With more carbon dioxide available in the air, trees have been growing faster. In North America and Europe, forests are climbing up the sides of mountains. A 1999 study of 35 species of nonmigrating butterflies in North America and Europe found that 63 percent of them shifted their ranges northward over the twentieth century. Even ticks are responding to the warmer winters and marching toward this pole.

These sorts of migrations aren't too surprising when you look at the Ice Age history of North America: the retreats and advances of glaciers triggered ecological scrambles north and south. If the warming continues, entire forests will soon be on the move. The U.S. Department of Agriculture has run computer models to see what will happen to plants and animals in the United States if the climate keeps getting warmer. The conifers and hardwoods of New England move into Canada, while oak and hickory forests of the Midwest decline as they get replaced by the pine forests that move out of the South. In the West, the saguaro cactus might move out of the Southwest deserts, heading north all the way into Washington State.

But global warming won't merely rearrange nature's furniture. The plants and animals that live in cold climates—either in the far north or high in the mountains—have nowhere to shift their range to. Coral reefs, which are proving to be extremely vulnerable to warming in the ocean, cannot simply uproot themselves and head to cooler waters. As a result, global warming may help destroy most of the world's coral reefs in the next 20 years.

Even the species that have enough room on the map to shift their range north or south may actually have trouble surviving. Many of them will be trying to colonize land that is now broken up into farms, suburbs, and cities. It is difficult enough to spare land for nature reserves where endangered plants and animals live today; setting aside new space for them in the decades to come may be even harder. But if we don't, they may simply shift themselves right off an evolutionary cliff.

HUMANITY LEAVES ITS MARK

If these predictions hold true, the next few centuries will see another bout of mass extinctions, with well over half of all species disappearing. Given that we may have inherited a planet at the height of its diversity, it could be the biggest of all in terms of the total number lost.

In a few important ways, this extinction pulse will be different from past ones. An asteroid cannot change its own course, but humans can. How big the extinctions become depends on what humans do in the next hundred years. With habitats disappearing and fragmenting so quickly, conservation biologists have focused their attention on ways to save the most diversity with the least amount of effort. Diversity is not smoothly distributed over the globe, or even within the tropics. A few places, such as Madagascar, the Philippines, and Brazil's Atlantic forests, represent "biodiversity hot spots." The top 25 hot spots contain 44 percent of the diversity of plants and 35 percent of vertebrates. They also comprise only 1.4 percent of Earth's land surface. They are going to disappear quickly if they aren't conserved. On average, 88 percent of the original area of the hot spots has already been destroyed, and their human population is growing fast. These cradles of diversity demand our immediate attention and care.

If extinctions continue to accelerate, the world will become, in a matter of centuries, a homogenized place. While the majority of species with limited ranges continue to go extinct, a few rugged species will thrive. Over 90 percent of the world's agriculture is based on only 20 species of plants and 6 species of animals. As the human population continues to rise, the fortune of these species will go on rising with it. Invaders will continue to spread— zebra mussels, for example, are expected to colonize much of the United States in years to come as they move from one waterway to another. The destruction of forests and other habitats will harm most native species, but a

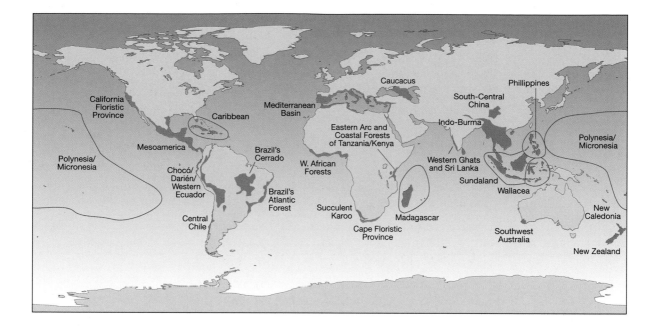

Much of Earth's biodiversity on land is concentrated in only 1.4 percent of the planet's land surface. These regional diversity hot spots are also home to some of the fastest-growing human populations.

few will prosper. In South American forests, hylid frogs can lay eggs in potholes and other temporary ponds, while wolf spiders can spin their webs on weeds. Today quillworts and other lycopsids are growing where tropical rain forests once stood, just as they did 250 million years ago.

"As long as humans are here and don't go extinct," says Ward, "that evolutionary faucet that you turn on after a mass extinction that creates new species—that'll never be turned on. As humans exist far into the future, I foresee a world in which biodiversity stays very low. And that to me is the tragedy."

It is possible that we ourselves will not escape these mass extinctions. We depend on wetlands to filter our water, on bees to pollinate our crops, on plants to build soils. And these plants and animals depend, in turn, on healthy ecosystems to survive. Biologists have run experiments in which they've altered the diversity in simple ecosystems such as grassland plots. With few species, ecosystems become more susceptible to droughts and other catastrophes. Humans may not be able to survive if the impoverished ecosystems we depend on collapse. Of course, humans are the most resourceful species on the planet, so we may find ways to survive even beyond such a disaster.

After past mass extinctions, life has recuperated and even rebounded. How it recovers from the current one depends partly on the destiny of the human race. Man-made global warming may end up being one of the most profound causes of extinction, but it cannot last forever. There is only so much oil and coal left—about 11 trillion tons. James Kasting, a climatologist at Penn State University, estimates that burning this much fossil fuel would increase atmospheric levels of carbon dioxide to about three times today's levels, raising temperatures between 3 and 10 degrees Celsius. It would take only a few centuries to use up these reserves. It will take hundreds of thousands of years for Earth to draw down the carbon dioxide to the levels it was at before the Industrial Revolution.

But long after the atmosphere recovers from its binge of carbon dioxide, and even after *Homo sapiens* is gone, the biological invaders we have sown around the world will keep regenerating, keep controlling the ecosystems that surround them. They will continue to frustrate the evolution of other plants and animals.

"Evolution has now entered a new mode," says Burney. "Something altogether new is happening, and it has to do with what humans do to the evolutionary process. And it's a very scary thing, because it's like we are taking evolution around a blind corner, something that nature hasn't dealt with before: species that can just hop a plane and wind up on the other side of the world; combinations of species that have never been combined before. It's a whole new ball game, and we don't know, really, where it will end."

PART THREE
Evolution's Dance

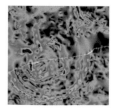

Coevolution

WEAVING THE WEB OF LIFE

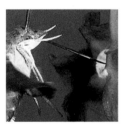

ONE OF THE MANY reasons we need to save endangered species from extinction is because we can find among them lessons about how evolution works. In the rapidly shrinking forests of Madagascar, one such species clings to existence: an extraordinary orchid named *Angraecum sesquipedale*. One of the petals on its pale white flower is shaped into a 16-inch-deep shaft, and nestled in the bottom of the shaft are a few drops of sweet nectar.

What could this deep nectary possibly be for? What evolutionary force created it? Wait long enough and the answer will arrive on the wing. A species of moth visits the orchid, slowing to a hovering stop over the flower. Its tongue, coiled up like a watch spring, begins to fill with blood, and as it does, the pressure forces it to straighten out. It grows to be 16 inches long, far longer than the moth's entire body. The moth snakes its tongue down the tube

Flowers depend on insects such as bumblebees to spread their pollen. In exchange, the flowers provide insects with nectar, oils, resins, or sometimes even the pollen itself.

until it reaches the sweet nectar. As it soaks up the nectar, it buries its face in the flower, and as it does, its forehead rubs against pollen grains. When the moth is finished with its drink, it curls up its tongue again and flies away with the pollen smeared over its head, to find another orchid to drink from. The pollen on its forehead brushes off on the new orchid, where it can fertilize that flower's eggs.

It may be hard to believe that a pair of species could be so tightly linked together, and yet nature is filled with such intimate partnerships, whether they are the beneficial ones between flowers and their pollinators or hostile ones, such as predators and the prey they hunt. Life consists for the most part of a web of interacting species, adapted to one another like a lock and key.

Partners such as the orchid and its moth did not spring into existence already linked together in the relationships they have today. They gradually evolved into closer and closer intimacy, and they continue to evolve today. Every generation of plants, for instance, adapts its defenses against insects, but the insects are also evolving ways to overcome their defenses at the same time.

Scientists are discovering that the way in which the evolution of one species drives the evolution of another—known as coevolution—is one of the most powerful forces shaping life. Coevolution can create anatomy as remark-able as a moth's 16-inch tongue. It is also responsible for creating much of life's diversity, as the spiraling coevolution of partners spawns millions of new species. Coevolution is also a fact that we ignore at our own peril. The crops we eat, the paper on which these words are printed—every plant we depend on is coevolving with intimate partners, both life-sustaining and life-destroying. If we alter their coevolutionary dance, we may have to pay a steep price.

SEXUAL GO-BETWEENS

The concept of coevolution dawned on Darwin in the 1830s as he pondered the mystery of how plants have sex. A typical flower grows both male and female sexual organs. The male pistils hold pollen grains, which enter the female anthers and fertilize their seeds. In order for plants to mate with other plants, they cannot pull up their roots and look for a partner. Somehow the pollen from one plant must get to the eggs of another. And not just any other plant: it has to reach another member of the same species.

For some plants, setting their pollen on the breeze is all that's necessary. But Darwin discovered that some plants use insects to carry the pollen for them. He watched bees come to scarlet kidney bean plants to drink the sweet nectar they produced. As a bee climbed up the petals of the bean's flowers, it

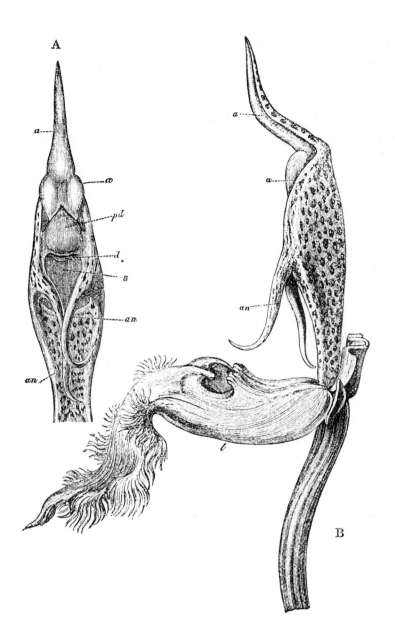

A drawing of the orchid *Catasetum saccatum* from Darwin's 1862 book on these extraordinary flowers. Darwin discovered that as pollinating insects walk along the petal toward the flower's nectar, they trip an antenna-shaped projection. The antenna acts like a trigger releasing a package of pollen that is flung on the insect's back.

invariably rubbed its back on the plant's pistils, picking up its pollen. And when the bee visited another plant, it unloaded the pollen on its female anthers. Darwin realized that flowers use bees for sex, giving them nectar for their services.

Another scientist might have been content with making such a remarkable discovery. But Darwin was never satisfied to understand nature as it was; he wanted to uncover clues of its history. In the case of flowers and their pollinating insects, he realized that evolution must have taken a complex path. Here Darwin was not dealing with a species adapting to physical conditions like the force of gravity or the viscosity of water. Here he was dealing with two species that were adapting to each other. While the force of gravity doesn't change, a species can change with every new generation.

In *Origin of Species,* Darwin sketched out one example of how coevolution might shape two species. Common clover is normally pollinated by bumblebees. But imagine that one day bumblebees suddenly went extinct. If the common red clover didn't get a new pollinating partner, it would be unable to reproduce and disappear as well.

Honeybees might fill the void. Normally, honeybees only pollinate a different form of clover, called incarnate clover. But some honeybees might start taking advantage of all the nectar in the common red clover that was now going to waste. At first the honeybees would have a difficult time, because their tongues aren't as long as those of bumblebees, so they wouldn't be able to get as much nectar. Any honeybee born with an unusually long tongue would be richly rewarded with lots of common red clover nectar, so natural selection might gradually lengthen the tongue of the honeybee.

Some pollinating insects have extraordinarily long tongues they use to drink the nectar of flowers held in extraordinarily long tubes. The shape of the tongues and tubes have coevolved.

Meanwhile, the common red clover might adapt to its new pollinator, the honeybee. A common red clover flower with a shape that made the efforts of a honeybee a little easier would be rewarded with the spread of its pollen. Gradually the red clover flower and the honeybee would evolve to suit each other.

"Thus I can understand," Darwin wrote, "how a flower and a bee might slowly become, either simultaneously or one after the other, modified and adapted in the most perfect manner to each other, by continued preservation of individuals presenting mutual and slightly favorable deviations of structure."

Not long after Darwin finished *Origin of Species,* he discovered just how drastically flowers and insects could affect one another. He began studying orchids, crouching in the fields around Down House to observe native species or studying exotic species at his greenhouse that he had sent from the tropics. In Darwin's time, most people considered orchids creations designed purely to please the human eye. But Darwin recognized that their shapes were not beauty for beauty's sake, but elaborate devices for luring insects into their sex lives.

As a mechanic might dismantle a car, Darwin figured out how the parts of an orchid worked together. One of the species that fascinated him was the South American orchid *Catasetum saccatum.* It keeps its pollen loaded on a disk attached to a flexible shoot, which is bent back so that the disk is lodged within the flower itself. There the disk remains, cocked like a crossbow. When an insect arrives at the orchid to drink its nectar, it has to land on a cup-shaped petal that extends horizontally out from the plant. To get to where the nectar is hidden, it has to walk across the petal, grazing its back against antennae hanging down overhead. The antennae are attached to the flexible shoot, and they act like a trigger, dislodging the shoot and letting it fling its pollen disk down against the back of the bee.

Moving the antennae, Darwin recognized, was the only way to set the pollen free. The specimens of *C. saccatum* that he studied had been delivered to him by train, and yet the jostling of the ride hadn't made the pollen explode. He poked at the orchids in various places with a quill, and nothing happened. "After trials made on fifteen flowers of three species," he later wrote, "I find that no moderate degree of violence on any part of the flower, except on the antennae, produces any effect." The orchid, Darwin realized, had coevolved with its insect visitors.

Darwin described these orchids, and many others, in a long-titled book: *The Various Contrivances by Which British and Foreign Orchids Are Fertilized by Insects, and on the Good Effects of Intercrossing.* Like *Origin of Species,* it was an argument for evolution, but it was subtler than its predecessor. Darwin guided the reader from orchid to orchid, showing how each flower's design represented an elaborate form of sex. Just as he had previously demonstrated how barnacles were highly evolved crustaceans, he now showed that orchids were highly evolved flowers. Evolution had stretched and twisted and transmogrified the parts of ordinary flowers to create the crossbows and other devices that orchids use to spread their pollen.

Darwin was so confident that coevolution created the shapes of orchids that he made a bold prediction in his book. Explorers had already discovered the Madagascar orchid *Angraecum sesquipedale,* with its 16-inch-long nectary. As outrageous as it might sound, Darwin predicted the existence of a suitably long-tongued insect on Madagascar. The orchid's pollen, he predicted, "would not be withdrawn until some huge moth, with a wonderfully long proboscis, tried to drain the last drop."

Darwin held out hope for that huge, wonderful moth even as decades passed without anyone discovering it. It was not until 1903 that entomologists reported the existence of just such an insect, which they named *Xanthopan morgani praedicta*—*praedicta* in honor of Darwin's prediction. Biologists now know of many other species of moths and flies with long tongues, which they use to drink the nectar of equally long-nectaried flowers. They can be found not just on Madagascar but in Brazil and South Africa as well. It's a lucky scientist whose most bizarre predictions get confirmed even once.

THE COEVOLUTIONARY MATRIX

Coevolution is far more powerful and widespread than Darwin ever imagined, even among the plants that inspired him. Scientists now recognize that the vast majority of flowering plants—290,000 species—depend on animals to spread their pollen (only 20,000 species can spread their pollen by wind or water). Instead of nectar, some of them get insects to spread their pollen by offering resins and oils the insects use to help build the walls of their hives. Tomatoes and certain other plants even offer up some of their own pollen. They typically keep their pollen in containers shaped like salt shakers, and when the insects land on a flower, they buzz their wings at a frequency that jostles the pollen free. The insects feast on the pollen, and in the process they get dusted with it as well.

While most of the animals that pollinate flowers are insects, about 1,200 species of vertebrates—mostly birds and bats—also do the job. Like insect pollinators, they have shaped the evolution of the plants they pollinate. The flowers that are pollinated by birds lure them with bright red petals (insects are blind to the color). Unlike fragrant orchids, bird-pollinated flowers are scentless, since birds have a poor sense of smell. They keep their nectar in long, wide tubes to suit the long, stiff beaks of birds. On the other hand, the plants that depend on bats to spread their pollen open their flowers at night, when bats leave their roosts in search of food. To make themselves easy to find, some bat flowers have evolved into cuplike shapes that can reflect and focus sound waves that bats use to echolocate. These acoustic mirrors catch the bats' attention and guide them to their meal.

Domesticated plants depend on pollinators just as their wild relatives do. Without them, an apple orchard would be fruitless, a cornfield cobless. But plants—both wild and domesticated—also depend on other coevolutionary partners to keep from starving to death. Plants use photosynthesis to turn carbon dioxide and water into organic carbon, but they have a harder time extracting nitrogen, phosphorus, and other nutrients from the soil. Fortunately, the roots of many species of plants are enmeshed in a vast filigree of fungus that can supply the nutrients they need.

The fungus produces enzymes that break down soil, allowing it to suck up phosphorus and other chemicals. It injects these nutrients into the plants, and in exchange, it draws out some of the organic carbon the plants create with photosynthesis. The fungus demands a steep cost for its services: about 15 percent of the organic carbon a tree creates in a year. But it's a price worth paying; without fungus, many plants become stunted and gaunt. Some species of fungus can kill soil nematodes and other plant enemies and may even make plants more resistant to drought and other disasters. It can store carbon it extracts from trees and then shuttle it through its web. If one tree it is connected to is suffering a carbon shortfall, the fungus can pump carbon into its roots. Forests, prairies, and soybean fields are not a collection of lonely individuals: they are just the visible tips of a huge coevolutionary matrix.

BIOCHEMICAL WARFARE

Coevolution can produce mutually rewarding friendships, but it can just as easily turn species into finely tuned enemies. The constant menace of a predator may drive animals to evolve faster legs or harder shells or better camouflage. In response, their predators are free to evolve faster legs of their own, or stronger jaws or stronger eyes. Predator and prey can thus get locked in a biological version of an arms race, each opponent developing some new adaptation, only to be outstripped by its enemy's evolution.

This arms race may produce brute strength or speed, but it can also create chemical warfare of exquisite sophistication. One of the best places to see it on display is in the wetlands and forests of the Pacific Northwest. There you can find the rough-skinned newt, an 8-inch-long amphibian with a brilliant orange belly. When one of these newts is attacked, it displays its belly, which a predator would do well to recognize as a warning. If it eats the newt, it will almost certainly die because the newt produces a nerve toxin powerful enough to kill 17 full-grown humans or 25,000 mice.

Since only a fraction of the poison made by a newt would be enough to kill most of its predators, it seems to be indulging in overkill. But there is one predator that remains a threat to even the most poisonous newts. Edmund Brodie Jr. of the University of Utah and his son Edmund III, a biologist at Indiana University, have discovered that the red-sided garter snake can eat rough-skinned newts without dying, thanks to a genetic resistance to the newts' poison.

In other predators (including other species of garter snakes), the poison blocks certain channels on the surface of nerve cells, jamming their communication and causing a fatal paralysis. But red-sided garter snakes have evolved nerve channels that cannot be completely blocked by the poison. The snakes may become immobilized for a few hours after they eat a newt, but they recuperate eventually. The menace of garter snakes has driven the evolution of more toxin in the newts, which has driven the evolution of more resistance in the snakes.

This arms race between snakes and newts doesn't move forward in some kind of simple, specieswide march. That's not how evolution works. Instead, the struggle between predator and prey plays out in hundreds of local populations. In places such as the San Francisco Bay area and the northern coast of Oregon, coevolution appears to be running at top speed, creating highly toxic

Father-and-son biologists Edmund Brodie Jr. and Edmund III (below) search for the fatally toxic rough-skinned newt (right). They are studying the coevolutionary pressure that has turned it into a killer.

newts and highly resistant snakes. But just as there are coevolutionary hot spots, there are also coevolutionary cold spots in their range. In the Olympic peninsula, for example, some populations of newts make almost no toxins at all, and the snakes that eat them have hardly any resistance.

The Brodies suspect that the unique circumstances of each place are setting coevolution's course. Evolving resistance to rough-skinned newts does not come without a price. The more resistant snakes are, it turns out, the slower they crawl. There's a trade-off between resistance and speed that the Brodies don't yet fully understand, but it puts resistant snakes at greater risk of getting attacked by birds and other predators. The coevolutionary cold spots

EVOLUTION

may be places where garter snakes are under heavy attack. On the other hand, the hot spots may represent places where garter snakes depend on the newts because other prey are rarer. Whatever the cause of these hot spots and cold spots, the genes that evolve in each of them spread across the ranges of newt and snake. Sometimes the cold spots will stop the arms race in its tracks, but in other cases the coevolutionary hot spots may drive an entire species to deadly extremes.

BEETLES VERSUS PLANTS: A 300-MILLION-YEAR WAR

The coevolutionary struggle between enemies may produce much more than exquisite poisons and sophisticated antidotes. It may be a driving force behind the diversity of life itself.

The possibility that coevolution could have such a profound effect was first raised in 1964 by two ecologists, Paul Ehrlich and Peter Raven. Ehrlich and Raven pointed out that while pollinating plants may be on friendly terms with insects, many wage all-out war. Insects make their living by chewing leaves, boring through wood, feasting on fruit, and otherwise ravaging plants. By eating leaves, the insects deprive them of photosynthesis. By nibbling on their roots, they cut off their supply of water and nutrients. If they chew away too much, they can kill plants outright. Even if insects limit themselves to a plant's seeds, they deprive it of its genetic legacy.

Plants have evolved physical and chemical defenses to ward off these hungry bugs. Holly plants have evolved leaves with sharp teeth along their edges that protect them from chewing insects. (If you clip off the teeth, bugs can quickly strip the leaves.) Some plants respond to insect bites by creating poisons in their tissues. Some defend themselves with a network of tubes in their leaves and stems full of sticky resin or gooey latex. When an insect chewing on the plant ruptures one of the tubes, a flood of resin or latex comes pouring out. It engulfs the insect, perhaps to entomb it in a lump of amber or trap it in a rubbery plug of latex. Other plants defend themselves by calling for help. In response to the bite of a caterpillar, they release chemicals that attract parasitic wasps. The wasps lay their larvae inside the attackers, devouring them from within.

Ehrlich and Raven suggested that any lineage of plants that happened to evolve a way to escape its pests might be able to grow aggressively and spread its range. It would become less likely to go extinct and more likely to branch off into new species. Over time, its descendants would be more diverse than other lineages of plants that hadn't managed to escape their pests.

To the insects, these new plants would be like a continent rising out of the ocean, just waiting for them to explore. But to reach it, they would have to evolve a way to overcome the plants' new defenses. Any lineage of insects that managed to find a way would be able to colonize the plants, and without any competition from other insects, they would thrive and diversify. This coevolution—a series of escapes and captures and fresh escapes—might account for the different levels of diversity among plants and among insects.

Ehrlich and Raven's hypothesis was alluring, but it was difficult to test. Scientists couldn't replay the past 300 million years and watch the coevolution of plants and their insect enemies play out. But by the early 1990s, a Harvard entomologist named Brian Farrell had figured out how to use the diversity of living plants and insects to see whether Ehrlich and Raven were right.

In his first test, he set out to see if the evolution of a new defense really does trigger an explosion of diversity in plants. As his test case, he chose the evolution of canals of latex or resin, which drown feeding insects. Farrell found that this defense arose independently in 16 lineages of plants. He counted up the number of species each lineage had given rise to and then compared it to the number of species in closely related lineages that lacked canals. In 13 out of the 16 cases, the groups with canals were far more diverse than their relatives without them. The gingko tree and conifers share a close common ancestor, for instance, but gingkoes, which lack canals, consist of a single species. Meanwhile, conifers evolved canals and have since produced 559 species, ranging from pines to yews. Daisies and dandelions belong to a diverse group of canal-bearing plants called Asterales, which has 22,000 species in its ranks. Its closest relative without canals, an obscure group called the Calyceraceae, has only 60. By evolving defensive canals, these plants have lived up to Ehrlich and Raven's prediction: they have thrived by fighting off their enemies.

Farrell then looked at the second part of Ehrlich and Raven's hypothesis: when insects manage to colonize these new lineages of plants, they experience bursts of diversity of their own. He chose a particularly diverse bunch of insects for his study: beetles. The biologist J. S. B. Haldane liked to say that if biology had taught him anything about the nature of the Creator, it was that he had "an inordinate fondness for beetles." While insects may be the most diverse group of animals, beetles, with 330,000 known species, are the most diverse group of insects. Farrell set out to figure out how they had become so diverse.

He first built himself an evolutionary tree of beetles. He found that the insects did not start out as plant eaters. The most primitive beetles had a diet that consisted of fungus and smaller insects. A few species of living beetles such as weevils carry on this primitive way of life today. But about 230 million years ago, a new branch of beetles shifted its diet to plants. Many of the plants we are most familiar with had not yet evolved. There were no flowering plants, for example. These early plant-eating beetles specialized on the plants that already existed, known as gymnosperms (the group that includes conifers, gingkoes, and sago palms). By feeding on gymnosperms, these new branches of beetles diversified into many more species than their fungus- and insect-eating cousins.

About 120 million years ago, flowering plants began to emerge and proved to be far more diverse than gymnosperms. As they diversified, five separate lineages of beetles managed to make the leap from gymnosperms to these new plants. Farrell found that in all five cases the beetles that shifted to flowering plants consistently became more diverse than their cousins who stayed behind, surviving on gymnosperms. In some cases the diversity of the new beetles has multiplied more than a thousand times. As Ehrlich and Raven had predicted, the newer beetles explored new ways of making a living on the plants. The beetles that feed on gymnosperms specialize on pinecones or similar seed-bearing structures. The beetles that moved to flowering plants began branching out to different kinds of food, such as bark, leaves, and roots.

Although no one has done such careful studies of other insects, the same pattern probably holds: coevolution with flowering plants was their secret to success. And like all good research, Farrell's work points to a new puzzle still to be solved: Why are there so many flowering plants? Again, at least part of the answer probably has to do with coevolution. As insects invented new ways to devour flowering plants, the plants were under intense pressure to find ways to hold them at bay. At the same time, some insects such as bees were coevolving more benign partnerships, helping flowering plants spread their pollen, which may have helped them fight against extinction and branch into more species. Diversity, scientists have found, begets diversity.

Beetles (a fossil of which is shown here) are the most diverse of all insects, with 330,000 species identified so far. Much of their success apparently stems from their colonization of flowering plants more than 100 million years ago.

MAN VERSUS BUG

Beetles and other insects have been coevolving with the plants they eat for hundreds of millions of years. In the past few thousand years, we humans have suddenly altered their coevolution by domesticating plants and trying to keep insects from eating them. For the most part, we've tried stopping insects without taking the reality of coevolution into consideration, with the result that much of our effort backfires. The swift development of resistance to pesticides is one of the most graphic cases of coevolution in action.

Humans began domesticating crops around 10,000 years ago. As the first farmers planted fields with crops like lentils and bulgur, insects fed on them just as they had on their wild ancestors. At first humans could do nothing but plead to the gods. Sometimes they would even take insects to court. In 1478 beetles were ravaging the crops around the Swiss city of Berne, and in response the mayor appointed a lawyer to go to an ecclesiastical court and demand punishment. "With grievous wrong, they do detriment to the ever-living God," they complained. A lawyer was appointed for the beetles, and defendant and plaintiff made their cases to the court. After hearing both sides, the bishop ruled in favor of the farmers, declaring that the beetles were incarnations of devils. "We charge and burden them with our curse," he declared, "and command them to be obedient and anathematize them in the name of the Father, the Son and the Holy Ghost, that they turn away from all fields, grounds, enclosures, seeds, fruits and produce, and depart."

The beetles did not depart. They went on eating their crops as before. It was then decided that the beetles were not devils but a punishment that God was visiting on farmers for their sins. Once the farmers gave a tithe to the Church from what little harvest they could manage, the beetles disappeared. Or perhaps when the beetles used up their food supply their population crashed naturally.

When court cases and invocations to the gods have failed, farmers have resorted to poisons. The Sumerians began putting sulfur on their crops 4,500 years ago; pitch and grease were popular in ancient Rome. Ancient farmers discovered that some plants contained substances that could help their crops ward off insects. The Greeks soaked their seeds in cucumber extract before planting them. In the 1600s, Europeans began to extract chemicals from plants such as tobacco that proved even more powerful than old potions. From an Armenian daisy in 1807 came pyrethrum, which farmers still use today.

At the same time that Europeans were discovering better pesticides, they were also building huge farms both at home and in their new colonies. For insects it was as if someone had laid out a vast banquet table. Epidemics of insects raced across entire countries. Farmers began resorting to harsher kinds of pesticides, such as cyanide, arsenic, antimony, and zinc; they mixed copper and lime into a concoction known as paris green. With the invention of airplanes and spray nozzles, entire farms could be blanketed with pesticides, and by 1934 American farmers were using 30 million pounds of sulfur, 7 million pounds of arsenic-based pesticides, and 4 million pounds of paris green.

Around 1870, a little fruit-eating insect arrived in San Jose, California, on some nursery stock shipped from China. The pest, which became known as the San Jose scale, quickly spread through the United States and Canada, killing orchard trees as it went. Farmers found that the best way to control the scale was to spray their orchards with a mixture of sulfur and lime. Within a few weeks of spraying a tree, the insect vanished completely.

Around the turn of the century, however, farmers began to notice that the sulfur-lime mixture wasn't working all that well. A handful of scales would survive a spraying and eventually rebound to their former numbers. In Clarkston Valley in Washington state, orchard growers became convinced that manufacturers were adulterating their pesticide. They built their own factory to guarantee a pure poison, which they drenched over their trees, yet the scale kept spreading uncontrollably. An entomologist named A. L. Melander inspected the trees and found scales living happily under a thick crust of dried spray.

Melander began to suspect that adulteration was not to blame. In 1912 he compared how effective the sprays were in different parts of Washington. In Yakima and Sunnyside, he found that sulfur-lime could wipe out every last scale on a tree, while in Clarkston between 4 and 13 percent survived. On the other hand, the Clarkston scales were annihilated by a different pesticide made from fuel oil, just as the insects in other parts of Washington were. In other words, the scales of Clarkston had a peculiar resistance to sulfur-lime.

Melander wondered why. He knew that if an individual insect eats small amounts of certain poisons, such as arsenic, it can build up an immunity. But San Jose scales bred so quickly that no single scale experienced more than a single spray of sulfur-lime, giving them no chance to develop immunity.

A radical idea occurred to Melander. Perhaps mutations made a few scales resistant to sulfur-lime. When a farmer sprayed his trees, these resistant scales survived, as well as a few nonresistant ones that hadn't received a fatal dose. The surviving scales would then breed, and the resistant genes would become more common in the following generations. Depending on the proportions of the survivors, the trees might become covered by resistant or nonresistant scales. In the Clarkston Valley region farmers had been using sulfur-lime longer than anywhere else in the Northwest and were desperately soaking their trees with the stuff. In the process, they were driving the evolution of more resistant scales.

Melander offered his ideas in 1914, but no one paid much attention to him; they were too busy discovering even more powerful pesticides. In 1939 the Swiss chemist Paul Muller found that a compound of chlorine and hydrocarbons could kill insects more effectively than any previous pesticide. Dubbed

DDT, it looked like a panacea. Cheap and easy to make, it could kill many species of insects and was stable enough to be stored for years. It could be used in small doses, and it didn't seem to pose any health risks to humans. Between 1941 and 1976, 4.5 million tons of DDT was produced—more than a pound for every man, woman, and child alive today. DDT was so powerful and cheap that farmers gave up old-fashioned ways of controlling pests, such as draining standing water or breeding resistant strains of crops.

DDT and similar pesticides created the delusion that pests could be not merely controlled but eradicated. Farmers began spraying pesticides on their crops as a matter of course, rather than to control outbreaks. Meanwhile, public health workers saw in DDT the hope of controlling mosquitoes, which spread diseases such as malaria. In his 1955 book, *Man's Mastery of Malaria,* Paul Russell of Rockefeller University promised that "for the first time it is economically feasible for nations, however underdeveloped and whatever the climate, to banish malaria completely from their borders."

DDT certainly saved a great many lives and crops, but even in its early days some scientists saw signs of its doom. In 1946, Swedish scientists discovered houseflies that could no longer be killed with DDT. Houseflies in other countries became resistant as well in later years, and soon other species could withstand it. Melander's warning was becoming a reality. By 1992 more than 500 species were resistant to DDT, and the number is still climbing. As DDT began to fail, farmers at first just applied more of it; when more no longer worked, they switched to newer pesticides, like malathion; when those started to fail, they looked for newer ones.

The quest to eradicate pests with DDT and similar poisons has been a colossal failure. Each year more than 2 million tons of pesticides are used in the United States alone. Americans use 20 times more pesticides today than they did in 1945, even though the newest pesticides are up to 100 times more toxic. And yet the fraction of crops lost to insects has risen from 7 percent to 13 percent—thanks in large part to the resistance insects have evolved.

The failure of DDT has been an unplanned experiment in evolution, as compelling as Darwin's finches or the guppies of Trinidad. As Ehrlich and Raven pointed out in 1964, plants have been producing natural pesticides for hundreds of millions of years, and insects have coevolved resistance to them. In the last few thousand years, insects encountered some new man-made poisons on the plants they've been eating, and they've been doing what they've always done: evolving their way around them. Coevolution has entered the age of humans.

The first time a pesticide is sprayed on a field, it kills most of the insects. A few insects survive because they don't get a fatal dose. Others survive because they have rare mutant genes that happen to protect them from the pesticide. These mutations might have arisen from time to time during the species' lifetime, but normally they put insects at a disadvantage. As the mutants were outcompeted, their mutation would disappear. The balance between new mutations and their eradication from the gene pool leaves a population of insects at any given moment with a few mutants in their ranks. At the moment that the pesticides arrive, the mutants suddenly are more fit than their counterparts.

There are many ways that mutant insects can resist pesticides. They may be born with thick cuticles covering their bodies, shielding them from the chemical. They may make a mutant protein that can cut a pesticide molecule into pieces. They may get irritable when they come into contact with the pesticide, flying away before they can receive a lethal dose.

After a spraying, the surviving insects are free from competitors. They may even be liberated from their parasites and predators, if the pesticide wipes them out as well. With few rival pests to eat their food, and without enemies to keep their numbers in check, they explode. As they mate with one another, or with surviving nonresistant insects, their mutant genes quickly spread. If farmers spray relentlessly enough, the nonresistant forms may disappear almost completely.

Insects coevolved with plants for hundreds of millions of years, developing resistance to their chemical defenses. In the past century they have quickly evolved resistance to pesticides used by farmers.

Pesticides are a clumsy substitute for coevolution. Plants and insects can evolve new attacks against each other from one generation to the next. But chemists need years to discover new pesticides, and as they search, the resistance insects evolve takes a heavy toll. Resistant insects require farmers to spend more money to buy new pesticides. Unlike the natural defenses that plants produce, pesticides may also kill earthworms and other subterranean creatures that are essential for creating new soil out of organic matter. Some pesticides kill bees and other pollinators. They can linger in the environment for years and travel for thousands of miles. Pesticides may also kill humans, directly poisoning farmhands, and there is some disturbing—although hotly disputed—evidence of a relationship between exposure to pesticides and some types of cancer.

Agribusiness has recently offered a solution to the pesticide crisis, in the form of genetically altered crops. Eight million hectares of farmland are now planted with crops carrying genes from bacteria that let them make their own pesticides. The genes come from *Bacillus thuringiensis,* a bacterium that lives in the soil and attacks butterflies and moths. In order to feed on its hosts, *B. thuringiensis* uses these genes to produce a protein that destroys an insect's gut cells. Cultures of the bacteria (known as Bt) have been kept since the 1960s, and the protein is sprayed just about everywhere, from organic farms to forests. It doesn't harm mammals, and it quickly breaks down in sunlight. Now biochemists have been able to insert Bt genes into plants such as cotton, corn, and potatoes, and these plants now produce Bt in their own tissues. Insects that attack the transgenic plants eat the Bt and die.

The Environmental Protection Agency is hoping that Bt-producing crops won't become another victim of coevolution. If a cotton farmer plants his entire farm with Bt-producing crops, the insects that live there will encounter vast expanses of plants all producing the same poison, and they'll evolve resistance. The EPA is mandating that farmers plant ordinary crops on 20 percent or more of their fields. These patches will become refuges where non-resistant insects can survive; they'll mate with resistant insects, the reasoning goes, and keep those resistance genes from becoming too common.

This approach depends on the cooperation of farmers, who will have to sacrifice some of the crops in their refuges to insects. But their grim experiences with pesticides will probably keep many of them from planting only Bt crops. If these crops succeed over the long term, it will be thanks to a proper understanding of coevolution. But if resistance does break out, farmers may have to buy new kinds of modified plants that produce a new kind of toxin—in other words, they will have to jump off the pesticide treadmill only to jump onto a transgenic treadmill.

Coevolution offers some hints about other ways to fight pests. Insects would be less harmful if farmers stopped growing vast carpets of monoculture. Growing a combination of different crops makes it harder for the specialist pests to build up the reproductive momentum they need to cause an outbreak. And consumers can help as well. When you shop for fruit, you probably pass over ones that are blemished. Fruit growers know this and go to great lengths to deliver spotless produce to supermarkets—a feat that requires using lots of pesticides to kill insects. In fact, a slightly blemished fruit is perfectly safe to eat. If consumers were more willing to buy less-than-perfect fruit, farmers could use substantially fewer pesticides. As a result, they would relax the evolutionary pressure on insects to resist the pesticides.

ANTS: THE FIRST FARMERS

We humans may pride ourselves on inventing agriculture, but we were not the first to do so. In one of the most extraordinary episodes of coevolution, one group of ants became mushroom farmers 50 million years ago. They remain hugely successful at agriculture today, and they have managed to avoid much of the grief we suffer with pests. We would do well to learn from them.

Fungus-growing ants live in tropical forests around the world. In many species, a caste of large ants marches out from the nest each day in search of trees and bushes. They climb onto the plants and chew off pieces of leaves, which they bring home in a little green parade. The big ants pass the leaves to a smaller caste, which tear the leaves into smaller pieces. They pass them to an even smaller caste, which chew them into smaller pieces, and so on, until the leaves have been transformed into a paste. The ants then spread the leaf paste like a fertilizer on carpets of fungus in their nest. The fungus can break down the tough tissue of the leaves and grow, and the ants can then harvest special nutrient-rich parts of the fungi. (Not all fungus-growing ants fertilize their fungus with leaves—many species search the forest floor for organic matter such as fallen flowers and seeds.)

The fungi that grow in the gardens of leaf-cutter ants have become completely dependent on their farmers. Free-living fungi propagate themselves by growing mushrooms filled with spores that are carried away by the wind. The garden fungi have lost the ability to sprout mushrooms. They are trapped in their ant nest, which they leave only when a young queen takes a bit of fungus in her mouth when she sets out to found a new colony.

The leaf-cutter ants get a huge benefit from caring for their fungus. Ants can't digest plant tissue, so most species cannot take advantage of the vast amounts of food that surround them. Leaf-cutter ants can let their garden fungus do the hard work of breaking down leaves for them. And thanks to their partnership, the ants have become one of the most powerful players in tropical forests, eating a fifth of all the leaves grown each year in some tropical forests.

To understand how such a remarkable partnership evolved, scientists study the evolutionary relationships between the ants and the fungus. Since the 200 species of ants known to farm fungus are all close relatives (there are no nonfarmers in their ranks), biologists have long assumed that their saga must have begun with a single primordial lineage of ants inventing agriculture. They then passed the secret down to their descendants in every new queen's mouth. As new species of ants emerged, their fungus would evolve into new species as well. If this was in fact the case, then the evolutionary tree of the fungus should perfectly mirror the evolutionary tree of the ants.

But the truth turned out otherwise. Since the early 1990s, Ulrich Mueller of the University of Texas and Ted Schultz from the Smithsonian Institution have been trekking through jungles around the world, gathering leaf-cutting ants and their fungi. Back at their lab, they and their colleagues have sequenced the ant and fungus genes and used them to work out their evolutionary relationships. The ants did not domesticate a single original fungus. Mueller and Schultz have discovered that they have tamed fungi at least six different times. Once these six different lineages of fungi were domesticated, they branched into new species as their ant farmers have evolved into new species of their own. But on many occasions fungus species have been exchanged between ant colonies.

Mueller is now studying how these shifts take place. "One possible scenario," he says, "is that pathogens wipe out entire gardens. Then the ants are forced to go to neighboring ants and steal a replacement, or temporarily join with them in one happy community. But occasionally we also see them invade a neighboring nest and wipe out the ants and take over their gardens."

Thanks to Mueller's work, ants now look more like human farmers than ever before. Our ancestors in China, Africa, Mexico, and the Middle East tamed a handful of plants and animals, a tiny fraction of the millions of wild species on Earth, just as ants domesticated a few of the hundreds of thousands of fungus species. As human cultures have made contact with one another, their crops have passed between them like fungal spores. The only major difference between us and the ants is that they stumbled across agriculture 50 million years before we did.

Leaf-cutter ants have to struggle with pests just as human farmers do. In the case of the ants, several species of fungi live as parasites on their garden fungus. It's possible for a spore of parasitic fungus to get into a garden and destroy it in a matter of days.

But Cameron Currie, who works with Mueller at the University of Texas, has discovered that ants use a fungicide to keep the parasitic fungus in check. The bodies of the ants are coated with a thin powdery layer of *Streptomyces* bacteria. The bacteria produce a compound that kills off the parasitic fungus while stimulating the growth of the garden fungus. Each of the 22 species of leaf-cutting ants that Currie has studied carries its own strain of *Streptomyces*.

Judging from the fact that all fungus-growing ant species that Currie has studied carry *Streptomyces* with them, it's possible that the very first ones 50

Biologists Ulrich Mueller and Ted Schultz search in Brazil for new species of leaf-cutting ants. They've discovered how ants have domesticated separate strains of fungi and borrowed them from other ant species.

A leaf-cutter ant carries home a fragment of a leaf to its nest, where it will be used to fertilize a garden of fungus, which the ant colony eats. This relationship between ant and fungus has been evolving for 50 million years.

million years ago used them as well. And yet in all that time, the parasitic fungus has not evolved any significant resistance to the fungicide. How is this possible, when we humans have inadvertently bred resistant pests in only a few decades? Currie and his colleagues are only just beginning to tackle that question, but they have a working hypothesis: when we use a pesticide, we isolate a single molecule and apply it to an insect. But *Streptomyces* is an entire living organism that can evolve new forms of fungicides in response to any resistance that emerges in the parasitic fungus. In other words, ants are using the laws of coevolution to their advantage, while we end up turning them against us.

EVOLUTION'S WIDOWS

Coevolution can marry species together, but extinctions can turn them into widows. If a species goes extinct, its partners may have to struggle to survive on their own. Sometimes the struggle proves too much, and they ultimately become extinct themselves.

According to Daniel Janzen, a University of Pennsylvania ecologist who works in the forests of Costa Rica, a number of trees in the New World have been evolutionary widows since the end of the Ice Age. Many species of plants disperse their seeds by growing fruits. The fruits attract animals with their sweet flesh, and the seeds have evolved tough coats that let them survive the passage through the animals' digestive tracts. The seeds can escape unharmed from the animals in their droppings, far away from the tree where they grew.

By getting away from their parent's tree, seeds have a better chance of surviving. Seeds that simply fall to the ground are more likely to be eaten by the beetles that linger under trees; even if the fallen should manage to sprout, they have to struggle under their parent's shade. And dispersing seeds is also good insurance against extinctions: if a hurricane should wipe out one stand of trees, their offspring a few miles away may be spared.

Just as orchids adapt themselves to particular kinds of pollinators, many plants adapt their fruits to attract certain kinds of animals. Some fruits with brightly colored skins catch the eyes of birds; the fruits that depend on bats and other nocturnal animals with poor color vision instead have rich aromas. But as Janzen has pointed out, some fruits make it difficult for any living animal to disperse their seeds. In Costa Rica, where Janzen works, the seed-dispersing animals include bats, squirrels, birds, and tapirs. None of them can eat the fruit of the tree *Cassia grandis*. *Cassia* produces fruit a foot and a half long, with seeds as big as cherries embedded in a fibrous pulp, encased by a woody hull. Because no living Costa Rican animal eats the *Cassia* fruit, it simply hangs on the tree until beetles drill their way into it and destroy most of the seeds.

The fruits that grow on *Cassia* and many other New World plants may be too big, hard, or fibrous for living animals to eat, but they would have been a perfect meal for giant ground sloths, camels, horses, and many other large mammals that went extinct 12,000 years ago. These animals had mouths big enough to take in the fruits, and teeth powerful enough to open them. For millions of years, Janzen argues, the giant fruits had coevolved with giant mammals. While other plants adapted their fruits for birds or bats, these plants depended on the megafauna.

Big mammals in the Old World still have this relationship with some plants—Sumatran rhinos, for example, feast on mangoes and pass their giant seeds in their dung. The hard shells and fibrous pulp of the Costa Rican fruits are a major obstacle to a small animal like a bird, but for a ground sloth it would be easy eating. A sloth would graze under the trees, sniffing for ripe fruits that had fallen to the ground. When it came across one, it would pop the entire fruit into its enormous mouth and crush the shell with its massive molars. While it chewed lazily on the pulp, the big seeds—which are usually coated in oil—would slip down into its giant gut. Later, after the sloth had shambled a few miles away, it would release the seeds in a giant plop of dung, where they would sprout into saplings.

The forests of Costa Rica were not unique during the Pleistocene: fruits were probably coevolving with giant mammals throughout the New World. Janzen has proposed that many other plants, such as avocados and papaya, are likewise widowed. When the giant mammals disappeared, these plants suffered a devastating loss. In some cases, their seeds could still be spread by smaller surviving animals, such as tapirs or seed-hoarding rodents, but their dispersal became far less reliable as more of their seeds were devoured by mice and insects. Many species became rare, as individual trees died out without being replaced.

The arrival of the Spaniards, bringing with them horses and cattle, went a small way toward restoring the Pleistocene ecology of the New World. These big mammals love to eat the widowed fruits. One Costa Rican fruit called the *jicaro* has a shell so hard that people use it for making ladles; Janzen has found that horses are the only animals living today in Costa Rica with a bite strong enough to crack it. They will do so happily, and after they've eaten the fruit inside, *jicaro* seeds survive the passage through their gut and sprout from their dung. Before the Spanish brought horses to Costa Rica, the *jicaro* was trapped in its own hard-shelled prison. While horses and cattle may ravage the New World wilderness in other ways—by trampling fragile soils and overgrazing grasslands—they can nudge plants like the *jicaro* a small way back to their Ice Age glory.

The extinction of the great mammals of the New World left behind trees that had long depended on them to spread their seeds. Except for what little help they've gotten from introduced livestock, Janzen argues, these widowed plants have been ebbing away over the past 12,000 years as their ranges shrank. Now, as the extinction rates accelerate, we may be creating a new generation of evolutionary widows.

These widows may come to include some of our own crops. Agriculture relies on pollinating insects for its existence, but humans have besieged pollinators over the past few centuries. Before Europeans arrived in North America, there were tens of thousands of pollinator species, including bees, wasps, and flies. The colonists brought with them honeybees from Europe, which they kept in managed hives. The honeybees competed with native bees for a limited supply of nectar, and their managed hives—with their reliable supply of honey—gave them a competitive edge. Bernd Heinrich of the University of Vermont has calculated that a single honeybee colony can wipe out 100 colonies of native bumblebees. An untold number of native pollinators have gone extinct, and many of the survivors are endangered.

Honeybees are now in decline as well. Pesticides have been ravaging their numbers, and parasitic mites recently introduced into the United States are killing off entire colonies. In 1947 there were 5.9 million honeybees in managed colonies, but by 1995 they had fallen to less than half that number, at 2.6 million. Feral honeybees have almost completely disappeared. If honeybees vanish, farmers will have to rely on the native species to pollinate their crops, but the natives may no longer be around.

It's easy to pretend that humans are the champions of the evolutionary race, that through some kind of superiority we have won Earth for ourselves. But in fact whatever success we enjoy depends on the balance between ourselves and the plants, animals, fungi, protozoa, and bacteria with which we have been coevolving. If anything, we are the most coevolved species that has ever existed, and depend more than any other species on the web of life.

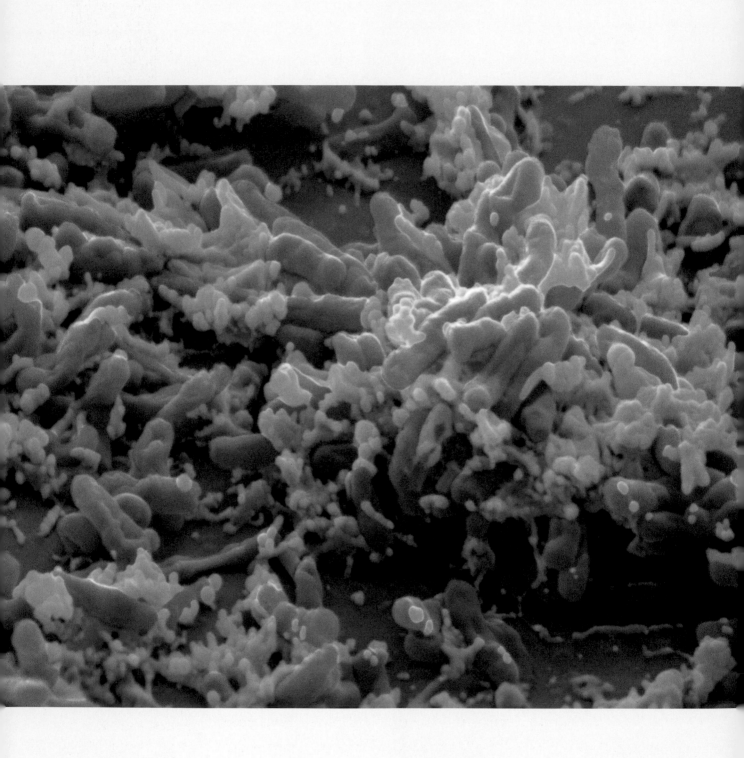

Doctor Darwin

DISEASE IN THE AGE OF
EVOLUTIONARY MEDICINE

ALEXANDER BIVELICH first came to the Tomsk Prison in central Russia in 1993. Convicted of theft, he was sentenced to three years. Two years into his term he began to cough up phlegm, and he spiked a fever. The prison doctors discovered a small infection in his left lung and diagnosed him with tuberculosis, a disease caused by the bacteria *Mycobacterium tuberculosis*. Bivelich might have picked it up by inhaling a droplet of sputum that had been hacked up by an infected prisoner, the bacteria settling into his own lungs. "I never thought I'd be infected," says Bivelich. "At first I did not believe what the doctors told me." But as the disease took hold in his body, he came to believe.

Tuberculosis ought to be an easily treated disease. It has been almost 60 years since Selman Waksmann of Rutgers University discovered that bacteria produce proteins that can kill *Mycobacterium*.

Tuberculosis is caused by the bacteria *Mycobacterium tuberculosis*, which scar the lungs of their victims.

Waksmann's drug joined a growing group of bacteria-killing antibiotics being discovered at the time. They were so lethal to bacteria that medical researchers thought infectious diseases such as tuberculosis would be eradicated within a few decades.

But *Mycobacterium* did not surrender Bivelich so easily. Prison doctors were able to treat him with antibiotics for a few months until he was released from jail in 1996. In 1998 Bivelich was back at Tomsk Prison, arrested for another theft. He had gotten no treatment for his TB on the outside, and when the prison doctors took another x-ray of his lungs, they found that the infection had spread during his freedom. Now his left lung was ravaged by lesions along with the right. They began giving him antibiotics again, but before long their tests revealed that the drugs were not stopping the spread of the bacteria. What were once wonder drugs were, for Bivelich, useless.

The prison doctors decided to switch Bivelich to a new regimen of antibiotics—powerful, expensive drugs hard to come by in Russia—and for a few months they managed to stabilize his health. But in time even these drugs proved useless. By July 2000, Bivelich's doctors were contemplating cutting out the diseased parts of his lungs. If the surgery and the drugs didn't stop the tuberculosis soon, he would probably die.

Bivelich's fate is not unusual in Russia. Drug-resistant strains of tuberculosis have emerged in Russia's crowded, filthy prisons, and today 100,000 prisoners carry strains of TB resistant to at least one antibiotic. Like Bivelich, many of these prisoners are petty criminals serving short terms. But thanks to tuberculosis, those short terms can become death sentences.

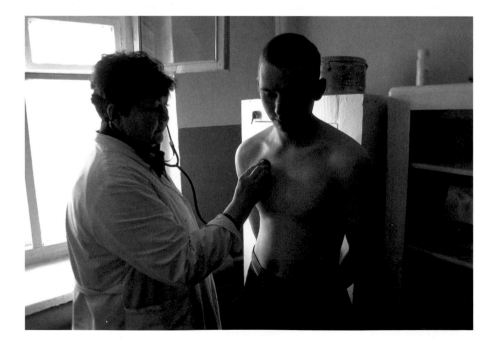

Resistant forms of tuberculosis are evolving in Russian prisons and threatening to become a global menace.

Bivelich is the victim of coevolution's dark side: the frightening speed at which parasites adapt to their hosts. Just as orchids adapt to bees, or fruit trees to the animals that spread their seeds, pathogens are forever evolving into new forms, hitting upon new ways to overrun their hosts' defenses. And just as many pesticides have lost their power to stop killing insects, drugs are becoming impotent in the face of mutating parasites. Drug-resistant forms of tuberculosis and other diseases are now evolving around the world, killing thousands. In the future, they have the potential to kill millions.

By understanding evolution, medical researchers may be able to find new ways to fight diseases. In some cases, by uncovering the evolutionary history of a disease—how a parasite first made humans its host and how humans evolved in response—they may find a cure. In other cases, scientists may even be able to harness the power of coevolution to tame the agents of disease.

THE TRIUMPH OF PARASITES

Wherever there is life, there are parasites. There are 10 billion viruses in every quart of seawater. There are parasitic flatworms that can live in the bladders of desert toads, which stay buried underground 11 months of the year; there are parasitic crustaceans that live only in the eye of the Greenland shark, which swims the icy darkness of the Arctic Ocean.

As much as we'd like to ignore parasites, they are among evolution's great success stories. They have probably existed in one form or another for billions of years. Biologists even suspect that certain viruses made out of RNA are actually survivors from the RNA world that predated DNA-based life. Judging from their abundance today, parasites have had a happy reign on Earth. In addition to viruses, many lineages of bacteria, protozoa, fungi, algae, plants, and animals have taken the parasitic path. By some estimates, four out of every five species are parasites.

Parasites and their host are fundamentally no different from beetles trying to devour the leaves of a tree. The parasites must consume their host to survive, and their host must in turn defend itself. These twin imperatives create a fierce coevolutionary struggle. Any adaptations that can keep a host disease-free will be favored by natural selection. Leaf-rolling caterpillars, for instance, fire their droppings out of an anal cannon, so that they don't end up creating a fragrant pile of frass that attracts parasitic wasps. Chimpanzees seek out foul-tasting, parasite-killing plants when they get intestinal worms. And when faced with an unbeatable parasite, some hosts try to make the best of a bad situation. When male fruit flies in the Sonora Desert are attacked by bloodsucking mites, they go into a mating frenzy in order to pass on as many of their genes as possible before they die.

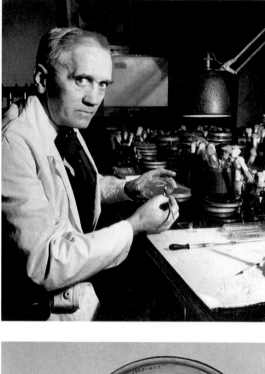

Parasites in turn have evolved ways around their host's defenses. Once a parasite gets inside its host, it has to escape the assault of immune cells, which blast it with poisons, suffocate it by plugging up its membrane channels, or just swallow it whole. Parasitic invaders use camouflage and subterfuge to survive. They may carry surface proteins that are exquisite forgeries of the proteins that our own bodies produce. Some of them use their mimicry to slip into cells through guarded passageways. Some parasites can jam the communication systems used by the immune system to spread the news of infection. Some can even send signals of their own that force immune cells to commit suicide. But as parasites evolve these ways around immune systems, hosts evolve new ways to kill the parasites, and the race rolls onward.

THE END OF A PANACEA

The coevolution between parasites and hosts has not faded into history's fog. It continues every day, and we humans are a subject of one of the newest experiments in host-parasite coevolution. We are trying to artificially improve our defense against bacteria with antibiotics, and it's becoming abundantly clear that we're in danger of losing this arms race.

When Selman Waksmann and his fellow scientists first discovered antibiotics, many people thought the war against infectious diseases was as good as won. But some researchers warned from the beginning that evolution might wipe away the miracles. Sir Alexander Fleming, the British microbiologist who discovered penicillin in 1928, was one of them. He ran an experiment in which he exposed bacteria to low levels of penicillin and gradually increased the exposure. In each successive generation, more and more of the bacteria could withstand the effects of the drug, and before long his petri dishes were swarming with bacteria that regular prescriptions of penicillin could not harm.

During World War II the American military closely guarded its stocks of penicillin, releasing only a few doses to civilian doctors whose patients were desperately ill. But after the war, pharmaceutical companies began selling the drug, even inventing a pill to take the place of injections. Fleming worried that doctors would prescribe it indiscriminately and, worst of all, people would be able to buy penicillin and take it on their own:

Sir Alexander Fleming (1881–1955) discovered in 1928 that a mold was killing the bacteria in his petri dishes by producing a compound called penicillin (bottom). Fleming rightly warned that bacteria could evolve resistance to it if the drug wasn't properly prescribed.

The greatest possibility of evil in self-medication is the use of too-small doses, so that, instead of clearing up the infection, the microbes are educated to resist penicillin and a host of penicillin-fast organisms is bred out which can be passed on to other individuals and perhaps from there to others until they reach someone who gets a septicemia or a pneumonia which penicillin cannot save.

In such a case the thoughtless person playing with penicillin treatment is morally responsible for the death of the man who finally succumbs to infection with the penicillin-resistant organism. I hope this evil can be averted.

Bacteria, microbiologists would later discover, are even more adept at coevolution than insects, able to alter their genetic makeup with staggering speed. Because they can divide several times an hour, bacteria can mutate quickly, stumbling across new formulas for resisting antibiotics. These mutations may create proteins that can destroy the drugs; some resistant bacteria are equipped with pumps in their membranes that can squirt the antibiotics out as quickly as they flow in. Normally, these mutants would not be favored by natural selection. But when faced with antibiotics, their offspring become successful.

Unlike insects, bacteria can acquire resistance genes not just from their parents but from the bacteria that surround them. Independent loops of DNA can shuttle from one microbe to another; they can slurp up the genes of dead bacteria and integrate some of them into their own DNA. Antibiotic-resistant bacteria can thus pass on their resistance genes not only to their descendants but to different species altogether.

A twenty-first-century Russian prison is a perfect laboratory for microbial evolution. Crime has soared since the fall of the Soviet Union, and the Russian courts are sending more and more people to prison—a million currently are jailed. But the prisons are in no shape to receive them. The prisoners receive a few cents' worth of food each day, leaving them malnourished and primed for infection. Then dozens of them are crammed into cells the size of living rooms. Those who are sick with tuberculosis can easily infect their prison mates with their coughs, letting the germs shuttle quickly from host to host, multiplying and mutating as they go.

Mycobacterium is a particularly tenacious bug, which can be destroyed only with a long course of antibiotics that usually lasts for months. If a patient doesn't take the full prescription of pills, the bacteria may be able to survive long enough for resistant strains to multiply. Russian prisons rarely provide a full course of antibiotics for their prisoners or make sure that they finish it. In their famished, undermedicated bodies, resistant bacteria spread easily.

When a person falls ill with resistant TB, doctors have to resort to much more expensive drugs, which can cost thousands of dollars. With so little money to spend on medicine, Russian prisons have no choice but to let new forms fester. The prison doctors have no illusions about curing their patients: they know that most of their patients will still be infectious when they are released from jail. The prisoners then carry resistant tuberculosis back to their hometowns to infect more people. By releasing sick prisoners back into the population at large, the government quintupled the tuberculosis rate in Russia between 1990 and 1996. It is now the leading contributor to increased mortality among young Russian men.

"All the strains that are in the Russian prisons will eventually come to our doorstep," says Barry Kreisworth, an epidemiologist with the Public Health Research Institute in New York City. In fact, Kreisworth has already detected some of the strains that evolved at the Tomsk Prison in immigrants arriving in New York City.

The Public Health Research Institute and other organizations are now trying to stop the spread of resistant tuberculosis in Russia and elsewhere by providing aggressive treatment with the most powerful antibiotics available. They hope to destroy resistant strains before they get a chance to evolve into new forms. The stakes are high in this gamble. If the bacteria continue to evolve, an unstoppable form of TB may emerge, one that is resistant to every known antibiotic.

The antibiotic crisis is out of hand not just in Russia but across the world. New strains of *E. coli, Streptococcus,* and other bacteria that can resist almost all antibiotics are emerging. Gonorrhea, once a harmless nuisance, has evolved into a life-threatening disease: in Southeast Asia, 98 percent of gonorrhea is now resistant to penicillin. In London doctors have isolated a remarkable strain of *Enterococcus* bacteria that has evolved to the point where it actually *depends* on the antibiotic vancomycin for its survival.

After 20 years of complacency, pharmaceutical companies are only now working on new antibiotics. It will take years until this next generation of drugs is ready; once they hit the market, no one knows how long they will stay effective against bacteria. In the meantime, we may face a frightening reversal of medical history. The risk of infection with unstoppable superbugs may make surgery as dangerous as it was during the Civil War.

Experts on antibiotic resistance are calling for global action. One way to cut down on the threat of resistant bacteria may be to stop encouraging their evolution. Antibiotics were introduced to the world as a panacea in the 1940s, and we still imagine that they can cure all things. As a result, they are prescribed far more often than they really need to be. (Many people, for example, think that antibiotics can kill viruses, when in fact they can only attack bacteria.) As a result, more than 25 million pounds of antibiotics are prescribed each year in the United States alone, of which a third to half is either inappropriately prescribed or just unnecessary.

Doctors need to do a better job of prescribing drugs, but patients have a duty to take their full courses of antibiotics so that the bacteria infecting them can't get a chance to breed resistance. Consumers have to resist the lure of antibiotic soaps and sprays, which encourage resistant bacteria to evolve. Meanwhile, the flow of cheaply made antibiotics sold over the counter in developing countries has to stop.

A number of scientists are also worried about the 20 million pounds of antibiotics that U.S. farmers feed to their livestock. Cows, chickens, and other animals are given antibiotics not to cure a particular outbreak but to keep them from getting sick in the first place. Farmers also find that antibiotics—for reasons still unknown—make animals grow faster. By pumping animals full of antibiotics, farmers are breeding resistant strains of *Salmonella* and other bacteria that can then attack people. In 1994 the FDA approved the use of antibiotics called quinolones in chickens, to prevent infections by intestinal bacteria called *Campylobacter jejuni*. Since then, quinolone-resistant *Campylobacter* cultures in humans have risen from 1 percent to 17 percent.

Bacteria are enjoying a strange new age. Never before in their long history have such a combination of molecules been used against them, and in such spectacular quantities. Genes for antibiotic resistance, once a burden, are now the secret to success. For our own survival, we have to bring this peculiar era to an end.

AIDS: EVOLUTION DAY BY DAY

Bacteria are not the only parasites that evolution has turned into a worldwide menace. Over the past few decades, the human immunodeficiency virus—the cause of acquired immunodeficiency syndrome, or AIDS—has evolved from obscurity into a global epidemic.

Viruses such as HIV are peculiar as parasites go. They are not alive in the sense that bacteria or humans are. They do not have a metabolism that allows them to extract energy from food and release wastes. They are simply small collections of DNA or RNA encased in a protein shell. When they invade cells, their genetic material commandeers the protein-making factories of their host. Their host cells make new copies of the virus, which burst out of the cell and search for new homes.

In their own way, viruses are just as ruthless as bacteria in their coevolution with their hosts. They do not have the cellular machinery that bacteria have to swap genes, but they can more than make up for this shortcoming by evolving on overdrive. The genome of HIV is only 9,000 base pairs, compared with 3 billion base pairs found in human DNA. But when a single virus infects a new human host and invades a white blood cell, it starts multiplying wildly. Within 24 hours, it has turned into a horde of viruses billions strong.

Almost as soon as the virus starts multiplying, our immune system starts recognizing infected white blood cells and destroying them, wiping out the viruses in the process. But despite the immune system's ability to kill HIV by the billions every day, HIV can survive these attacks for years. The secret to its longevity is its ability to evolve. The enzymes that HIV uses to make new copies of its genes are very sloppy, making one or two mistakes on average every time they duplicate the virus's genome. Among the many mutants that spring up, a few strains will turn out to be hard for the immune system to recognize. Because HIV replicates so quickly, these resistant viruses quickly become the dominant strains in a person's body. It takes time for our immune system to shift its attack toward the new strain, and once it does, the viruses evolve even newer forms that escape the immune system yet again.

Virus and host remain balanced in this equilibrium for years, poised between population explosions and implosions. Without an HIV test, infected people have no way of knowing that a coevolutionary struggle is raging under their own skin. Only when HIV has destroyed the immune system, allowing other parasites to invade and causing full-blown AIDS, does the virus make itself known.

Drugs now exist that can interfere with the enzymes HIV uses to replicate itself, and they can slow the progress of AIDS. Yet even though anti-HIV drugs have been available only for a few years, the virus's mutational over-drive is already threatening to make them useless. Just as the virus can evolve around the latest wave of attacks from the immune system, it can mutate into forms that cannot be harmed by drugs. Even one or two mutations are enough to overcome these drugs. In a matter of weeks, a patient's load of HIV can bounce back up to the level it was at before the treatment began.

Switching to a different drug makes it possible to kill off most of these resistant viruses, but among the survivors a new mutant may emerge that can resist the new treatment. Doctors therefore favor giving their patients cocktails of several drugs at once. A virus may be able to evolve resistance to a single drug with one or two mutations, but it is much less likely to escape several other drugs at the same time. Yet even under attack from drug cocktails, multiple-drug-resistant HIV is emerging.

IN SEARCH OF THE ORIGIN OF AIDS

No one knows if drug cocktails will ever become a truly practical cure for AIDS. At best, they only keep the virus in check, and they cost tens of thousands of

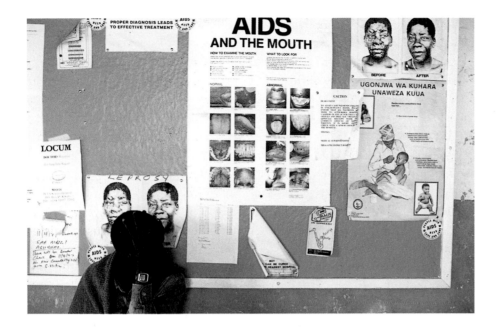

AIDS was unknown 20 years ago. Today, it infects 36 million people. Sub-Saharan Africa has been hit particularly hard, with 25.3 million infected people.

dollars every year, which puts them beyond the reach of the vast majority of AIDS victims. In the hopes of finding new treatments, other researchers are studying the history of the virus. Hidden in its past they may uncover a cure that they would not otherwise find.

When AIDS was first recognized as a disease of its own, it seemed to come out of nowhere. In the early 1980s American gay men started coming down with bizarre diseases that a healthy immune system could easily suppress. Their immune systems had collapsed, and soon researchers in France and the United States discovered that HIV was responsible. It was a dangerous virus, they found, but a fragile one. A cold virus can travel through the air and cling to fingers and lips. But HIV needs to be helped from the bloodstream of an old host to a new one, through sex, shared hypodermic needles, or contaminated blood transfusions.

By the end of the 1980s, researchers recognized that they had a global plague on their hands. But AIDS is like no other epidemic. During the Black Death of the 1300s, for example, Europeans who came down with bubonic plague usually were dead in a few days; HIV can take 10 or 15 years to do its damage. Because of its slow, masked course, HIV spread quietly during much of the 1980s, slipping from one unsuspecting victim to the next. By 2000, 36 million people were suffering from AIDS, and 21.8 million people had already died of the disease. Sub-Saharan Africa has suffered the heaviest blow, with 25.3 million people now infected with AIDS.

Where did HIV come from? There are hardly any clues from before the virus became epidemic in the 1980s. (The oldest known sample of HIV comes from the blood of a patient in Zaire in 1959.) But scientists can go back in

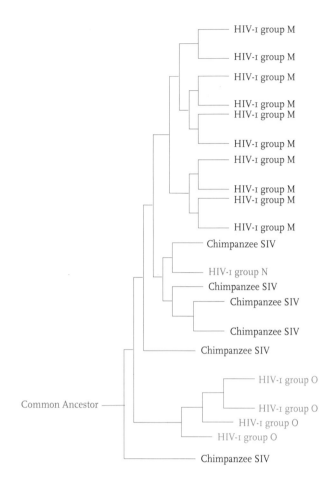

This evolutionary tree represents the scientific consensus on the origin of HIV-1 viruses, the most important cause of AIDS. The trunk of the tree at left represents an ancestral virus that gave rise to strains that infected chimpanzees (known as SIV). Over time, new species branched off and infected humans, becoming HIV.

time by looking at the genetic code of today's HIV and building an evolutionary tree for the virus.

HIV belongs to a class of slow-breeding viruses known as lentiviruses (*lentos* is the Latin word for "slow"). Cats, both wild and domesticated, get feline immunodeficiency virus; cows get bovine immunodeficiency virus. And most significantly, primates get simian immunodeficiency viruses (SIVs), which resemble HIV. Unlike humans, though, most primates never seem to get sick from their infections. The viruses may have once been as lethal to them as HIV is to humans now, but natural selection has preserved only the resistant primates.

Scientists have found evidence that the HIV epidemic is the result of SIVs jumping from primates into humans several times. HIV exists in many different strains, which are classed as either HIV-1, the form found in most parts of the world, or HIV-2, which is limited to West Africa. In 1989 virologist Vanessa Hirsch of Georgetown University and her colleagues found that HIV-2 is more similar to SIV from the sooty mangabey, a monkey from West Africa, than it is to HIV-1. Likewise, the SIV of the sooty mangabey is more like HIV-2 than the SIV of other monkeys. Sooty mangabeys are kept as pets in West Africa and hunted for food; Hirsch suggested that HIV-2 originated as people were scratched by the monkeys and picked up infected mangabey blood through open wounds.

Although HIV-1 is much more common, it wasn't until 1999 that a clear picture of its history emerged. Beatrice Hahn of the University of Alabama in Birmingham and her colleagues discovered that SIVs found in chimpanzees were the closest known relatives of HIV-1. And not just any chimps—the viruses most like HIV-1 all came from a single subspecies of chimpanzee: *Pan troglodytes troglodytes,* which lives in Gabon, Cameroon, and surrounding countries in equatorial West Africa. From this one subspecies of chimp, Hahn concluded, strains of HIV-1 had evolved at least three separate times.

Hahn and her colleagues are now putting together a picture of how HIV got its start. It's only a hypothesis, certainly, but one that has been bolstered as more evidence has come to light. The ancestors of sooty mangabeys and chimpanzees carried the ancestors of HIV for hundreds of thousands of years. Hunters sometimes contracted these viruses—along with many others—when they killed and butchered monkeys and chimps. But the ancestors of HIV, poorly adapted for life in humans, had little chance of establishing themselves in a new host. Even if they could survive in the body of a single hunter, they couldn't spread very far. They only rarely came into contact with hunters, and the hunters themselves lived in remote villages without much contact with the outside world. The virus was likely to die off before it could be transmitted to other hosts.

The dramatic changes that came to West Africa in the twentieth century finally unleashed HIV on our species. Cities sprang up; railroads were built into the interior; loggers moved deeper into the forests; people were forced to move to plantations to work. The market for bush meat grew, and with it the contact between hunters and primate blood. With people moving quickly across the countryside on buses and trains, a virus could easily spread from its first human contacts to new hosts.

The diversity of HIV in the people of equatorial West Africa is immense compared to the rest of the world, and that, Hahn argues, is the result of many separate leaps the virus has made from primates into humans there. HIV-2 has jumped from sooty mangabeys to humans as many as six times, and HIV-1 has crossed from *Pan t. troglodytes* at least three times. Most of these leaps were dead ends. Only two of the six HIV-2 strains have made any headway in humans, while the global AIDS epidemic is due mainly to a single strain of HIV-1. As West Africa came into closer contact with the rest of the world, the virus spread to Europe, the United States, and elsewhere.

For now, this hypothesis awaits more tests. Hahn's tree of HIV is based on viruses from only six chimps; as more data comes in, researchers may have to rearrange the branches of the tree. But it is no easy task finding viruses in wild chimps, and the job gets harder every day: the trade in chimpanzee meat that may have triggered the AIDS epidemic is driving *Pan t. troglodytes* rapidly toward extinction.

These endangered chimpanzees may carry the first chapters of the biography of AIDS inside them. They are infected with the closest known relatives of HIV-I, and yet their immune systems are able to hold the viruses in check. Because their virus is so closely related to HIV, the adaptations they have evolved against it may be the secret to a human cure. If they should disappear, clues to a cure may disappear with them.

"Our hospitals are filled with incurable infectious diseases such as AIDS—well, these same diseases have affected animals, but they haven't had the benefit of an emergency room," says Stephen O'Brien, a virologist at the National Cancer Institute. "All they've had is natural selection. If we look through their genomes and discover how they deal with these same onslaughts, we'll be much better equipped to come up with therapies for humans."

SAVED BY THE BLACK DEATH?

O'Brien studies evolution in search of a different sort of weapon against HIV. Humans have evolved in response to parasites in the past, and it's possible

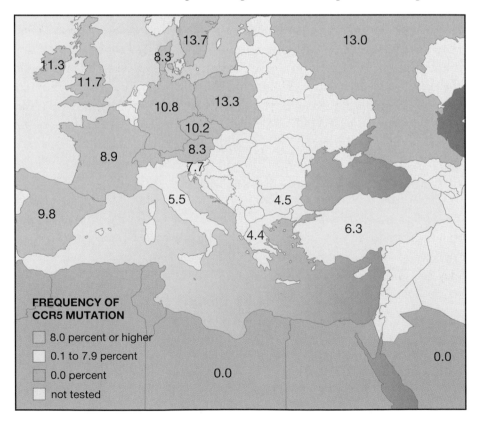

A mutation to the CCR5 cell receptor provides protection from HIV. It is most common in northern Europe and rare outside the continent. Researchers suspect that it may have been raised to such high levels by the Black Death and other epidemics of bubonic plague.

that some of those adaptations may be protecting some people from HIV today.

Since 1985 O'Brien has collected samples of blood from people who are at high risk for being exposed to HIV, such as homosexuals and intravenous drug users. He analyzed their DNA, comparing the genes of people who have become infected with HIV to those who haven't, in the hopes of finding mutations that might defend humans against the virus.

By the mid-1990s, after collecting more than 10,000 samples, O'Brien and his team were starting to flag. "We were beginning to lose our enthusiasm. We went through several hundred genes, one after another, and every one of them gave the same answer—that there was no effect." But in 1996 that finally changed. In that year, several different teams of scientists discovered that in order to enter a white blood cell, HIV has to pry open a receptor on the cell's surface called CCR5. O'Brien's team turned back to their thousands of samples and looked for mutations to the gene that makes the CCR5 receptor.

The Black Death claimed somewhere between a quarter and a third of all Europeans between 1347 and 1350.

"We were stunned," says O'Brien. They came across a mutation to CCR5: some people were missing a 32-base section of the gene. The mutation ruined the protein that the CCR5 gene is supposed to create. As a result, people who carry two copies of the mutant CCR5 gene have no CCR5 receptors on the surface of their cells (people who carry only one mutant copy of the gene have fewer receptors than average). O'Brien discovered a striking correlation between the mutation and HIV infection: people who carried two copies of the mutant CCR5 gene almost never became infected. "It was the first serious genetic effect we were able to discover," says O'Brien, "and it was a big one."

Without CCR5 receptors, HIV's doorway into white blood cells is bricked up. As a result, people who carry two mutant copies of CCR5 can be repeatedly exposed to HIV and yet resist it completely. Those who carry only a single copy of the mutant gene make fewer CCR5 receptors than normal; while they may get infected by HIV, the mutation slows down the onset of full-blown AIDS for two or three years.

O'Brien's team was also surprised to see who carries the CCR5 mutation. It is relatively common in Europe, with 20 percent of the population carrying one or two copies. It's most common of all in Sweden, and as you move towards southeast Europe, the frequency of carriers tapers off. Only a small fraction of Greeks carry it; an even smaller number of Central Asians do. From the rest of the world, this particular mutation is missing altogether.

The only way the CCR5 mutation could have reached such high levels was if it was somehow valuable to the ancestors of northern Europeans and had been favored by natural selection. "It would have been a breathtaking selective pressure," says O'Brien, "and the only thing that would fit would be some infectious disease epidemic that was killing off thousands, if not millions, of individuals, and favoring those who carried it."

Whatever sort of event favored the gene among Europeans happened 700 years ago, according to O'Brien. He was able to determine its age by examining the DNA that surrounds the CCR5 mutation. Over time, variations have emerged, and O'Brien used that variability to estimate how long ago the gene arose. And it turns out that 700 years ago, something *was* putting Europe under massive natural selection: the Black Death.

The Black Death, which claimed over a quarter of all Europeans between 1347 and 1350, was only the most deadly of a long string of epidemics of bubonic plague that struck the continent for centuries. The plague acted like a human pesticide: any mutations that could help Europeans survive would become far more common in later generations. O'Brien suspects that CCR5 was just such a fortunate mutation, and with each plague outbreak its frequency ratcheted upward.

The bubonic plague is caused by *Yersinia pestis,* a bacterium that can live inside rats and spread to humans through the bite of a flea. Like HIV, *Yersinia* binds to white blood cells. Rather than invade the cells, it injects toxins into the cells to stun the immune system, allowing the bacteria to multiply without being attacked. No one knows exactly how *Yersinia* binds to white blood cells. O'Brien and his team are finding out. If his hypothesis is right, *Yersinia* needs to use CCR5. Europeans who were born without CCR5, he proposes, were protected during the Black Death; today, some of their descendants are protected from HIV.

If the CCR5 mutation does provide resistance to bubonic plague, it might turn out to be an extraordinary exaptation. Thanks to the gruesome natural selection of the Black Death, some Europeans may now be protected from a virus that depends on the same cell receptors. The fact that the AIDS epidemic has been far more destructive in Africa and Southeast Asia than in Europe or the United States might be due in part to the different evolutionary histories of the continents. O'Brien hopes that ultimately the benefits of the CCR5 mutation can be translated into a treatment for HIV. If medical researchers can invent a drug that can block normal CCR5 receptors, they may make people immune from HIV without causing dangerous side effects.

Even if the evolutionary research of scientists like O'Brien and Hahn should lead to a cure for AIDS, there may be many more new diseases to grapple with in the future. The AIDS epidemic came about as nine different primate lentiviruses leaped from primates to humans. There are 24 other known primate lentiviruses, all related to HIV, that may be poised to make the jump as well. The modern world, with its frustrating mix of wealth and poverty, of intercontinental airplane flights and secondhand needles, is primed for their entry.

PLAGUE TAMERS

With so many diseases emerging, doctors may have to try a new way of controlling parasites: by taming them. When a disease-causing parasite invades a host, it faces a trade-off. On the one hand, it can breed like mad in a person's body, feeding on its host's tissues and spewing poisons until its host dies. While it may make many trillions of copies of itself in the process, it may risk extinction if it kills its host before it can infect a new one. On the other

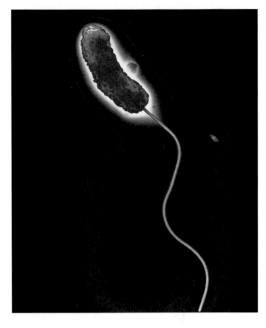

Cholera (caused by the bacteria *Vibrio cholerae*, above) can cause fatal epidemics. Amherst College biologist Paul Ewald (left) has found that poor water supplies make the bacteria evolve into more deadly forms.

hand, a parasite might take a much gentler approach, multiplying so slowly that its hosts don't even realize that they are sick. It may have a much better chance of spreading on a fork or a handshake, because it keeps its hosts alive long enough to transmit it. But if it lives alongside a more aggressive strain that can also breed faster, it may get outcompeted and driven to extinction.

Paul Ewald, a biologist at Amherst College, has been exploring how different parasites negotiate these trade-offs. As a rule, he finds that if a parasite depends on a mobile host to get transmitted, it will be gentle. Rhinoviruses,

A painting inspired by the 1912 cholera epidemic during the Balkan Wars.

which cause colds, can be transmitted only by sneezes or skin contact, so they depend on healthy hosts who can mingle with other people. "So, not surprisingly, the rhinovirus is one of the mildest viruses that we know about," says Ewald. "In fact, nobody has ever been known to die from a rhinovirus, and that's not true for almost any other disease organism of humans."

On the other hand, if a parasite doesn't depend on the health of its host to get to a new one, it can afford to be more vicious. Malaria, for example, uses mosquitoes as its ferry, and its brutal fevers often leave its victims bedridden.

Not every pathogen follows this rule, Ewald points out. Smallpox, for example, has no vector like a mosquito to carry it to a new host, so it has to find new hosts on its own. And yet it is one of the most lethal diseases known. It can afford to be virulent because, unlike cold viruses and other benign diseases, it can survive outside a host for a decade, waiting for a person to pick it up. When it does get into a new host, it breeds madly until the host dies, and then waits for its next chance.

All of these parasites are continually evolving in response to their surroundings, and Ewald has predicted that if it becomes easier or harder for parasites to spread, they will adapt. He has tested his prediction with several different diseases, including cholera. Cholera spreads by releasing toxins that give its hosts diarrhea, allowing it to escape their body. Another person may then pick up the bacteria in a bathroom, handle some food, and infect someone else. On the other hand, cholera can also spread if sewage contaminates supplies of drinking water. The first route depends on a healthy host that can come into contact with other people; the second depends only on bad water supplies. According to Ewald's theory, cholera should evolve to be more toxic in places where the water supply is contaminated.

That's exactly what Ewald found during a 1991 cholera outbreak in South America. "Cholera came into Peru and then quickly, within a couple of years, spread all throughout South and Central America," he explains. "When the organism invaded countries with clean water supplies, it dropped in its harmfulness." In Chile, a country with clean water, it evolved into a mild form; in Ecuador, where water supplies are much worse, it became more dangerous.

Instead of eradicating diseases, Ewald argues, we may have more luck trying to domesticate them. It wouldn't be the first time we've domesticated natural enemies. "Wolves have been harmful to us throughout our evolutionary history," says Ewald, "but we've been living with some wolves which have evolved into dogs. Instead of harming us, they now actually help us. I think we can do the same thing with these disease organisms."

Domesticating parasites is not as hard as it may sound. To domesticate *Plasmodium,* the parasite that causes malaria, people may need only to put screens in their windows. Unable to sail through open windows, the mosquitoes that transmit *Plasmodium* cannot bite as many people in a given night, slowing down the transmission rate. If a strain of *Plasmodium* is adapted to kill its hosts quickly, window screens will put it at an evolutionary disadvantage, because its hosts will die before it can infect someone else. Gentler strains will outcompete the harsher ones, and fewer people will die of malaria.

When it comes to disease, evolution has been working against us for millennia. It's about time we harnessed its powers.

Passion's Logic
THE EVOLUTION OF SEX

Lᴵғᴇ ɪѕ ᴀ ᴅᴀɴᴄᴇ of partners—cold viruses and their sniffly hosts, orchids and the insects that pollinate them, garter snakes and poisonous rough-skinned newts. But no list of life's dance partners would be complete without Male and Female. For the vast majority of animal species, the dance between the sexes is essential to their existence.

As vital as sex may be, it is a glorious, glittering puzzle. Why do peacocks drag around such grand tails—but not peahens? Why is it that when Australian redback spiders mate, the male hurls himself onto the female's poisonous fangs, becoming a meal for her at the end of the act? Why do ant nests contain thousands of sterile female workers, all serving a fertile queen? Why do males always have small, mobile sperm, while females have giant, immobile eggs? Why are there males and females at all?

Male bighorn sheep fight for the opportunity to mate with females.

The answers are to be found in evolution. Sex, biologists now suspect, is itself an evolutionary adaptation. It gives sexual organisms a competitive edge over ones that reproduce without males and females. But while sex may benefit both males and females, it creates a conflict of interest between them. The best reproductive strategy for a male is not the same as the female's. Over countless generations, this conflict gradually shapes animals in many ways, from their anatomy to their behavior. And the conflict doesn't end once a male and female have mated. Within the womb and within the family, the struggle continues, until it has shaped even the societies of animals.

Evolutionary biologists have found that the peacock's tail, sterile ants, and suicidal spiders can all make eminent sense once they recognize the conflict between the sexes. And their success at understanding how animals are shaped by sex naturally raises a thorny question: Are some parts of human psychology the result of the evolutionary pressure of sex as well?

WHY SEX?

The question of why we have sex is not one that occurs to most people. We have it because we want children, or because it feels good, or both. But many organisms can reproduce without sex. Bacteria and many protozoa can simply divide themselves in two without the help of a partner. Asexual animals are rare, but they exist. Some species of whiptail lizards in the western United States, for instance, have no males. One female will mount another female, bite her neck, wrap around her like a reptilian doughnut, and otherwise mimic what a male lizard does while mating. Herpetologists suspect that the mounting female makes the other female ovulate. But she needs no sperm to fertilize her eggs. They simply start dividing and growing into embryos. As these clones begin to develop, their mothers return the favor to their pseudo-mates by playing the part of the male. When the lizards give birth, they produce only females, all of which are identical to their mothers.

Sex is not only unnecessary, but it ought to be a recipe for evolutionary disaster. For one thing, it is an inefficient way to reproduce. In a population of asexual whiptail lizards, every new lizard can bear baby lizards of her own; in a population that reproduces sexually, only half of them can. If sexual and asexual members of a species are living side by side, the asexuals should quickly swamp the sexuals with their explosive birthrate. And sex carries other costs as well. When males compete for females by locking horns or singing, they're using up a huge amount of energy, and sometimes even putting themselves at risk of being attacked by a predator. "The cost of sex is immense," says Robert Vrijenhoek of the Monterey Bay Aquarium Research Institute.

By all rights, any group of animals that evolves sexual reproduction should be promptly outcompeted by nonsexual ones. And yet sex reigns. Peacocks show no sign of evolving away their tails; new generations of redback spider males are throwing themselves into the jaws of death just as their fathers did. Meanwhile, only a fraction of 1 percent of all vertebrates reproduce asexually like the virgin whiptail lizards.

Why is sex a success, despite all its disadvantages? Scientists have recently been gathering support for a surprising hypothesis: sex fights off parasites. Parasites take a tremendous toll on their hosts, and any adaptation that helps hosts escape them may become hugely successful. In the 1970s, a number of biologists began building simple mathematical models of the coevolution between parasites and their hosts that suggested it moves in circles, like a deadly merry-go-round.

Imagine a pond of fishes that reproduce by cloning. Each fish is an identical copy of its mother, but the fishes are not all carbon copies of one another. A mutation may arise in one fish and be passed down to her descendants. They will form a strain that can be distinguished from other strains by unique mutations.

Now suppose a fatal parasite invades the pond. The parasite mutates as it spreads, forming strains of its own. Some strains of the parasite carry mutations that make them good at attacking certain strains of fish. The strain that can attack the most common fish has the most hosts to attack, and it soon becomes the most common of all parasite strains. The other parasite strains, limited to fewer hosts, dwindle to low levels.

But the parasites undermine their own success. They thrive so intensely in their strain of fish (call it Fish A) that they kill their hosts faster than they can reproduce. The population of Fish A crashes, and as it disappears, its parasites have a harder time finding new hosts to infect. Their numbers crash as well.

This attack on Fish A gives the rarer strains of fishes an evolutionary edge. Unburdened by parasites, their numbers swell. Eventually another fish strain becomes most common—call it Fish B. As it grows more successful, it becomes fertile ground for the rare parasites that are best adapted to them. They begin to multiply and catch up with the explosion of their host. Now Fish B crashes, to be replaced by Fish C, and so on.

Biologists call this model of evolution the Red Queen hypothesis. The name refers to the character in Lewis Carroll's *Through the Looking Glass* who takes Alice on a long run that actually brings them nowhere. "Now, *here,* you see, it takes all the running you can do, to keep in the same place," the Red Queen declared. Hosts and parasites experience a huge amount of evolution, but it doesn't produce any long-term change in either of them. It's as if they're evolving in place.

The Red Queen and Alice run as hard as they can just to stay in place— much like the ongoing coevolution of parasites and their hosts.

William Hamilton, an Oxford biologist, proposed in the early 1980s that sex might bring an advantage to animals struggling through this Red Queen race, because it makes it harder for the parasites to adapt to them. A sexual animal is not a clone of its mother; it carries a combination of genes from both its mother and father. Nor is it a simple blend of its parents' genes. As cells divide into eggs or sperm, each pair of chromosomes wraps around each other and swaps genes. Thanks to that sexual dance, the genes of a male and a female can be shuffled into billions of different combinations in their offspring.

As a result, sexual fish don't evolve into distinct strains; their genes scatter throughout the pond's population, mingling with the genes of other fish. Fish genes that have lost their protective powers against the parasite can be stored away in the DNA of fish that also carry more effective ones. These obsolete genes may later be able to provide more protection against new parasite strains, and they'll be able to spread once more through the pond's populations. Parasites can still attack fish that reproduce sexually, but they cannot force them into boom-and-bust cycles as dramatic as the ones suffered by their clonal cousins.

The ups and downs through which parasites drive the asexual fish may make their genes deteriorate. For any given gene, some fish will have defective versions and others will have defect-free copies. Each time the fish go through a crash, there's a chance that some of the defect-free fish will die, taking their unblemished genes with them. After enough crashes, the defect-free genes may disappear completely.

Once a perfect version of a gene disappears from the fish population, it's unlikely to return. The only way evolution can repair a defective gene is for a mutation to change the faulty part of its sequence. But mutations strike a gene at random, anywhere along its sequence. It's far more likely that a mutation will just cause more harm to the gene. The Red Queen phenomenon among asexual fish renders more of their genes defective as time goes by. But sexual fish mix their genes together in every generation, so that defect-free genes rarely disappear for good. The overall quality of their DNA remains high and may even make them more fit than the asexual ones. Their good genes may give them more stamina or the ability to draw more energy out of the insects they eat. Even though they breed more slowly, their resistance to parasites may give them an evolutionary edge over asexual fish.

That was the hypothesis, at any rate, and while it looked promising in theoretical models, scientists needed to test it in the real world. In the 1970s Robert Vrijenhoek discovered one natural experiment among topminnows that live in Mexican ponds and streams. These topminnows sometimes mate with a closely related species, creating a hybrid fish with three copies of genes rather than two. The hybrids are always female, and they always reproduce by cloning rather than by mating. In order to trigger their eggs to grow, they need to get sperm from a male fish, but they don't actually incorporate his genes into their eggs.

Vrijenhoek and his colleagues have studied the topminnows in several different ponds and streams, and each one gives a different confirmation of the Red Queen hypothesis. Many topminnows are infected with parasitic flukes, which form black cysts in their flesh. In one pond, Vrijenhoek found that the hybrid clones had many more cysts than sexual topminnows. In other words, the clones provided an easier target than the sexual fish, because the parasites could adapt to their immune systems faster. In a second pond where two strains of clones lived, the more common strain was subject to more infections—just as the Red Queen hypothesis predicts.

The topminnows in a third pond seemed at first to contradict the Red Queen hypothesis: the sexual fish were more vulnerable than the clones. But as Vrijenhoek studied that pond more closely, it turned out to be an even stronger confirmation. It had dried up in a drought a few years earlier, and after it had returned, only a few fish recolonized it. As a result, the sexual fish were highly inbred and were thus deprived of the genetic variety that represents the advantage of sex. Vrijenhoek and his colleagues added some more sexual topminnows to the pond to boost the diversity of its DNA. Within two years the sexual fish were immune to the parasites, which had switched to attacking the clones.

SPERM AND EGG

The advantages of sex have allowed it to come into existence dozens of times, in many separate lineages of animals, plants, red algae, and other eukaryotes. The first sexual animals probably just sprayed their sex cells (called gametes) into an ocean current and let them struggle to find each other. Although sex has evolved independently numerous times, most gametes look pretty much the same: the egg is big and immobile, while the sperm is a small swimmer. When the sperm fuses to the egg, it unloads only the DNA in its nucleus; its mitochondria and other organelles are blocked from entering.

This arrangement is popular because it works so well. David Dusenbery, a biologist at the Georgia Institute of Technology, has identified the advantages by building a mathematical model of gametes struggling to find each other. In his model both gametes can swim around or remain stationary; they can be the same size or different. Dusenbery finds that gametes are a bit like two people lost in a giant forest at night. If both of them wander around, they will be unlikely to find each other. It is better for one of them to remain motionless and send out signals to the other one.

In the case of people, the signals might be shouts; for gametes, they are powerful odors known as pheromones. The louder people shout, the easier they are to hear. For gametes, shouting louder means producing more pheromones. Dusenbery finds any increase in the size of a gamete makes it able to make many more pheromones, extending the range of its communication. And in fact, it is eggs that send out pheromones to attract sperm, not the other way around.

Of course, an entire search party fanning out through a forest will do a better job of finding someone than a single person. Likewise, one way to increase the odds of contact is to use many sperm instead of one to search for the egg. According to Dusenbery, evolution would favor any mutation that would make a species' eggs bigger or its sperm more plentiful. It would be able to use less energy to reproduce successfully, because its gametes would do a better job of finding each other; it could survive in places where less efficient forms couldn't reproduce.

By evolving to bigger sizes, eggs not only could spread pheromones more effectively, but they could store more energy that they would need to fuel their cell division once they were fertilized. The more energy an egg could provide, the less the sperm would need to bring with them. They could become even smaller and more numerous, increasing their chances of fertilizing eggs and propagating their genes. And in the face of shrinking sperm carrying fewer resources, natural selection favored eggs that could supply even more. Over time sperm evolved into little more than mobile crates of genes, while eggs became giant, energy-rich cells.

As the big-egg, little-sperm arrangement evolved, it created a huge imbalance between the sexes. A single man can produce enough sperm in his life to make every woman on the planet pregnant many times over. But each woman ovulates only once a month, and like other mammals, she has to carry her baby inside her for months and nurse it when it's born. Each birth puts her at risk of dying from complications, and nursing makes her burn tens of thousands of extra calories. The vast reproductive potential of men has to fit through the narrow bottleneck of womanhood.

While a single male of a species may be able to fertilize every female, there are other males who wouldn't mind doing the same. In many species, this conflict leads to battles among the males. Exactly what sort of battle evolution produces depends on the species in question and its ecology. Northern elephant seals will slam their 2,000-pound bodies against one another, spraying blood and foam, in order to be the sole mate for a harem of dozens of females. Musk ox bulls will ram their thick horns against one another on the Arctic tundra, and 1 out of every 10 will die of a fractured skull. Even male beetles and flies have evolved antlers of their own that they use to battle for the right to sex.

FEMALE CHOICE

Competition between males was well known to nineteenth-century naturalists, Darwin included. It fit into his theory of evolution without much trouble: if males competed with one another for females, the winners would mate most often. If having a slightly thicker skull gave the edge to the winners, more males of the next generation would have thick skulls. A pair of lumps might make the skull even more effective in fighting, and so those lumps might evolve into horns.

But Darwin wondered what the females were doing during these battles. Were they simply waiting passively to be possessed by the winner of a match? The notion of passive females might appeal to some Victorian gentlemen, but Darwin recognized there was a problem with it: it could not account for species in which males do not go *mano a mano*.

Consider the peacock and his splendid tail. "The sight of a feather in a peacock's tail, whenever I gaze at it, makes me sick!" Darwin once said. The great fan of iridescent eyes is not essential to *Pavo cristatus*—the females survive perfectly well without one. A male can't use its tail to bludgeon another male into submission. In fact, it's a burden, since it weighs down the peacock when it tries to escape from a fox. Yet despite its drawbacks, male peacocks grow a new set of tail feathers every year to replace the ones they shed at the end of the previous year.

"Darwin had a real problem with peacocks, because they seemed to go against his theory of evolution by natural selection," says Marion Petrie, a biologist at the University of Newcastle-upon-Tyne. "He thought about it a lot and it was several years before he produced his explanation for why peacock trains might have developed. And he had a special term for the process that might have caused the development of the train, which he called sexual selection."

Peahens prefer peacocks with bigger tails, which reflect the quality of their genes.

During the mating season, peacocks will assemble together in groups called leks, drawing females to them with their cries. As soon as a female comes into view, a male will raise his tail and make it shiver. Darwin proposed that peahens judge peacocks by their tails. They find certain kinds of tails attractive and choose to mate with their owners. Whether their choice was based on aesthetics or on some desirable qualities in the peacock, Darwin didn't say. In either case, by choosing among males, peahens behave like pigeon breeders who select certain pigeons for qualities that natural selection would have ignored in the wild. A fantail pigeon pleases a breeder's eye; a resplendent peacock pleases a peahen. With each generation, the preference of females brings more reproductive success to males with pleasing traits. In time, Darwin argued, the choices of females could have created something as extravagant as a peacock's tail.

Sexual selection, as Darwin dubbed this new force in evolution, didn't win many admirers. Alfred Wallace thought good old natural selection was sufficient. Female birds were drab because they spent their time in nests, he claimed, where they needed to be camouflaged for protection. Extravagant colors might be the normal state of feathers, and male birds—with less need of camouflage—were not subject to the dulling force of natural selection.

For decades most biologists continued to doubt that females had much say in the matter of sex. Only in the past 20 years or so have researchers performed experiments that can show whether females have preferences. It turns out that they have very strong ones—strong enough to drive the evolution of the peacock's tail.

Marion Petrie, for example, demonstrated that peahens have very clear tastes when it comes to peacocks. "Lots of females, where they've got free choice amongst several males, will actually approach one male, and he will get a high proportion of females in the population," says Petrie. And peahens, Petrie has shown, choose those fortunate males on the basis of their tails. Elaborate tails with lots of eyes are more attractive to peahens than less ornamented ones. Peacocks have on average 150 eyes on their tails. By trimming only a few eyes from a peacock's tail, Petrie could significantly reduce his chances of getting picked. A peacock with fewer than 130 eyes rarely mated at all.

Other biologists have shown that the females of many other species also have strong preferences in their choice of mates. Hens like roosters with big bright combs; female swordtail fish like males with long tails; female crickets like males with the most complicated calls. Since these displays are hereditary, sexual selection could indeed have driven their evolution. And since long tails and bright colors and loud calls all make heavy demands on male animals, there must be a limit to how extravagant they can become. If they pose too much of a survival cost to males, natural selection will put a ceiling on their evolution.

Darwin was always a bit evasive about one fundamental question of sexual selection: Why did a female prefer a particular kind of tail or comb? He simply said that she found it attractive. In 1930 the British geneticist Ronald Fisher rephrased the idea in a more formal way: if female birds find long tails attractive, short-tailed male birds will have a harder time finding mates. A female who chooses a long-tailed mate will presumably have long-tailed sons, and her male offspring will have a better chance of finding mates. Mothers, in other words, just want their sons to look sexy.

But a growing number of scientists now believe that females are not being arbitrary when they pick their mates. They are actually attracted to displays that can reveal a male's genetic potential.

Females have fewer chances to pass their genes on to the next generation than males, and so evolution frequently makes them much more cautious about their choice of a mate. One powerful threat to a female's offspring are parasites. Even if a female has good genes for protecting herself from diseases, she will dilute their power in her offspring if she mates with a male with a weak genome.

A female animal cannot send her suitor's genes to a lab for analysis, but she can detect clues of his fitness in the way he looks or acts. To sing loudly or grow bright feathers, a male can't be too weakened by his fight with parasites. Exactly what sort of display the males of a species will evolve as a way to impress their females depends on the peculiarities of the species itself. Primates are the only mammals with good color vision, which may be the reason why some primate species are the only mammals who use brilliant reds and

Rooster combs, like peacock tails, function as a way for males to attract females.

blues in their sexual displays. But whatever form a display takes, it has to represent a sincere sacrifice. If females can be fooled by false displays, their offspring will not inherit good genes from their fathers. The attraction of the false display will evolve away into oblivion.

A rooster's comb doesn't physically burden the bird like a peacock's tail does, but it is a sincere sacrifice nonetheless. The comb, like many other male displays, needs testosterone to trigger its growth. But testosterone also lowers the rooster's immune system. To grow a comb, a rooster must put himself at greater risk of getting sick. Only truly strong roosters can squander their immune systems this way.

Another sincere form of advertising is symmetry. As an embryo develops, it can be buffeted by different kinds of stress. Its mother may not be able to find enough food while she is pregnant, for example, and so the embryo will not have the energy it needs to grow properly. Some animals are genetically predisposed to withstand these assaults and grow up healthy. But in other animals, the choreography of embryonic development gets thrown into chaos by the stress. As a result, they may grow up to be infertile or susceptible to diseases. A female looking for a mate would do well to avoid males with this instability.

Developmental instability leaves its mark on the visible symmetry of an animal's body. For the most part, an animal's body is a pair of mirror images. The same intricate network of genes that builds its left side needs to perform precisely the same job on the right. If the development of an animal is disrupted somehow, the exact symmetry of its body may be thrown off. An antelope's horns may grow to different lengths; a peacock may grow different numbers of eyes on each side of its train. Symmetry may be a badge of fitness.

Researchers are now testing whether females use male displays to judge their genes, and many of their results support the idea. Female crickets prefer males whose songs have extra syllables in them, and the length of a cricket song is a reliable indication of a strong resistance to parasites. Female barn swallows prefer males with long tail feathers that are symmetrical, and length and symmetry are both reliable clues that it is healthy. Marion Petrie has shown that peacocks with bigger tails are more likely to survive than peacocks with smaller tails, and those survival rates get passed on to their offspring.

One of the best ways to test an evolutionary hypothesis is to find an exception that proves the rule. Not every animal species has males competing over display-judging females. In a few species the roles of the sexes have been partially reversed. The female pipefish (*Syngnathus typhle*) places her eggs in a pouch in the male's body, and the male essentially becomes pregnant. For

Female barn swallows prefer males with long, symmetrical tail feathers.

several weeks he carries the eggs, supplying them with nutrients and oxygen from his own blood. In a single breeding season, each female can make enough eggs for two males to carry, creating a fierce competition among the females for the limited number of available males. As a result it is the male pipefish, not the females, who choose their mates, preferring big, ornamented females over small, plain ones.

When animals choose mates they don't make conscious decisions. Peahens do not count the eyes on a peacock tail and think to themselves, "Only 130 eyes? Not good enough. Next!" Peahens probably experience a complicated chain of biochemical reactions at the sight of a sumptuous peacock tail that leads them to mate with its owner. Such is the case for most sorts of adaptive behavior: although they are based only on instinct, they can carry out a sophisticated strategy for survival.

BATTLE OF THE SPERM

Once a male has earned a female's attention and has successfully mated, he doesn't automatically become a father. His sperm still have to struggle through her reproductive system to find an egg to fertilize. And quite often, his sperm won't be alone: they will be competing with the sperm of other males that the female has mated with.

It may seem odd that a female would go to great lengths to choose a male and then end up mating with another one. But nothing is simple when it comes to sex. Sometimes a female's choice gets overridden by big males who grab her and forcibly mate with her. In other cases, a female who has chosen one male may encounter a better specimen and mate with him as well. Hens, for example, prefer to mate with dominant roosters, but a subdominant male can sometimes mount a hen before a dominant rooster can scare him off. The hens don't treasure these flings. If a subdominant male mates with a hen, she is likely to squirt out his semen. That increases the odds that when the dominant male mates with her later, his sperm will fertilize her eggs, producing stronger chicks.

Promiscuity is rampant in the animal world, even in species that generations of scientists had been convinced were utterly faithful. Around 90 percent of all bird species live monogamously, for instance, with a male and female joining together for a year or even a lifetime, building nests together and working together to raise chicks. Monogamy is a matter of survival: without the help of both parents, the helpless chicks may not live to adulthood. But when ornithologists began to sample the DNA of chicks in the 1980s, they found that in many species, some of the birds did not have the father's genes. In most species, a few percent of chicks were illegitimate; in some, the percentage was as high as 55 percent.

A monogamous female bird doesn't cheat on her mate randomly. Female barn swallows, for example, judge the quality of males in part by the length of their tail feathers. Females who pair up with short-tailed males for the season are much more likely to cheat on their partners than females who pair up with long-tailed males. Female swallows have only a limited amount of time to choose a male companion for the breeding season, so they can't wait forever to find the perfect partner. But they can offset the shortcomings of their partner by mating with more desirable males that come visiting. And they can enjoy the help of their partner in raising the chicks, even if he is not their father.

A male thus faces a quandary. Despite all his efforts to court a female, he has no guarantee that his sperm will fertilize her eggs. She may carry the sperm of another male in her, or she may later mate with another male. As a result, males in many species have evolved ways to compete in utero.

One way is to make a lot of sperm. The scramble of sperm inside a female is like a lottery: the more tickets the males buy, the higher their chances of winning. Among primate species, for example, the average size of their testicles is directly proportional to the average number of partners the females mate with. The more intense the competition, the more sperm a male primate produces.

A sneakier way to win a lottery is to destroy the tickets of the other contestants. Male fruit flies have poisonous semen that disables the sperm of pre-

The males of some species of damselflies scrape the sperm from previous matings out of females.

vious suitors inside a female. Male black-winged damselflies have penis-like organs covered in spines; before they deposit their own sperm inside a female, they use the spines like a scrub brush to clean out the sperm of other males. They can get rid of 90 to 100 percent of another male's sperm, giving their own a much better chance of fertilizing eggs. The male dragonfly *Hadrothemis defecta* uses inflatable horns on his penis-like organ to push the sperm of other males deep into recesses in the female's body. Only then does he place his own sperm closest to her eggs.

Yet another way for males to win the lottery is to keep other males from buying tickets in the first place. In addition to their sperm poison, male fruit flies have chemicals in their semen that decrease a female's libido. With less interest in mating, she will be less likely to receive the sperm of other males. Among Sierra dome spiders, the females attract potential mates by scenting their webs with a male-attracting pheromone. Once a male locates the female, he destroys her web so that other males will have a harder time finding her.

For the males of some species, the best way to raise the odds that their sperm will succeed is to commit suicide. The male Australian redback spider routinely sacrifices himself for sex. He begins his courtship by plucking the strands of the female's web, transmitting a love song of sorts that may last for hours. If she doesn't chase him away—or if another male already with her doesn't do it for her—he approaches. She looms over him, her body weighing 100 times more than his own. To any animal in his position death could come at any moment: her bite is as deadly as that of her relative, the black widow.

The male redback crawls onto her belly. He extends an appendage that sprouts from his head. Known as a palp, it looks like a miniature boxing glove. On its tip is a long coiled tube, which he threads into the female's body. He then starts to pump sperm through the palp, into her. Suddenly, using his palp as a fulcrum, he swings his body up from the female's abdomen and flips over onto his back, landing on her fangs. She begins to chew on his abdomen and injects venom into his body, which begins to digest his innards to goo. She dines slowly, as the male continues to inseminate her, and after a few minutes the male pulls away from her. He retreats a few centimeters away and grooms himself for 10 minutes or so. Even as his body is disintegrating from within, he returns for more, inserting a second palp into her body and performing another flip. The female resumes her meal, biting deeper into his body. The mating may take half an hour; by then the male is barely alive, and when he withdraws his second palp, the female weaves a shroud of silk around him. There is no escape for him this time. After a few more minutes of feeding, the female has reduced him to a mummified husk.

Maydianne Andrade, a biologist at the University of Toronto, has studied the suicide of the redback spider to see whether it represents an evolutionary adaptation. Not all male redbacks get devoured, she has found. Only hungry females devour their mates, and as a result a third of males survive their deadly somersault. That discrepancy gave Andrade the opportunity to measure the reproductive success of cannibalism.

Female redbacks seem to be in control of how long copulation will last. In cases where they don't devour their mate, Andrade has found that sex lasts on average 11 minutes. But when a female chooses to eat a male, copulation can last 25 minutes. As she feasts on her mate's body, his palps can continue inseminating her. By offering himself as a meal, a male can stretch out the sex. As a result, a cannibalized male delivers more sperm, fertilizing twice as many eggs as a male that survives. And once the male is dead, the female tends to spurn new suitors—perhaps because she is sated with sperm or with food. In either case, the chances of another male injecting her with competing sperm go down, giving the dead male a better chance of having fertilized her eggs.

These sorts of benefits are apparently more valuable to male redback spiders than their own lives. They have few chances to mate for a number of reasons—they have short life spans, and their sperm-delivering palps snap off during sex, rendering them sterile. So these spiders make their one chance count.

CHEMICAL WARFARE OF THE SEXES

The struggle for sexual success is a continually shifting battle, one that carries on with every generation. It's hard to catch the struggle in action, but thanks to some brilliant experiments, scientists have been able to see a few glimpses. William Rice, a biologist at the University of California at Santa Barbara, has studied the chemical warfare that male fruit flies use to help their sperm compete against other sperm.

Not only does a male fruit fly's semen disable the sperm of other males and wipe out a female's libido, it even speeds up her egg-laying schedule. By making her lay her eggs sooner after mating, he reduces the time in which she might mate with other males. The chemicals that male fruit flies use to alter their mates are poisonous to the female. They don't kill her immediately, but the more often she mates, the shorter her life span becomes. It doesn't matter to the male fruit fly that his mate dies young. Since male fruit flies don't take care of their offspring, his only evolutionary interest is to produce more fertilized eggs.

The arsenal that male fruit flies use has the same effect on females as the pesticides farmers use to kill them: it makes them evolve. Just as pesticides trigger the evolution of resistance, female fruit flies have evolved ways to neutralize the poisons in semen. And the evolution of female defenses has encouraged the semen of the male flies to become more toxic.

In 1996 Rice was able to document this deadly pas de deux. He maintained a large breeding population of fruit flies, and thanks to some peculiarities in fruit fly genetics, he was able to manipulate the flies so that their descendants were all males and only inherited the genes of their fathers. These new male clones then mated with a fresh supply of females from Rice's breeding population, producing the next generation of males.

Each new batch of females was unfamiliar with the chemical warfare of Rice's males, so they had no opportunity to evolve defensive adaptations. Meanwhile, the males that produced more toxic semen could manipulate the females more effectively and father more flies. Forty-one generations later, Rice had created a race of supermales who mated more often and more successfully with females than their ancestors. Their success cost the females dearly: as the males' semen became more poisonous, their mates died at a much younger age.

Rice found still more evidence for this sexual arms race by forcing the flies to declare a cease-fire. In 1999 he ran an experiment in which he paired off males and females into monogamous couples. Instead of competing with other males, these males could mate only with the partner Rice gave them. When their eggs hatched, Rice again sorted the new flies into monogamous pairs. Since the monogamous males faced no competition, the poisonous chemicals in their semen no longer offered any evolutionary benefit. And once the males abandoned their poisons, the females had no incentive to evolve antidotes to them. After 47 generations, Rice found that the monogamous males had become significantly less harmful to their mates and the females were less resistant to toxins in semen.

Rice's fruit flies were able to enjoy a more tranquil life, but only because he forced them into it. On their own, they couldn't have found their way to a truce. Any male fly that can evict another male's sperm will spread his genes more successfully than a one-mate male. And any female who can defend herself will be favored as well. Evolution does not have the foresight of a biologist in a lab, so love among the fruit flies is truly blind.

TUG OF WAR, IN UTERO

Even after the mating is over and an egg has been fertilized, mothers and fathers can use evolutionary tactics to increase their success. In mammals such as humans, a fertilized egg implants itself in the mother's uterus and starts growing a placenta. The placenta pushes blood vessels into the mother's body to draw in blood and nutrients. A growing embryo needs a huge amount of energy, which can put a dangerous drain on the mother's resources. If a mother lets an embryo grow too fast, she may cause herself grievous harm, threatening her future fertility or maybe even her life. Evolution should therefore favor mothers that can keep their babies in check.

But fathers have a different evolutionary agenda. A fast-growing, healthy baby is an unalloyed good for the father. After all, the rate at which the baby grows can't threaten his own health or his own ability to have more children.

David Haig, a Harvard biologist, has proposed that the conflicting interests of mother and father are served by the genes they give to the baby. Maternally inherited genes do different things than paternally inherited ones. Take, for example, a gene called insulin-like growth factor 2 (IGF2). The protein that this gene produces stimulates the embryo to draw in more nutrients from its mother. In experiments on pregnant mice, researchers have shown that the mother's copy of the IGF2 gene is silent, while the father's copy is active. It turns out that mice also carry another gene that makes proteins whose job it is to destroy IGF2 proteins. And the mother's copy of this IGF2 destroyer is active, while the father's is shut down.

In other words, the father's genes are trying to speed up the growth of the mouse embryos while the mother's genes are trying to slow them down. You can actually see the effects of this struggle in experiments where scientists shut down the paternal or maternal genes. If they shut down the father's copy of IGF2, mice are born only 60 percent of their normal size. But if the maternal copy of the gene that destroys IGF2 is shut down, the mouse is born 20 percent heavier. If Haig is right, we are all the compromise between our parents' competing interests.

MATERNAL INVESTMENTS

A father's reach can only extend so far. There are some ways that mothers can take control of the fate of the embryos they are carrying without any interference from their mates. They can invest different amounts of energy into their eggs, depending on how desirable the father is. Female mallard ducks, for example, will lay bigger eggs for high-dominance males than low-dominance ones.

In some species, mothers can increase their reproductive success by determining the sex of their child. The most adept gender controller is the Seychelles warbler. These Indian Ocean birds live in pairs, each on its own territory. On the 70-acre Cousin Island, there is not enough land for new warblers to be guaranteed their own territories. As a result, young females will sometimes stay at home with their parents rather than search for a mate of their own. They help build nests, defend territory, incubate eggs, and feed newborn chicks. Seychelles warblers are a help to their parents when there's enough food for everyone. But if a warbler family must eke out a living on a low-quality territory without much food, daughters become more of a burden than a help.

In 1997 Jan Komdeur, a researcher at the time at the University of Groningen in the Netherlands, compared the eggs that were laid by birds living on low- and high-quality territories. On high-quality territories, he found that for every male chick the warblers produced, they produced six females. On low-quality territories, they produced only one female for every three males.

These sex ratios, Komdeur discovered, are not locked in by the genes of individual warblers. The birds can, in effect, decide how many sons and daughters they will have. Komdeur proved this by transferring some warbler pairs from Cousin Island to two other islands in the Seychelles that are uninhabited by the birds. He chose warblers that had been stuck with low-quality territories on Cousin Island and were producing mostly sons, putting them on territories where they could find plenty of food. As soon as the warblers began breeding in their new homes, they started producing mostly daughters.

The evolutionary logic to their strategy is clear. When food is scarce, it is better to make a lot of males. They will leave the nest as soon as they can in search of mates and new territories, leaving the parents to raise new fledglings on what little food they can find. (The males may not find a new territory, and die, but that's a risk worth taking.) When times are good, daughters make good helpers, and so a warbler mother somehow alters the balance of males and females. Exactly how the warblers choose between sons and daughters, no one yet knows, but that doesn't take away from Komdeur's discovery that they can indeed make the choice.

DARWINIAN FAMILY LIFE

Once an animal is born, it may grow up in a big family or find itself an orphan. Among mayflies, both parents are dead by the time their eggs hatch. Among black bears, the mother takes care of her cubs for a year while the father offers no help at all. Among barn swallows, the father will work just as hard as the mother to bring food to his fledglings until they're ready to fly. And elephants may live with brothers, sisters, aunts, uncles, and grandmothers in clans that endure for decades.

Raising children can be just as crucial to an animal's reproductive success as finding a mate. If a male dung beetle has sex with thousands of female dung beetles but all his children die within a week of hatching, all his sexual conquests have been, evolutionarily speaking, for nothing. In many species, mothers and fathers work together to raise children. But their conflict of interest can threaten their family bond. Males who spend their time raising another male's babies are less likely to pass on their own genes. As a result, males in some species have gotten wise to their cheating partners. Andrew Dixon at the University of Leicester observed how much care reed buntings took in feeding and protecting their young. When he sampled the DNA of the birds to see which were actually related, Dixon found that when fewer fledglings in a nest belonged to the father, he put less effort into bringing food to the nest.

But in many species, ranging from mice to langur monkeys to dolphins, males do not simply neglect young animals that don't belong to them. Sometimes they turn homicidal. This disturbing behavior has been particularly well studied in lions. A pride of lions consists of a dozen or more lionesses and up to four males, along with their cubs. When the male cubs grow to maturity, they are driven away by the older males. Together with other outcasts, they search for another pride where the males seem weak. They fight with the residents, and if the residents run away, they take over the pride. Cubs are in grave danger at that moment: the new males are likely to take them in their jaws and crush them to death. Of all the lion cubs who die in their first year, adult male lions kill one out of four.

What looks like cruelty to us humans, a number of zoologists argue, is evolution's logic at work once again. A male lion's ultimate goal in taking over a harem is to father cubs of his own. Nursing prevents a lioness from coming into estrus, so the presence of cubs in a pride means a male lion may have to wait for months before mating with their mothers. Given that a male lion may be overthrown in a couple years—and his own cubs possibly killed if they are too young at the time—he has no time to waste as a stepfather.

Lionesses do what they can to keep their cubs alive. The roar of alien males will make them stand up, snarl, and gather together to put up a fight.

The more lionesses defending a cub, the better its chances of surviving. That may be why lionesses form prides to begin with, as a way to fight against infanticide.

But lionesses are not always able to defend their cubs, and when new males take over they try to start families again. After a male lion has killed a lioness's cubs she comes into estrus and soon becomes sexually insatiable. The dominant male lion of the pride will mount a lioness in estrus nearly a hundred times a day, and after a day or two he will become exhausted. Now the lower-ranking male lions step in and mate with her for a few days more. When she gives birth four months later, the paternity of her cubs is a mystery. That uncertainty may explain why male lions don't kill cubs in their own pride. There's always a chance that the cub they kill is their own—a disaster for their own evolutionary legacy.

When the first reports of infanticide in animals emerged in the 1970s, many researchers were skeptical. Surely, they said, these brutal males must be pathological. Besides, how could they possibly know which infants were their offspring and which weren't? But more cases kept pouring in. Stephen Emlen, a Cornell University ornithologist, found a particularly elegant way to test the infanticide hypothesis. In 1987 he was studying jacanas, a Panamanian bird that is, like the pipefish, a species in which the usual sex roles are switched. Male jacanas sit on eggs and raise the young, while the females rove around their territories, mating with many males and fighting off intruding females. Sometimes the attacker may drive away the resident female jacana and take over her males.

When male lions take over a pride of lionesses, they may kill the resident cubs. Without a nursing cub, a lioness becomes sexually receptive again and can bear the new male's own offspring.

If a male lion taking over a new pride should benefit by killing cubs, Emlen reasoned, then a female jacana taking over a harem of males would benefit by killing their chicks. Emlen needed to shoot some jacanas to get their DNA, and he decided to choose two females whose male partners were caring for nests of babies. He shot one of the females one night, and by the next morning a new female was on its territory, pecking at the chicks and hurling them to the ground until they were dead. The male jacana looked on helplessly. Within hours she was soliciting him to mate, and he mounted her. The next night Emlen shot a second female, and the following morning the same violence played itself out again.

"If we're thinking about individuals leaving genes to the next generation," says Emlen, "this makes sense, despite its grisliness."

FOR THE GOOD OF THE GENE

Infanticidal lions, adulterous swallows, and sex-ratio-skewing warblers may give the impression that animal life is little more than sexual selfishness. And yet in many cases, evolution has produced animals that have given up their own struggle for sex altogether.

William Hamilton was intrigued by this paradox, particularly in the case of bees and other social insects. In a honeybee hive, there is a single queen, a few males, and 20,000 to 40,000 female workers. The workers cannot reproduce on their own; they spend their lives gathering nectar, keeping the hive in good working order, and feeding the queen's larvae. They will defend their hive against attackers, dying in the process. In terms of evolution, it seems like a mass suicide.

But Hamilton suggested that thanks to the peculiar genetics of honeybees and related insect species, the workers are actually working for the long-term benefit of their genes. A queen makes sons and daughters in distinctly different ways. Males start out as unfertilized eggs, which divide and develop into full-grown insects without any sperm. Because they don't receive any DNA from a father, male honeybees have only one copy of each gene. On the other hand, a queen mates with one of her male consorts and uses a standard Mendelian shuffle to create daughters, each with two copies of each gene.

Sister bees thus have a remarkably close bond—closer than human sisters. Human sisters inherit one of their father's two genes, and one of their mother's pair. Since the chance of inheriting any particular gene is thus 50 percent, human sisters share, on average, half of their genes. Sister bees, on the other hand, inherit identical genes from their father, because male bees have only a single set to give. Combined with their mother's DNA, sister bees thus share on average three-quarters of their genes. If a female bee were to have a daughter of her own, she would share only half of her genes with her offspring, the rest of them coming from her mate. A female honeybee thus has more in common with her sisters than her daughters.

Under these circumstances, Hamilton argued, it's not surprising that worker bees forgo their own reproduction to work for the good of the hive. The larvae that they help to raise are so closely related to them that they can spread their genes more quickly than if they tried mating themselves.

With one swift jab, Hamilton punched through the paradox of altruism that had puzzled biologists since Darwin's day. If evolution consists of a competition among individuals for survival and reproduction, it makes little sense to help others. Perhaps, some researchers suggested, animals acted selflessly for the good of the species or at least the good of a group. But that kind of altruism simply didn't square with what biologists knew about how genes spread over time.

Hamilton looked at the question of altruism from the gene's point of view. Altruism might not benefit the altruistic individual, but it might be a good way to make more copies of the altruistic individual's gene. It raises an animal's fitness, but not because it raises the animal's own chances of reproducing. Hamilton called this indirect benefit of altruism "inclusive fitness."

Hamilton's rule of inclusive fitness has been confirmed splendidly. Not only are female workers more closely related to their sisters than to their own daughters, but they are more closely related to one another than to their own brothers. Brothers don't receive any of the paternal genes that their sisters inherit from their father, and they share only half of their mother's genes. Thus while a sister shares 75 percent of the genes of her sisters, she shares only 25 percent of her brother's genes. In other words, she is three times more closely related to her sisters than to her brothers. This difference is reflected in the ratio of brothers and sisters in a nest. In the colonies of many social insects, the ratio of females to males is 3 to 1. It is the worker females, not the queen, who set that ratio. They take worse care of the male larvae than the female ones.

But inclusive fitness should produce the 3-to-1 ratio only in nests where the queen mates once and uses that same sperm to create her entire colony. If a queen should mate with another male and use his sperm as well, the sisters will not share their father's genes. Liselotte Sundstrom, a Finnish entomologist, discovered colonies of *Formica* ants in which the queens mate either once or multiple times. In the single-father colonies, she found that the 3-to-1 ratio of larvae reigns. But in the multiple-father colonies, the ratio is close to 1-to-1. Since sisters are no more closely related to one another than to their brothers, there is no incentive to the workers to favor one sex over the other.

THE GENEROSITY OF PEACOCKS

Hamilton's inclusive fitness may help explain family life not just in ant colonies but among birds and mammals as well. When Marion Petrie began studying peacocks, she was curious not just about their tails but about their leks. Why, she wondered, did peacocks gather together in groups in order to strut before peahens? The less successful males would invariably be passed over for the males with the most resplendent feathers. Wouldn't it be better for males to look for females on their own so that they wouldn't suffer by comparison?

At an English zoo called Whipsnade Park 200 peacocks roam freely over the grounds. In 1991 Petrie took 8 Whipsnade peacocks to a farm more than 100 miles away, where she penned each of them with 4 peahens. She collected their eggs each day and hatched the chicks in an incubator. She put rings on their legs for identification, and then mixed them together with chicks from other pens. A year later, she brought 96 of the young peacocks (12 from each of the 8 fathers) back to Whipsnade.

In a peacock's fourth year he chooses the spot where he begins to display his tail. In 1997 Petrie watched the peacocks she had returned to Whipsnade as they gathered to form their leks. She checked their rings and looked in her charts to see who their parents were. She was surprised to find that brothers and half brothers stayed much closer together than did unrelated peacocks. A peacock's nearest neighbor was five times likelier to be related than would be predicted by chance. Somehow, even though they had no knowledge of their parents and had never had a chance to get to know their siblings, they had found one another at Whipsnade.

As a family affair, a peacock lek makes sense. A peacock shares many of the same genes as his brother, so if his brother reproduces successfully, the genes they share get carried forward. For some peacocks, it may be more rewarding to help their brother find a mate than to search for one themselves. If a female mates with any of them, their collective genetic legacy is the winner.

Inclusive fitness may even help explain some of nature's most intricate soap operas. In Kenya, for example, a bird called the white-fronted bee-eater lives in what scientists once thought was a sort of utopian commune. They form giant colonies of up to 300 individuals, their nests consisting of holes in a mud cliff packed in as densely as apartments. The adult birds seemed to early ornithologists to live in peaceful monogamy. When the children were grown they often stayed on to help their parents tend to younger brothers and sisters. Sometimes they even helped their neighbors.

In the 1970s Stephen Emlen began visiting the bee-eaters to figure out just how altruistic the birds really were. He and his colleagues spent years watching the nearly identical birds flitting among the nests or darting away for food and returning not long afterward. They built up genealogies of families, which they confirmed by testing the DNA of the birds. What they found was that what looked like simple altruism was actually a complicated family intrigue.

Bee-eaters, Emlen discovered, do not live in simple two-parent nuclear families. They form big multigenerational clans of up to 17 birds, including parents, grandparents, uncles, aunts, cousins, nieces, and nephews. An extended family occupies groups of neighboring nests, and relatives will spend a lot of time visiting one another. If a predator kills a nest of chicks, a son who was helping to raise them would move into a nearby nest to continue helping there. He is not helping a stranger, but an uncle, perhaps, or a sister. Since these relatives share some of his own genes, it makes sense for him to help them if he cannot help his immediate family. And helping, Emlen and his colleagues discovered, can make a huge difference to the success of a nest. The addition of a helper to a nest can double its productivity.

Bee-eaters live in a web of family conspiracies. Female bee-eaters do in fact visit the nests of strangers, but not to help raise chicks; instead, they try to lay their eggs in their nests. If the strangers don't realize that the egg doesn't belong to them, they will spend the effort raising the chick, effort that the real mother can use for raising even more chicks in her own nest. During the time after a mother has laid her eggs and before they have hatched, families are perpetually on their guard for these intruders. But their own daughters, Emlen discovered, will try to sneak their own eggs into their mother's nest. Emlen was surprised that the daughters even had eggs to lay, given that they were still living with their family and hadn't yet been paired off with a male bee-eater. But he discovered that daughters sometimes fly miles away from their nests to consort with males of other bee-eater colonies.

The parents conspire as well. If one of their sons pairs up with a female and tries to set up his own nest, his father pays him so many social visits that he can't start his own family. The son will then be more likely to come back to his parents' nest to help raise more siblings. What looks like a utopia is actually a swirling cauldron of inclusive-fitness-raising behaviors.

SEXUAL POLITICS OF THE CHIMPANZEES

Emlen and other researchers have undermined many cases of so-called altruism. In only a few species, it now seems, do animals help out strangers without any regard for blood ties. Vampire bats are one. They spend each night searching for animals whose blood they can drink. But if they fail, they can return to their cave and beg for some blood from an unrelated bat that had a more successful night. Robert Trivers, a Rutgers University anthropologist, has dubbed cases like these "reciprocal altruism." Evolution can favor reciprocal altruism, Trivers argues, because in the long run two unrelated animals that help each other may enhance their chances of survival more than if they behave more selfishly. Vampire bats burn up their food quickly, so if they go two or three days without blood they will starve. Donating blood to an unrelated bat may be a sacrifice, but it is also an insurance policy.

Reciprocal altruism may be particularly likely to evolve in species with big brains. If you have the mental capacity to recognize individuals and keep a scorecard of who has been good to you and who has been taking advantage of your kindness, you can use reciprocal altruism to your advantage. So it shouldn't come as a surprise that some of the best evidence for animals helping strangers comes from our closest relatives—chimpanzees and bonobos (a separate ape species sometimes known as the pygmy chimpanzee).

Chimpanzees cooperate with unrelated chimps, do favors for them, and sometimes even make sacrifices for them. They may join together on hunting expeditions, looking for duikers or colobus monkeys and sharing their kills. Reciprocal altruism may help chimpanzees gain social power—two subordinate males, for example, can make an alliance to overthrow the top male in their group. And chimpanzees don't just hand out favors blindly. They keep track of their kindness, and if they are betrayed they will cut off their generosity or even punish a cheating chimp.

In chimpanzee society, males can take advantage of reciprocal altruism while females cannot. Whereas males spend their whole lives in the band where they were born, female chimps leave when they reach adulthood. Once a female chimp joins another group, the demands of raising children keep her from establishing long-term relationships with the new chimps. Lugging a nursing baby makes it impossible for her to keep up with the group as it searches for fruit. Because baby chimps depend on their mothers for up to four years, a female chimpanzee may end up spending 70 percent of her adult life away from her group.

As a result, male chimps have all the power. They bond with other males, making alliances that help them climb their way up through the chimp hierarchy. Reciprocal altruism also lets male chimps cope with their unstable supply of food. Chimpanzees rely mainly on fruit, which requires them to travel continually to find ripe trees. To supplement their diet, males can hunt together for meat and share the spoils, and they can also form raiding parties to attack smaller groups of chimps, taking over their fruit trees.

Meanwhile the females, who never get a chance to form alliances and enjoy the benefits of reciprocal altruism, can't gain the power that male chimps have. If a group of chimpanzees comes across food, the females invariably wait while the males have their fill. Male chimpanzees also inflict violence on females. They will hit them to coerce them into having sex, and

when a new female shows up in a group with a baby, the male chimps may kill it. "Chimpanzee society is horridly patriarchal, and horridly brutal," says Richard Wrangham, a primatologist at Harvard University.

As in other species, female chimpanzees do not suffer passively. They do what they can to protect their babies and find good mates. Compared to other apes, young female chimpanzees take a long time to reach sexual maturity; some primatologists have suggested that the delay is a strategy for reducing the chances that they will come into a new group bearing an infant that may get killed.

Once a female chimpanzee does reach sexual maturity, she uses sex to protect her babies. Each time she becomes sexually receptive, her genitals swell and turn pink, and she approaches all the males of her group. Dominant males tend to have the most sex with her, but they cannot keep the females from mating with others. On average, a female chimp will have sex 138 times with 13 different males for every infant she gives birth to. Yet the signal of her swollen genitals is misleading: she is actually fertile for only a short time. As a result, about 90 percent of the times a female chimp has sex she actually cannot conceive. As with lions, female chimpanzees may be having sex with many males in order to defend against the infanticidal instinct of males, by making it harder for males to determine the paternity of their babies.

Primatologist Richard Wrangham (below) has helped uncover the social life of chimpanzees, a male-dominated life in which female chimps have to cope with violent attacks and even infanticide.

LOVE, NOT WAR

The evolutionary conflict between the sexes sometimes leads to male-on-female violence, as in the case of chimpanzees, but it doesn't have to. If the conditions are right, apes may evolve a tranquil existence, in which sex becomes far more than a matter of the survival of genes. It becomes a tool for keeping the peace.

The peaceful apes in question are the bonobos. Bonobos are relatively new to science; scientists first recognized they are different from chimpanzees only 70 years ago. In 1929 a German anatomist was studying the skull of a juvenile chimpanzee in a Belgian colonial museum when he realized that it actually belonged to a small adult of a different species. Bonobos, which live south of the Zaire River in the Democratic Republic of Congo, are not only smaller than ordinary chimpanzees but more slender, with long legs and narrow shoulders. Their lips are reddish and their ears are small and black. Their faces are flatter than a chimp's, and they have long, fine black hair neatly parted in the middle.

Bonobos, also known as pygmy chimpanzees, diverged from chimpanzees 2 to 3 million years ago. In that time, the two species have evolved very different social lives. While chimps live in male-dominated societies, females are in control in bonobo society.

The differences between chimps and bonobos are more than just anatomical. During World War II the Allied forces bombed the German city of Hellabrun. One of the zoos in the city had a colony of chimpanzees, which were not affected by the terrific sound of the explosions. Another zoo nearby kept a colony of bonobos, and all of them died of fright. A few years later two German primatologists studying bonobos at Hellabrun noticed that their sex lives were quite unlike that of chimpanzees. They wrote that chimpanzees mated *more canum* (like dogs) while bonobos did so *more hominum* (like people). Unlike any other primate except humans, bonobos had sex face-to-face.

The German primatologists were ignored by other ape experts, and it wasn't until the 1970s that a new generation of scientists rediscovered that bonobos were dramatically different from chimpanzees. As with chimpanzees, male bonobos stay in the community where they are born and a female must leave to find a new community when she reaches adulthood. But when she arrives she doesn't face a gang of bullying males ready to kill her baby and force her into sex. In bonobo society, the females dominate. If you toss a bunch of bananas into their midst, the females eat first, and the males wait their turn. If a male bonobo tries to attack a female, he's liable to be stormed by a pack of angry females. They've been known to pin an offending male to the ground as one of them gives his testicles a painful bite. Male bonobos live in a hierarchy of their own, but it is the sons of the dominant females who rank high; among themselves, males form hardly any bonds at all.

A female bonobo joining a new community also enters a perpetual orgy. While female chimps have swollen genitals for 5 percent of their adult life, female bonobos are sexually receptive 50 percent of the time. Their sex lives start early: young bonobos start trying to mate long before they can possibly conceive. And bonobo sex is not just heterosexual. Young males will fence with their penises or give oral sex to each other. Females, meanwhile, specialize in rubbing their genitals together until they reach orgasm (what primatologists call the "g-g rub").

Among bonobos, sex is not just for reproduction, or even for protecting babies against angry males. It is a social tool. A new female will work her way into a bonobo community by approaching a resident female and giving her lots of sexual satisfaction. This favor wins her an alliance, and as she makes more of them, she can make her way toward the core of the community.

Sex can also defuse the tensions that build up in bonobo society. When bonobos come across food—be it a fruit tree or a termite nest—they start screaming in excitement. But rather than fighting over the food as chimpanzees might, the bonobos proceed to have sex. Likewise, if a male has a fit of jealousy and chases another male away from a female bonobo, the two males may later reunite for some scrotal rubbing. Sex keeps the underlying competition from escalating into all-out war. "The chimpanzee resolves sexual issues with power; the bonobo resolves power issues with sex," writes Frans de Waal, a primatologist at Emory University, in his book *Bonobo: The Forgotten Ape.*

Chimps and bonobos share a common ancestor that primatologists estimate lived 2 or 3 million years ago. Richard Wrangham and his colleagues have proposed that the difference stems from where the two primates live. Bonobos live in humid jungles, where the supply of fruit is much more reliable year-round than the open forests where chimpanzees often live. And even if bonobos should run out of fruit, they can turn to the herbs that grow in abundance in their forests.

Thanks to the abundance of food, bonobo groups don't have to move as quickly as chimps do. Female bonobos can keep up even when they have babies in tow. With enough food for everyone, the females don't compete with one another, and can form long-term alliances. By cooperating, the female bonobos can keep the males in line. As a result, infanticide is unknown in bonobo society. Because males are peaceful in their own groups, they don't wage war against other groups. When two groups of bonobos meet up, they have sex rather than fight.

"It looks as though a relatively simple change in the feeding ecology is responsible for this dramatic difference in sexual behavior," says Wrangham.

For female bonobos the benefits of this social structure are clear: they start becoming pregnant several years earlier than chimpanzees and can have more offspring. The difference, researchers suspect, lies in the fact that a female chimp has to cope with the threat of infanticide. Bonobo females, thanks to their social power, no longer have to worry about it.

Alliances, betrayals, deception, trust, jealousy, adultery, motherly affection, suicidal love—it all sounds rather human. When biologists talk about divorce among birds or adultery among mice, these words always bear invisible quotation marks around them. Nevertheless, we humans are animals— males with abundant sperm, females with scarce eggs—and our ancestors were subject to evolution just as much as any pipefish or jacana. Could it be that inclusive fitness, reciprocal altruism, and conflicts between males and females have something to do with the way we act, or even the way we think?

Drop this question at a bar full of biologists and prepare to dodge the flying pint glasses. Why are humans such a tender subject? To understand the murkiness of the matter, we first have to understand where humanity came from.

PART FOUR

Humanity's Place in Evolution and Evolution's Place in Humanity

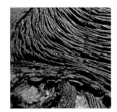

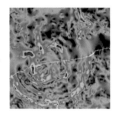

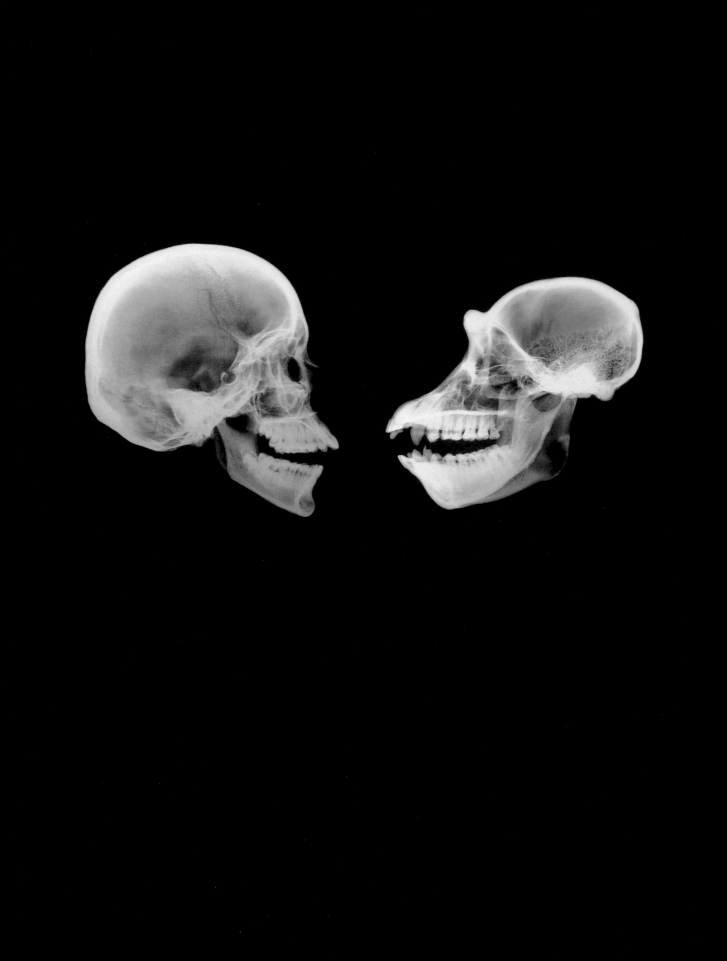

The Gossiping Ape

THE SOCIAL ROOTS
OF HUMAN EVOLUTION

THE MOST OBVIOUS way to distinguish us humans from the millions of other species on Earth is by looking at the things we make. Even from thousands of miles away, our creations would be obvious to an alien naturalist passing by on an interstellar *Beagle:* the swarm of satellites, space stations, and bits of space junk orbiting the planet; the signs we leave on Earth's face, from the Great Wall of China to the constellation of city lights glowing at night; the stream of transmissions—phone calls, cartoon shows, and the rest of our great telecommunication babble—beaming out into space.

Technology may be an obvious hallmark of humanity, but it's not the only one. Compared to other animals, we are a supremely social species. We live in a global network of nations, alliances, tribes, clubs, friendships, corporations, leagues, unions, and secret societies.

Relative to their body size, humans have brains that are more than twice as big as those of chimps. The challenges of hominid social life may have driven the expansion of our brains over the last 5 million years.

A 700,000-year-old hand
axe from Tanzania.

Our social nature would be much harder for our passing alien naturalist to detect, and yet the invisible links that bind us together are no less important to human nature than the highways or cities we may build for ourselves.

When we look back through time at the evolution of our own species, we are in the same plight as the alien naturalist. We can see vestiges of the technology of our ancestors; we can even touch them. As early as 2.5 million years ago, our ancestors were chipping away at stones to create blades for cutting the flesh from carcasses. By 1.5 million years ago, they were making powerful stone hand axes that they may have used not just for flesh-cutting, but for crafting other tools, like digging sticks. By 400,000 years ago, the oldest spears appear, and the technological record becomes more and more dense as it approaches the present day. No other animal has left behind signs of technology in the 4 billion years of life's history on Earth. But while you can wrap your fingers around a million-year-old hand axe, you cannot touch the society of the person who crafted it or the experiences he or she had living in it.

Yet as difficult as it is to glimpse the social evolution of humans, scientists suspect that it was a crucial factor in the rise of our species—perhaps *the* crucial factor. Our chimplike ancestors had chimplike social lives, but 5 million years ago they branched away from other apes and began to explore a new ecological niche on the savannas of East Africa where their social lives became far more complex. Much of what makes humans special—our big brains, our intelligence, even our gift of language and our ability to use tools—may have evolved as a result. At the same time, the competition for mates and the struggle for reproductive success among these hominid ancestors of ours may have left their mark on our psychology, shaping our capacity for love, jealousy, and all our other emotions.

DARWIN'S AFRICAN GUESS

As Darwin was putting together his theory of natural selection, he couldn't help but wonder how humans had come to be. There were no million-year-old hand axes yet known for him to examine; in fact, before the late 1850s there were no recognized fossils of ancient humans whatsoever. He sometimes jotted his thoughts in a notebook, but he did not dare make them public. In 1857, two years before Darwin published *Origin of Species,* Wallace asked him in a letter if he would discuss the origin of mankind in the book. Darwin replied, "I think I shall avoid the whole subject, as so surrounded with prejudices, though I fully admit that it is the highest and most interesting problem for the naturalist."

His silence was purely strategic. Humans must have evolved, like any other animal. But Darwin didn't delve into that ramification of his theory, hoping that he could get a fair hearing. Yet as cautious as Darwin was in writing *Origin of Species,* many of his readers immediately wondered where humans fit into his theory. Making the question all the more pressing, explorers were returning from the jungles of Africa at the time with chimpanzees and gorillas. Huxley and other biologists examined them and showed that they were even more like humans than orangutans were. In 1860, Darwin wrote to Wallace to say that he had changed his mind: he would write an essay on man.

It would take Darwin 11 years to finish it. In the interim he was bogged down by new editions of *Origin of Species* and his book on orchids; a book on the domestication of animals and plants exploded into a two-volume monster; he fell sick for months at a stretch. But through all those distractions, the pressure to speak about human evolution only grew. How could natural selection spontaneously produce human beings in all their wonder, with their ability to speak and reason, to love and explore? Even Wallace gave up. He decided that our oversized brains were far more powerful than necessary—we could easily survive with minds slightly more advanced than an ape's. The creation of humans must, he concluded, be the work of divine intervention.

Darwin did not agree, and in 1871 he finally set forth his argument for human evolution in *The Descent of Man and Selection in Relation to Sex*. It was a hodgepodge of a book. Darwin used a few chapters to introduce readers to the theory of sexual selection, which he thought was responsible for the differences between human races. (Even Darwin came up with a few clunkers.) In a book that was supposed to be the story of human evolution, Darwin spent hundreds of pages detailing how sexual selection might work on other animals. But he also managed to include evidence suggesting that humans had evolved into their current form from apelike ancestors.

By the time Darwin got around to writing *The Descent of Man,* only a few hints of our antiquity had emerged, and they were ambiguous ones at that. In 1856 a miner in the Neander Valley in Germany unearthed pieces of a skeleton, which was dubbed Neanderthal Man. Its brow was massive and low, which raised the question of whether it was a separate species or—as Huxley claimed—was at one extreme of human variation. Other scientists had found not fossils but tools—flints and stone scrapers—alongside the fossils of extinct hyenas in England and France. They spoke of humanity's great antiquity but could say little more.

Because the fossils and tools shed so little light on human evolution, Darwin instead compared humans to great apes. Bone for bone, they are almost identical. As human embryos develop, they pass through virtually identical stages as gorillas or chimps. Only relatively late in their development do they start to diverge, taking on different proportions. These similarities, Darwin argued, were signs that apes and humans descended from some ancient common ancestor. After our ancestors diverged to a branch of their own, they gradually evolved all the traits that make us uniquely human. Since humans are so similar to gorillas and chimpanzees, and gorillas and chimpanzees both live in Africa, Darwin made a guess as to the land of our origins: "It is somewhat more probable that our early progenitors lived on the African continent than elsewhere."

In 1871 Darwin's readers may have thought he was firing a scientific shot in the dark. But 130 years later he has been vindicated by a wealth of evidence. Researchers now know that the similarity between the genes of humans and African apes is just as striking as their anatomy. In 1999 an international group of scientists offered an evolutionary tree of humans (shown on the next page), based on the most extensive study of our genes to date. Humanity forms a little tuft nestled alongside the chimpanzee lineages. Their tree demonstrates that, genetically speaking, we are practically a subspecies of chimp.

By gauging the rate at which our genes mutate, scientists estimate that the last common ancestor of chimpanzees and humans lived 5 million years ago. Since Darwin's day, paleoanthropologists have discovered many fossils of ancient humans, as well as a dozen other human-like species (known as hominids). These fossils show that human evolution was marked by five great transitions. The first, which began about 5 million years ago, gradually pushed our ancestors out onto the African savannas. The second saw the invention of the first stone tools about 2.5 million years ago, and the third came a million years later, as crude blades were transformed into massive hand axes. Half a million years ago, our ancestors went through a fourth transition, mastering fire and becoming more adept at making spears and other tools. And finally, 50,000 years ago, humans began leaving behind signs of truly modern minds—paintings on cave walls, carved jewelry, intricate weapons, and elaborate burials.

The oldest and most chimplike fossil of a hominid was discovered in the early 1990s by a team of scientists working in Ethiopia. There they unearthed a collection of teeth, bits of a skull, and some arm bones dating back 4.4 million years. The fossils were apelike but had some features that were more like humans than chimps. When its mouth closed, some of the skull's upper and lower teeth fit together in a human-like way. Its spine contacted the bottom of its skull, as our own spine does. (In chimpanzees and other apes, the point of

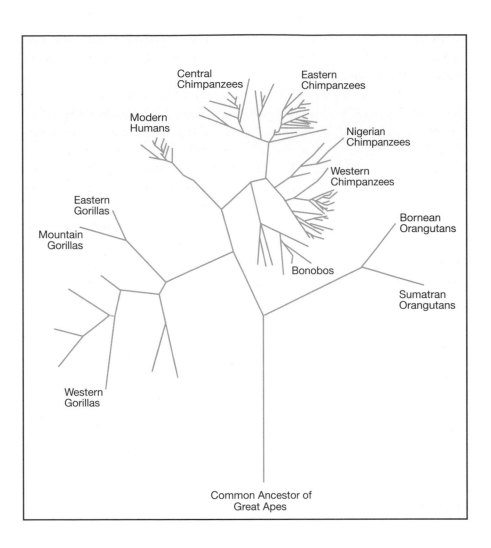

This evolutionary tree, based on DNA, shows how an ancestral ape gave rise to orangutans, gorillas, chimpanzees, and humans. The length of the branches reflects how far the genes of each population of apes have diverged from their relatives. Humans, the tree reveals, are barely distinguishable on a genetic level from bonobos and chimpanzees.

contact is closer to the back of the head.) But at the same time, the Ethiopian creature had some distinctly chimpish traits. It had massive canine teeth, as chimps do, covered by only a thin layer of tooth enamel. It wouldn't have been able to eat much meat or tough plants; it presumably ate only soft fruits and tender leaves, as chimps do today.

We have met this kind of strange mixing of traits before—in the walking whales, the fish with legs and toes, the invertebrates with glimmerings of the vertebrate brain. This Ethiopian creature, known as *Ardipithecus ramidus*, is not a missing link between man and chimp, but it lies on a branch close to the split between our ancestors and theirs.

While *A. ramidus* remains the oldest known hominid, other scientists have found fossils of several other hominid species dating back well over 3 million years, all of them in East Africa. On the shores of Lake Turkana in Kenya, paleoanthropologist Maeve Leakey discovered a 4.2-million-year-old hominid,

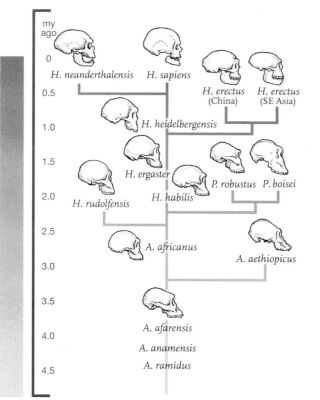

my
ago

0
H. neanderthalensis   *H. sapiens*

0.5
H. erectus (China)   *H. erectus* (SE Asia)

1.0
H. heidelbergensis

1.5
H. ergaster
P. robustus   *P. boisei*

2.0
H. rudolfensis   *H. habilis*

2.5

3.0
A. africanus
A. aethiopicus

3.5

4.0
A. afarensis

A. anamensis

4.5
A. ramidus

By comparing fossils of extinct hominids, researchers have drawn this evolutionary tree of humanity's ancestry. *Ardapithecus ramidus,* near the base of the tree, evolved soon after the hominid lineage branched from that of chimpanzees. Over the course of millions of years, over a dozen species of hominids evolved—many of whom lived side by side. But by 30,000 years ago, only a single hominid species survived—our own.

which she dubbed *Australopithecus anamensis.* In Ethiopia, Kenya, and Tanzania, several teams of scientists have dug up a species called *A. afarensis,* which endured from about 3.9 to 3 million years ago. (This is the most famous early hominid, including in its ranks a specimen named Lucy, the nearly complete skeleton of a female *A. afarensis* discovered by Donald Johanson.) Other fragments of ancient hominids have also been found in the same region of East Africa, and may turn out to be species in their own right.

These early hominids lived during tumultuous times. A cooling global climate was turning the wide carpet of jungles across sub-Saharan Africa into a ragged quilt of patchy forests and open woodlands. Chimpanzees and hominids appear to have adapted to the change in very different ways. The chimps clung to the dense forests that survived the climate change in central and western Africa. The hominids, meanwhile, adapted to the more open habitats of East Africa.

As the climate cooled, the bodies of our ancestors changed. Their toes became less like fingers. Their legs grew longer. They held their heads and backs more upright. Kevin Hunt of Indiana University has proposed that these changes occurred as hominids shifted to a new sort of diet. Earlier hominids may have climbed into trees in their jungle home to find food, much as chimps do today. But as their forest habitat became less dense, Hunt proposes that our ancestors began to gather the fruit that hung from low trees. By standing on two legs, an early hominid could brace itself by holding on to a branch with one arm as it grabbed fruit with the other. These changes in diet also altered the way hominids walked. The first hominids probably walked slowly on all fours, using the knuckles of their hands to support their weight as chimps still do. But with longer legs, hominids began to move around bipedally without the help of knuckles.

Walking upright was one of the biggest changes that our ancestors underwent, but the first bipedal hominids could hardly stroll as we do today. An average human walking at a comfortable pace travels about 3 miles per hour. With their short legs, early hominids would have had to run to match that speed. They were forced to walk more slowly, and as a result they covered only a short distance each day. Early hominids may have walked only from one tree to the next, sometimes picking the lowest fruit while standing on

the ground, and sometimes climbing into the trees using their long arms and curved fingers to grip the branches. (They probably also scrambled off the ground from time to time to get away from saber-toothed cats and other predators.)

As one millennium followed another, hominids spread out across a broader range. New hominid species emerged, leaving their fossils as far north as Chad and as far south as South Africa. And by 2.5 million years ago, they were leaving behind something altogether new in the fossil record: stone tools.

Hominids made tools by banging rocks together to chip off their edges. In the process, they created simple blades, which they could use to chop or scrape. Hominids are not the only apes who make and use tools. Orangutans will strip off branches to probe for honey or termites inside trees. Chimps are even more versatile: they can use sticks as probes; they can also place nuts on a rock and smash them with a second rock, like a blacksmith pounding on an anvil. They can use leaves like sponges to soak up water, or as umbrellas in the rain, or as a dry seat on mud. But the tools that hominids invented 2.5 million years ago were beyond the abilities of their ape relatives.

The limits of chimps were put on display by Nicholas Toth of Indiana University in the early 1990s. He tried training a clever captive bonobo

"Lucy" is the nearly complete skeleton of a 3-million-year-old member of *Australopithecus afarensis* (reconstructed at right with a male *A. afarensis*).

Chimpanzees can use tools; here a chimp digs for termites with a stick. But they are unable to fashion stone tools in the way hominids began to 2.5 million years ago.

named Kanzi to make stone tools. For months Kanzi banged rocks together, but in the end he made no progress. Part of his trouble was that his thumbs didn't have the range of motion of humans and some other hominids. As a result he couldn't deliver the precise blows required to shape a rock. But just as importantly, his brain couldn't master all the variables involved in chipping a stone with another stone—how much force to use, where to hit one stone against the other, and so on. By 2.5 million years ago, however, our ancestors had figured all this out.

As with bipedality, toolmaking may have had its origins in climate changes. Between 3 and 2 million years ago, East and South Africa turned much drier than in the past, and grasslands replaced many of the old woodlands. Hominids evolved an even more upright stance, which may have served as an adaptation for surviving in hot open habitats. The tropical rays of the sun couldn't hit as much of their bodies if they stood, and an upright body could be cooled more by passing breezes. Antelopes and gazelles, well-adapted for the new habitats, spread across the savanna, and some of their fossils show signs of being hacked open or scraped clean of meat. Hominids may have been following these mammals across the savannas, eating their flesh, either by scavenging carcasses left behind by lions and other predators or by scaring the animals off their kills.

When the oldest stone tools were first chipped, there were at least four species of hominids alive in Africa. The most likely candidates for the original toolmakers were the first members of our own genus, *Homo*. The earliest known *Homo* first appear in the fossil record about 2.5 million years ago, around the age of the oldest known stone tools. They're different from other hominids in some striking ways. They have opposable thumbs, and big brains. Judging from the cavities in their skulls, the brains of early *Homo* were 50 percent bigger than the earliest hominids, relative to their body size.

With stone tools, hominids could add much more meat to their diet, even if they didn't have the jaws of a hyena or the claws of a lion. The evolution of big brains raced on, and in a few hundred thousand years, hominid brains were double the size of a chimp's, housed in long-legged bodies that could reach 6 feet in height. All traces of tree climbing were now gone. These hominids, known as *Homo ergaster,* were the first to warrant the title of human beings. Like modern humans, they had a wanderlust, and before long they had left Africa altogether. By 1.7 million years ago *H. ergaster* had reached what is now the Republic of Georgia near the Caspian Sea, where they left behind skulls and tools.

But the tools that these Georgians used—the standard chipped rocks that hominids had used for at least 800,000 years—were about to become obsolete. The hominids that remained in Africa took another technological leap around 1.5 million years ago, inventing hand axes. These new tools took far more skill to make than the earlier models. In order to make a hand axe, a hominid had to flake a rock on both sides, giving it a much sharper edge. Whoever made them didn't just bang rocks until they could be used to cut. They had a particular tool in mind.

Homo ergaster, a 1.7-million-year-old hominid, is considered the earliest species to warrant the term *human.*

The invention of hand axes and other new stone tools allowed the African hominids to fuel their hungry minds. Brain tissue demands 22 times more energy than an equivalent piece of muscle at rest, and hominids now had gigantic brains to feed. Hominids may have used their new tools to get more meat, butchering bigger animals with tougher hides. But if living hunter-gatherers are any indication, these hominids did not survive on meat alone.

Even today, hunter-gatherers with much more sophisticated weapons like poison-tipped arrows do not catch enough game to feed their families. Kristen Hawkes, an anthropologist at the University of Utah, has studied the diet of the Hadza, a group that lives on the East African savannas. While they occasionally eat a gazelle or some other big animal, they depend on roots and tubers for a steady supply of calories. Hawkes has proposed that early *Homo* women began to use the improved stone tools 1.5 million years ago to craft digging sticks with which they could excavate roots.

It was not long after the invention of these new tools that even bigger waves of hominid migration surged out of Africa. Around 1 million years ago, hominids began moving into Asia and Europe, bringing their tools with them. By 800,000 years ago hominids spanned the Old World, from Spain in the west to Indonesia in the east. But these hominids did not move farther than about 50 degrees north—the southern edge of England. It would be hundreds of thousands of years before they would push farther north. As Hawkes has pointed out, 50 degrees marks the line beyond which the weather is too cold for many tubers to grow. If hominids were an army that marched on its stomach, they had to grind to a halt at that barrier.

TOOLS AND ALLIANCES

Without tools, hominids could not have spread so far. But what evolutionary force made it possible for those hominids to make their tools in the first place? Throughout hominid evolution, their brain expanded in fits and starts. As they gained bigger and more complex brains, hominids presumably became able to handle the challenge of making tools. That still leaves a question unresolved: How did these big brains arise?

The answer may have its roots in the social lives of apes and monkeys. Compared to other animals, primates are a particularly social bunch, spending their lives in bands, making alliances, and struggling up the social hierarchy. Primates are keenly aware of the shifting reality of their social world—often even more than of the physical world. Vervet monkeys, for example, are terrified of pythons, but they can't recognize a fresh python track. On the other hand, they can keep track of the genealogy and history of their band. If two vervets get in a fight, their relatives will bear the grudge and may harass one another days later.

In some cases, primates have such a keen awareness of their fellow primates that they can deceive them. Andrew Whiten, a primatologist at St. Andrews University in Scotland, once watched a chacma baboon named Paul sneak up on an adult female named Mel. Mel was digging into the ground to get her hands on a nutritious plant bulb. Paul looked around and saw that there were no other baboons in sight. Suddenly he let out a yell, and within seconds his mother came running. Assuming that Mel was harassing her son, she drove Mel off a small cliff. Paul then took Mel's bulb for himself.

And of all the nonhuman primates, the most deceptive and crafty are our closest relatives, the great apes. "It's as if the apes have been reading Machiavelli," says Whiten. "They're very concerned to climb up the social hierarchy, and make the right allies to enable them to do that. But at the same time if the occasion is right, just as Machiavelli would have advised, they'll deceive those friends and ditch them."

Neuroscientists once assumed a keen social intelligence was nothing special in itself. They thought that the brain was a general-purpose information processor, and it used the same strategies to solve any problem it encountered, social, physical, or otherwise. But evidence now suggests that our brains do not work in a general-purpose way. They appear to be a collection of modules—distinct networks of neurons, each of which is dedicated to solving a particular task.

To see a module at work, look at the optical illusion on this page. Even though this picture consists of three circles with wedges cut out of them, you see a triangle. That's because in the visual centers of your brain, there's a module whose job is to perceive edges on objects, even when those edges aren't completely visible. Instead of a random collection of lines, your brain can recognize them as the boundaries of an object. The signals carried from your eyes to your brain are massaged by many different modules, each dedicated to its own image-processing job, and once they're done, your brain merges the information together into a three-dimensional image of the world you see.

You didn't have to go to school to develop these visual modules; they began to form when you were an embryo and matured as you began using your eyes. Many biologists see them as adaptations created by natural selection—organs as distinct as an elephant's trunk or a bird's beak—which evolved as solutions to problems our ancestors regularly faced. Visual modules may have evolved as our distant primate ancestors struggled to recognize their favorite fruits or navigate through trees. Instead of building up a new picture of the world 60 times a second from scratch, our brains use modules to extract the chunks of information that really matter.

Just as we see with special modules, we may use other modules for perceiving our social world. Simon Baron-Cohen, a psychologist at Cambridge University, has traced the outlines of these social-intelligence modules by studying people with different brain disorders. One group of subjects has a condition known as Williams syndrome. They typically have low IQs between 50 and 70, they may have trouble telling their left from their right, and they can't perform simple addition. And yet people with Williams syndrome often turn out to be gifted musicians and voracious readers. They are fascinated by other people and are remarkably empathetic.

To study the social intelligence of people with Williams syndrome, Baron-Cohen invented a test. He looked through magazines for pictures of particularly expressive faces and cut out a strip that included their eyes. He then showed these strips to people with Williams syndrome and asked them to tell him what the people in the pictures were feeling, based only on their eyes. The answers his subjects offered tended to be the same as those of a control group of normal

Our brains contain many modules for carrying out specialized mental tasks, such as filling in the edges of objects we see. The optical illusion above reveals these edge-building-neurons at work.

adults. Although people with Williams syndrome may be brain-damaged, that damage did not harm their ability to look through the windows of the soul.

Baron-Cohen got a mirror-image result when he performed the same tests on autistic children. Autism does not automatically produce a low IQ; in rare cases autistic people may even be brilliant. But they consistently find it difficult to grasp the rules of society or understand what other people are thinking and feeling. When Baron-Cohen had autistic subjects look at pictures of eyes, they failed miserably at guessing the states of mind of the people in the pictures. Something about their brains keeps them from putting themselves in other people's places.

Baron-Cohen's work may expose the outlines of the modules that give us social intelligence. If these social-intelligence modules are damaged—as in the case of autistic people—other forms of intelligence may escape unharmed. People with Williams syndrome demonstrate that brain damage can harm some forms of intelligence while leaving social intelligence untouched.

The evolution of social intelligence may have been one of the most important factors in the rise of humans, not to mention the evolution of primates in general. The evidence is in the weight of primate brains. Robin Dunbar, a psychologist at the University of Liverpool, has compared the size of brains—particularly the neocortex, the outermost layer of the brain, in which high-level thought processes take place. Some primates, such as lemurs, have a relatively small neocortex for their body size, while other primates, such as baboons and chimps, have big ones. Dunbar discovered a striking pattern: the size of a primate's neocortex is tightly correlated with the average size of the groups in which it lives. The bigger the group, the bigger the neocortex.

When primates live in larger groups, Dunbar concludes, it puts bigger demands on their social intelligence. They need to keep track of their alliances and grudges, of relatives and acquaintances. Mutations that produce a bigger, more powerful neocortex are favored in these species because they make it possible for a primate to enhance its social intelligence. Not surprisingly, primates with a bigger neocortex engage in more deception than primates with smaller ones.

If we humans follow the primate rule—which only makes sense, given that we are primates—then the evolution of social intelligence must have played a crucial part in the development of our extraordinarily big brains.

EVOLVING A THEORY OF MIND

The earliest hominids were distinctly chimplike—in the shape of their body, in the sorts of habitats they made their home, and even in the size of their brain. Their social life was probably chimplike as well, demanding a social intelligence on a par with that of living chimps. To explore that potential connection, scientists have tested chimpanzees to see just how much they understand about their fellow chimps. Does their Machiavellian behavior come from a capacity to understand that other chimpanzees have a mind like their own? Do they have what psychologists would call a "theory of mind"?

Studies of chimpanzees suggest that they have only the rudiments of one. They know, for instance, what their fellow chimps can and cannot see. Brian Hare, a primatologist at Harvard, and his colleagues demonstrated this with a series of experiments with dominant and subordinant chimps. Whenever chimps come into conflict over food, the dominant ones win. In Hare's experiments, a dominant chimp and a subordinate one would be released at the same time into a cage from opposite sides. Hare placed two pieces of fruit in the cage, and the subordinate chimp could tell that the dominant chimp could see only one of the pieces of fruit, because her view of the second one was blocked by a piece of PVC tubing. Knowing this, the subordinate chimp consistently went for the hidden fruit, so as to avoid a confrontation with the dominant chimp over the visible one.

"We're increasingly realizing that chimpanzees do have the beginnings of the elements of a theory of mind," comments Harvard chimp expert Richard Wrangham. "So we now know that a chimp can look at another chimp, see what it sees, and then base its own strategies on what that other chimp has seen. We don't know of any other species, other than humans and chimpanzees, that can do that."

But chimpanzees don't seem to be able to put themselves fully in the minds of other chimps. Daniel Povinelli, a primatologist at the University of Southwestern Louisiana, performed an experiment comparing the social intelligence of chimps with 2-year-old children. He had his subjects gesture to one of two observers to ask for a piece of food—one observer wore a gag, the other a blindfold. The 2-year-old children understood that the blindfolded person could not see the gesture and asked the gagged person instead. The chimps, on the other hand, were just as likely to gesture to the blindfolded person as the gagged person.

Hare's experiment shows that chimpanzees understand some basic facts about seeing—that barriers, for example, can prevent other chimpanzees from seeing things. But Povinelli's experiment suggests that chimpanzees don't understand the full scope of vision—that there is a mind on the other side of those eyes perceiving the images that enter it.

These findings suggest that the common ancestor of chimps and humans could not actually conceive of the other members of its species as having minds of their own, capable of thinking as they did. They did not, in other words, have a theory of mind. Our hominid ancestors must have evolved one only after they split from chimpanzees 5 million years ago.

Andrew Whiten and Robin Dunbar have both argued that hominids began evolving a theory of mind as our ancestors gradually shifted from living in dense jungles to open woodlands and finally to savannas. They began coming in regular contact with big, dangerous predators like lions and leopards, and they could no longer hop into trees for safety. These hominids had to stick together in even bigger numbers than their ancestors had. Life in bigger groups would encourage the evolution of more social intelligence, which required bigger brains. In the process, hominids evolved into mind readers. By looking at the eyes of their fellow hominids, they could tell not only what they could or could not see but what they were thinking. They could read body language and reflect on the past actions of other people. In the process, hominids began to do a better job of deceiving one another, making alliances, and keeping track of one another.

Whiten suspects that once this sort of social evolution got started, it spiraled out of control. Any individual hominid born with a keener theory of mind would be able to deceive the rest of his or her band and ultimately might enjoy more reproductive success. "Now that creates a selection pressure in everybody else to be better at detecting deception," says Whiten. "And really, detecting deception means having a better notion of what's going on in the other's mind. That's mind reading."

Hominid evolution may have become a feedback loop of ever-increasing social intelligence, producing our ever-expanding brains. Ultimately, this evolutionary spiral transformed hominid society itself. It eventually became too hard for a dominant male to enforce a hierarchy in his band because his subordinates had become too clever. Hominid society shifted from a chimplike hierarchy to an egalitarian structure. Each individual used his or her theory of mind to keep track of everyone else, making sure no one cheated the group or tried to dominate it.

Only when hominids began to live in an egalitarian society, Whiten argues, could they fully take advantage of the hunter-gatherer way of life. Men could work together to plan a hunt, and they could leave the women and children behind without being paralyzed by suspicion. Likewise, women could organize expeditions of their own to find tubers and other plants. With tools and cooperation, hominids carved themselves a new ecological niche in the savanna.

"A theory of mind makes us as sublime as we are because we can feel for others so much," says Whiten. "At the same time, it allows us to be that much more sneaky than any other species on the planet."

PLEISTOCENE PASSIONS

For well over a million years, our hominid ancestors survived on the African savannas, scavenging or hunting meat and gathering plants. It was during that vast preamble to modern life that our ancestors first came to depend on tools for their existence and, according to Dunbar, the first time that they lived in complex societies, understanding their fellow hominids with a theory of mind.

In that world, natural selection would have favored certain behaviors and abilities. Some of them were basic survival skills—the ability to make a stone tool, perhaps, or powerful eyesight for spotting game at great distances. Other behaviors helped them find mates. The same powerful evolutionary forces that created the peacock's tail and infanticidal lions were presumably at work on our Pleistocene ancestors.

If hominid behavior was sculpted by the demands of sex and family, are we still ruled by those Pleistocene passions? Few questions in evolution have triggered so much debate, so much anger and rancor. A number of scientists argue that we are ruled by these passions; they even go so far as to claim that we can dissect them and discover their original adaptive value. Opponents say that human behavior has been unmoored from its evolutionary pier: any attempt to pin an emotion or custom in living humans to some adaptation on the African savanna a million years ago is pure scientific hubris. This debate is about more than nature versus nurture: it gets to the very heart of how we should go about trying to understand the evolutionary past.

Edward O. Wilson of Harvard opened up this psychological can of worms in his landmark 1975 book *Sociobiology*. Most of his book surveyed the great success that scientists were having in using evolutionary theory to understand the social lives of animals. What at first seemed like evolutionary paradoxes turned out, after some careful study, to make sense. Sterile worker ants can help pass on their own genes because they are closely related. When a male lion takes over a pride and kills the cubs, he brings their mothers into estrus so that they can bear his own offspring.

"Let us now consider man," Wilson wrote in the opening of the last chapter of *Sociobiology*, "as though we were zoologists from another planet completing a catalog of social species on Earth." Human beings are primates living in big societies. They descend from hominids who probably evolved reciprocal altruism and food sharing. Barter, exchange, and favors became a crucial part of early human society, just as did deception and subterfuge. Males and females had specific roles in these early societies, the males killing game and the females raising children and gathering plants. Sexual selection, Wilson speculated, must have helped drive human evolution. "Aggressiveness was constrained and the old forms of primate dominance replaced by complex social skills," he wrote. "Young males found it profitable to fit into the group, controlling their sexuality and aggression and awaiting their turn at leadership."

By trying to turn psychology into evolutionary biology, Wilson created a sensation. *Sociobiology* became a best-seller and inspired a front-page article in *The New York Times*. Human behavior, the newspaper declared, "may be as much a product of evolution as is the structure of the hand or the size of the brain." But Wilson inspired a lot of antipathy as well. Much of it came from the academic left, which accused Wilson of using science to justify the status quo, to apologize for all the inequalities of modern life. Protestors stormed scientific conferences where Wilson spoke, chanted their anti-sociobiology slogans; once they even dumped water on his head.

More restrained critics argued that humans didn't fit into sociobiology's straitjacket. Inclusive fitness might account for how a sterile ant cares for a queen's offspring but not for the many forms that human families can take. Consider the Nuer people of the Sudan. They treat barren women as men and allow them to marry other women who are impregnated by other men. The children that these wives give birth to are treated by the Nuer as the barren woman's. Such a family comes into being out of cultural traditions, not out of a genetic imperative.

Today many anthropologists still raise this sort of objection to sociobiology. But beginning in the 1980s, some dissenters found that their own data was consistent with it. Kristen Hawkes was one of them. She launched her anthropological career among the Binumarien people of the New Guinea highlands, studying the ways in which kinship influenced the way they behaved. She studied the categories that the Binumarien used to classify relatives, for example, and the ways in which relatives helped one another. Only when she returned to the United States did Hawkes start to think seriously about the possibility that evolution might influence human culture, and she decided to use her data on the Binumarien to test sociobiology.

If Wilson was right, the Binumarien should make clear distinctions between people of different levels of genetic relatedness. After all, to boost their inclusive fitness, they should help a brother before they help a cousin. But Hawkes found that the Binumarien language did not make sociobiological distinctions possible. Two men that Westerners would call cousins, for example, are called brothers among Binumarien. (Western society has its own slippery vocabulary: an uncle may be your parent's brother—and share on average a quarter of your genes—or the husband of your aunt, sharing no genes at all.)

It might seem as if Hawkes had found a case against sociobiology—that the relative closeness of relatives didn't make much difference to how a Binumarien thought of them. But underneath the surface of their family terms, it appears that inclusive fitness is at work. For food, the Binumarien raise pigs and grow crops such as sweet potatoes in their gardens. All adults in the community have gardens of their own, so any gardening help they might give to someone else means that they have less time to tend their own garden. Regardless of the terms they may use for their kin, Hawkes found that they spent more time working in the gardens of people who are genetically closer to them than those who were more distantly related.

As some anthropologists warmed to sociobiology, sociobiology itself grew more nuanced in the 1980s. Its advocates no longer argued that genes created behavior in a deterministic way; instead, they showed how genes regulate the way in which animals make unconscious decisions about mating and raising their offspring. These adaptive strategies—"decision rules," as Stephen Emlen has called them—let animals behave in different ways in different circumstances.

Emlen himself has shown just how complex decision rules can be among the bee-eater birds he studies in Kenya. A young female bee-eater can choose to breed at the start of her first breeding season, help a breeding pair at her nest burrow, or sit out the season altogether. If an unpaired dominant older male courts her, she almost always leaves her family and home territory to join him in nesting in a different part of the colony, particularly if he has a group of helpers to assist in feeding the offspring they will produce. But if young subordinate males are the only mates around for the choosing, the young female bee-eater will refuse to join them, because young males come with few helpers and will be harassed by their father to come back to their nest to raise their younger siblings.

Emlen has shown that the forces of evolution can create a subtle, flexible strategy in a humble bird, an animal with a tiny brain hardly capable of much thought. Why couldn't equally complex—and unconscious—decision rules have evolved in hominids as well?

The new sociobiologists also began to focus on the evolutionary pressures our ancestors experienced on the African savanna. For more than a million years our ancestors were living in the same place—African grasslands—and survived there in the same way—as small bands of hunter-gatherers. They killed or scavenged game with stone tools, and the rest of their food they got by digging up tubers and collecting other plants. They had to find mates and

Bee-eaters have to make many choices about whom to mate with and how to raise their offspring. Researchers have found that these unconscious decisions are influenced by a drive to replicate their genes. Some evolutionary biologists claim that the same can be said about humans.

raise children under the same conditions. Over time both their bodies and their minds adapted to this way of life. As part of that adaptation, our ancestors might have evolved mental modules well-suited to life on the savanna. They used the modules like the blades on a Swiss army knife, each well-adapted for its own job in a world of hunter-gatherers.

For many contemporary sociobiologists, those days on the savanna are gone, but in terms of evolution they're not forgotten. Industrialized civilization has existed only for a couple of centuries, and it has been a matter of only a few thousand years since humans shifted from hunting and gathering to agriculture. That's a fraction of 1 percent of the entire span of hominid evolution. Although our lives may now be very different, there hasn't been enough time for natural selection to change our psychology very much.

Looking at ourselves this way may help us understand why we are better at some mental tasks than at others. This approach to the mind, called evolutionary psychology, is championed by the husband-and-wife team of Leda Cosmides and John Tooby, a psychologist and an anthropologist, respectively, at the University of California at Santa Barbara. Cosmides and Tooby have used this approach to interpret some peculiar results from psychological experiments. In one case, Cosmides updated a classic psychological experiment in logic called the Wason Test. Imagine that someone lays four cards in front of you. They read Z, 3, E, and 4. You are told that there are also symbols on the other sides of the cards, and that there's a general rule that a vowel card always has an even number on the other side. Which card or cards do you need to turn over to see if this rule holds?

The answer is that you need to check E and 3. (You don't need to check the 4 card, because even if it has a consonant on the other side, the rule isn't broken.) People generally do very badly with the Wason Test. But Cosmides showed that their success rises dramatically if it is translated into social terms. Say someone sets down another four cards in front of you that read 18, Coke, 25, and Rheingold. The cards have the ages of people in a bar on one side and the drinks they have ordered on the other. Which cards would you have to turn over to find out if any of them were breaking the law by drinking alcohol under the age of 21?

The correct answer is 18 and Rheingold. Almost everyone who takes this form of the test gets it right, even though it is based on precisely the same underlying logic as the original Wason Test. Cosmides and Tooby argue that we are so good at this form of the test because we are skilled at keeping track of social complexities. Our ancestors evolved a module for sensing cheaters, because in a band of hunter-gatherers who had to share meat, tools, and other valuable items, people would benefit from being able to figure out if any individual was trying to take advantage of the rest.

Evolutionary psychologists argue that our ancestors would have also needed particularly powerful modules for our behavior with the opposite sex and with our children. It's with them, after all, that our ultimate reproductive success rests. In this way, we are no different from other animals. A peahen depends on evolved decision rules to choose a mate, but she doesn't actually work out an equation of the costs and benefits of her choice in her head. Instead, the things she sees, smells, and experiences trigger her to do certain things. People, likewise, don't fall in love according to cold, rational calculations of genetic benefits. But according to evolutionary psychologists, feelings like love and lust and jealousy are triggered by adaptations in our brain.

What, for example, do people find attractive in others? That's the first step, after all, in choosing a mate. Many animals show a strong liking for symmetry, possibly because it's a reliable clue to good health. It may also be a good clue in humans as well. In one study, David Waynforth of the University of Mexico measured the symmetry of men's faces in Belize and found that the men with more asymmetrical faces were more likely to suffer from serious diseases.

The symmetry or asymmetry of a face is often too subtle for people to consciously recognize, but it does influence how we judge people's attractiveness. David Perrett of St. Andrews University used a computer to manipulate photographs of faces, creating ones that were more or less symmetrical. He then showed his subjects a group of original pictures and their doctored versions, asking them to rate how attractive they were. Perrett's subjects tended to pick the symmetrical over the asymmetrical faces.

Faces may not have been the only clues that our Pleistocene ancestors used to choose potential mates. When girls reach puberty, they take on certain features that signal that they are becoming fertile. Their hips widen because they

Men tend to find a hip-to-waist ratio between 60 and 70 percent to be most attractive in women (shown here as the third and fourth from the left). Is this a cultural fluke or an adaptation for finding good mates that evolved in the brains of hominids a million years ago?

are putting on fat that can act as a reserve of energy during pregnancy. Fertile women have waist measurements that are between 67 percent and 80 percent of their hips. Men, children, and women who are postmenopausal have more closely proportioned waists that are 80 percent to 95 percent the size of their hips. A low hip-to-waist ratio correlates with youth, health, and fertility.

We appear to have a finely tuned awareness of these proportions. Devendra Singh, a psychologist at the University of Texas, has surveyed men and women of different ages and cultures, showing them pictures of women with different hip-to-waist ratios. As a rule, a ratio of 60 to 70 percent is considered attractive. Even as tastes in women's appearance change, this ratio holds up. Singh has found that while *Playboy* models and pageant queens have gotten thinner over the years, their hip-to-waist ratio has remained the same. This enduring taste may have emerged more than a million years ago as a way for men to choose mates who were more likely to bear children.

NOT SO HAPPILY EVER AFTER

Male and female animals have some inevitable conflicts of interest. A male can theoretically have thousands of offspring in his lifetime, while a female has far fewer opportunities and has to use more energy for each one. Our Pleistocene grandmothers and grandfathers would have experienced the same evolutionary conflict. While the men didn't necessarily have to pay a heavy cost for a child, the women faced some inescapable burdens. They had to be pregnant for nine months, risked death from various complications along the way, and burned tens of thousands of extra calories as they nursed their child.

As a result of that conflict, men and women evolved an attraction to different kinds of qualities in their mates—qualities we still look for today. David Buss, a psychologist at the University of Texas, has conducted a long-term survey of thousands of men and women from 37 different cultures, from Hawaii to Nigeria, asking them to rank the qualities that are most important in choosing someone to date or marry. Buss finds that in general, women tend to prefer older men and men tend to prefer younger women. Men ranked physical attractiveness higher than women did, while women valued good earning potential in potential husbands. Buss argues that these universal patterns expose adaptations that evolved in our Pleistocene ancestors: women were attracted to men who would be able to give support for their children, while men were more interested in finding a fertile, healthy mate.

Evolutionary psychologists argue that these conflicts of interest make men and women behave differently. Surveys confirm what many people already suspect, that men are far more willing to have sex than women. Men express a desire for four times as many sex partners in their lifetime; they have more than twice as many sexual fantasies. Men let less time elapse before seeking sex with a new partner, and they're more willing to consent to sex with a total stranger.

But just because Pleistocene women evolved to be choosy about their mates doesn't mean that they were perfectly faithful. As we saw in the last chapter, the animal world is rife with females who cheat on their partners. Among monogamous birds, for example, females will sometimes mate with a visiting male, and her cuckolded partner has to raise the chicks. A female bird puts herself at risk when she cheats on her partner, because he may abandon the nest. But the rewards may justify the danger, since she may be able to find a male with better genes than her partner to father her chicks. Pleistocene women might have evolved some similar decision rules about infidelity, which women today have inherited.

Ian Penton-Voak of St. Andrews University has surveyed women about what kind of face they find attractive on a man. He used a computer to generate "feminized" and "masculinized" male faces, and found that when women are ovulating they prefer masculine faces. The features that make a face masculine—a brow ridge, a jutting jaw, strong cheekbones, for example—may act like a peacock's tail, advertising a man's good genes. They are produced by testosterone, and testosterone depresses a man's immune system. That trade-off may make a masculine face a costly display—one that can be used only by a man with a strong immune system.

Penton-Voak suggests that a woman's keener attraction to a masculine man when she's most likely to conceive may be an adaptation for snagging good genes for her children. When she is ovulating, a woman may be more likely to have an affair with such a man, but during the rest of her menstrual cycle she may become more interested in the man who is helping to raise her children.

With these sorts of conflicts of interests driving the evolution of our ancestors, some of our uglier emotions may actually have originated as useful adaptations. David Buss has proposed that jealousy, far from being a pathology, was one such mechanism. There's no obvious signal that lets people know that their mate is cheating on them. Men cannot even tell when women are ovulating; unlike other primates, women do not get genital swellings. Under these uncertain conditions, jealousy makes ample evolutionary sense according to Buss. A "jealousy module" in the brain could let a person stay alert to the subtlest clues of betrayal that a purely rational mind might dismiss. ("Is that a new cologne I smell?") If those signs build up to a certain threshold, the jealousy module would trigger a reaction that could avert the threat—or cut the person's losses.

Buss marshals a number of experiments and surveys as evidence for the adaptiveness of jealousy. If you hook electrodes to a man's forehead, you can

In *The Cruel Sister* by John Faed (1851) a woman suspects her lover is attracted to her sister. According to some psychologists her jealousy is actually an evolutionary adaptation.

measure the stress he experiences when you ask him to think of his romantic partner cheating on him. Men experience more stress thinking about their partners having sex with another man than thinking about them becoming emotionally attached to someone else. (The thought of sexual betrayal makes their heart beat five extra beats a minute, on a par with the effects of drinking three cups of coffee.)

Women, on the other hand, tend to show the opposite reaction, experiencing more stress at the thought of emotional abandonment. Surveys have shown a similar pattern not only in the United States but in Europe, Korea, and Japan. For men, Buss concludes, sexual betrayal is a bigger threat to their reproductive success; for women, emotional betrayal may signal that a man will abandon her and no longer provide for her children.

If Buss is right, evolutionary psychology offers a better way to cope with jealousy than conventional psychology. Therapists typically treat jealousy as something unnatural, that can be eliminated by boosting self-esteem or desensitizing patients to the thought of their spouse cheating on them. Buss does not condone the ugly side of jealousy—the violence of wife beaters and stalkers—but he argues that pretending jealousy can simply go away is pointless. Instead, he suggests, people should use jealousy to strengthen relationships rather than to destroy them. A flash of jealousy can prevent us from taking a relationship for granted.

The psychological adaptations of our ancestors do not doom us to unhappiness, Buss and other evolutionary psychologists argue; we simply have to acknowledge their reality and work around them. Today, for example, stepparents are expected to treat their stepchildren no differently from biological children. That's an unrealistic expectation according to evolutionary psychologists. They argue that parental love, which encourages us to make enormous sacrifices for our children, is yet another adaptation designed for ensuring the survival of our genes. If that's true, then stepparents should have a much harder time feeling the full depth of parental love for children who are not their own.

There are some pretty chilling statistics to back up this hypothesis. In a conflict between stepparents and stepchildren, there isn't a biological bond to reduce the tension and conflicts are thus more likely to spiral out of control. It turns out that being a stepchild is the strongest risk factor for child abuse yet found. And a child is 40 to 100 times more likely to be killed by a stepparent than by a biological parent. Stepparents are not inherently evil; they simply haven't developed the same depths of patience and tolerance that parents have. And this, evolutionary psychologists argue, points a way to reduce the risks of conflict: stepparents need to be aware that they have to overcome some obstacles to having a happy family that a biological parent doesn't encounter.

MODULE OR MIRAGE?

The new generation of sociobiologists is attracting critics of its own—including a number of evolutionary biologists. They argue that the sociobiologists are too eager to draw certain conclusions from their data, and that in some cases they are not understanding how evolution actually works.

Take, for example, a troublesome book published in 2000 called *A Natural History of Rape*. Two biologists, Randy Thornhill and Craig Palmer, proposed that rape is an adaptation—a way for men who would otherwise have little access to women to increase their reproductive success. Forced sex is not unique to humans; it has been documented among certain species of mammals, birds, insects, and other animals. Thornhill himself has shown that it is a regular part of the scorpion fly's mating strategies. Some scorpion fly males woo females by hoarding dead insects that they like to eat, driving away other scorpion fly males who try to steal the carcasses. Other males secrete saliva onto leaves and wait for females to come by and eat it. And others simply grab females and force them to copulate.

Thornhill found that the biggest scorpion flies were those that hoarded dead insects and attracted the most females. Medium-sized scorpion flies made do with their salivary gifts, which won them fewer mates, and the smallest males attacked. But each scorpion fly can use any of these strategies if the conditions are right. If the biggest males disappear, the medium males can hoard insects and the small ones start drooling.

Thornhill and Palmer argue that our ancestors might have incorporated rape into their sexual strategies as well, as a tactic to use when other means fail. They point to evidence that rape victims tend to be in prime reproductive years—suggesting that reproduction is at the top of the rapist's unconscious agenda. Female rape victims of reproductive age fight back more against their attackers than women of other ages because, Thornhill and Palmer claimed, they have more to lose in terms of reproduction. Thornhill and Palmer also claim that surveys show reproductive-aged women are more traumatized by the experience than other women. They are "mourning" the loss of their ability to choose their mate through normal courtship.

A Natural History of Rape was the subject of a scathing review in the journal *Nature*. Two evolutionary biologists—Jerry Coyne of the University of Chicago and Andrew Berry of Harvard—picked apart the evidence in the book. Girls under 11—too young to reproduce—made up only 15 percent of the population but 29 percent of the victims in a 1992 survey, a percentage far higher than you'd expect according to the book's hypothesis. The authors claimed that this number was so high because American girls are experiencing their first menstrual period at earlier ages than in previous generations, which "contributes to the enhanced sexual attractiveness of some females under 12." Coyne and Berry weren't impressed. "In the end," they charged, "the hopelessness of this special pleading merely draws attention to the failure of the data to support the authors' hypothesis."

And the fact that reproductive-age women fight back says nothing about evolution: they are also much stronger than little girls and old women. "In exclusively championing their preferred explanation of a phenomenon, even when it is less plausible than alternatives, the authors reveal their true colours. *A Natural History of Rape* is advocacy, not science," Coyne and Berry wrote. "In keeping with the traditions established early in the evolution of sociobiology, Thornhill and Palmer's evidence comes down to a series of untestable 'just-so stories.'"

Coyne and Berry were referring to the title of Rudyard Kipling's 1902 book of children's tales, which tell how the leopard got its spots, the camel its hump, and the rhinoceros its skin. The irritation that Coyne and Berry express about evolutionary psychology is common among biologists. They know just how easy it is to make up a story about the evolution of adaptations, and how hard it is to figure out what anything in nature is really for.

To document real adaptations, biologists use every tool they can possibly find, testing for every possible alternative explanation they can think of. If they can run experiments, they will. When an adaptation—say, deep tubes for holding nectar in flowers—is found on many different species, scientists construct their evolutionary tree and trace the rise of the adaptation from species to species.

The human brain is far more complex than a flower, and researchers have fewer ways to study its evolution. Chimps and other apes can offer a glimpse as to what our ancestors may have been like 5 million years ago, but after that, we evolved in a unique direction. We cannot put 100 *Homo erectus* in some fenced-off compound and run experiments to see who will be attracted to whom.

Instead, evolutionary psychologists often rely on surveys. But their samples, usually a few dozen American undergraduates—mostly white, mostly affluent—can hardly be expected to represent the universal human condition. Some evolutionary psychologists appreciate this problem and try to replicate their results in other countries. But even then they may be too eager to jump to cosmic conclusions. In his book *The Dangerous Passion*, David Buss writes: "People from the United States and Germany give roughly equivalent responses, revealing a large sex difference in the desire for love to accompany sex—a desire that transcends cultures." Compared to the Binumarien of New Guinea or African pygmies, the differences between Americans and Germans are hardly transcendent.

Any particular human behavior may be created or shaped by culture, and even if it has some genetic basis, it may not actually be an adaptation at all. This has been the chief complaint of Stephen Jay Gould, who has been a critic of sociobiology ever since the publication of Wilson's book. Like Coyne and Berry, he sees evolutionary psychologists falling victim to a trap that all biologists have to take care to avoid. An eager search for an adaptational explanation, Gould argues, may blind biologists to the fact that they are dealing with an exaptation—something that has been borrowed from its original function for a new one. Birds now use feathers for flight, but feathers first appeared on dinosaurs that couldn't fly. They probably used them for insulation or as sexual displays to other dinosaurs.

Gould even claims that some things that seem like adaptations may have come into existence for no particular function at all. In a classic 1979 paper, Gould and a fellow Harvard biologist, Richard Lewontin, explained how this could happen with an analogy: the domed roof of the basilica of Saint Mark's in Venice. The dome sits on four arches that are joined at right angles. Because the tops of the arches are rounded, there is a triangular space at each corner. Three centuries after the dome was built, these spaces—known as spandrels or pendentives—were covered with mosaics.

It would be absurd to say that the architects designed the spandrels so that they could contain triangular mosaics. It would be absurd to say that the

The domed roof of the basilica of Saint Mark's in Venice has become a metaphor in the sociobiology debate. The triangular spaces between the arches beneath the dome are nothing more than a by-product of the overall design of the building. Critics of sociobiology argue that many features of the human brain may also be evolutionary by-products rather than the direct result of natural selection.

spaces were designed for anything at all. If you want to put a dome on four arches, spandrels are automatically part of the deal. You may later put the spandrels to some use, but that use has nothing to do with its basic design.

Evolution deals in spandrels as well, Gould and Lewontin argued. For a simple example, consider snail shells. All snails grow their shells around an axis, with the result that an empty column forms in the middle. In some snail species, the column fills with minerals, but in many the column remains empty. A few species use their open column as a chamber for brooding their eggs. Now, if a biologist was in high storytelling spirits, he or she might say that this chamber is an adaptation that evolved for brooding eggs, and might praise the cleverness of its design by pointing to how it is lodged at the center of the shell. But the fact is that the column has no adaptive function at all. It's just a matter of geometry.

Gould accuses evolutionary psychologists of mistaking spandrels in the human brain for precise adaptations. He is perfectly happy to grant that the human brain got larger as it adapted to life on the African savannas. But that increased size and complexity gave our ancestors a flexibility that allowed them to figure out how to kill a Cape buffalo or determine when a tuber was in season. It could be reconfigured to read, write, or fly planes. But we do not carry any of these particular abilities in some hardwired form in our brains. "The human brain must be bursting with spandrels that are essential to human nature and vital to our self-understanding but that arose as nonadaptations, and are therefore outside the compass of evolutionary psychology," Gould declares.

The debate over evolutionary psychology won't be resolved any time soon. It is a vital matter that gets at the very heart of human nature, and just how powerful an effect natural selection can have on it. But it can be rancorous and sometimes get mean-spirited. Evolutionary psychologists sometimes insinuate that their critics are dewy-eyed utopians, and their critics attack evolutionary psychologists as rabid conservatives who want to pretend that capitalism and sexism are hardwired into our brains. These sorts of insults are not just beside the point; very often they're flat-out wrong. Robert Trivers, who first came up with the idea of reciprocal altruism, is not a conservative; he has described himself as a liberal who was happy to find that his research suggested a biological basis for fairness and justice. And the anthropologist Sarah Blaffer Hrdy, who first showed how significant infanticide can be in animal societies, uses her work in sociobiology to offer a feminist perspective on evolution: that females are not coy, passive creatures as once thought, but active contestants in the evolutionary arena.

As hard as it may sometimes get, it's important to stay focused on the science, or the lack thereof, in evolutionary psychology. The weight of the scientific evidence will ultimately determine whether it stands or falls.

TOWARD LANGUAGE

The glacial monotony of hominid life began to break up about half a million years ago. The tools that humans left behind started to show signs of change. Instead of hacking a stone into a single axe, humans learned how to make a number of blades from a single rock. A hand axe made by a Kenyan 700,000 years ago didn't look much different than one made in China or Europe. But starting 500,000 years ago, regional styles emerged. New sorts of technology became more common. Humans learned how to make javelin-like spears, and they learned how to make reliable fires. And as in the past, the rise of new tools was reflected in the expansion of human brains. For about the next 400,000 years, human brains would grow at an extraordinary rate, until 100,000 years ago, when they reached their present size.

According to Robin Dunbar's work on primate brains, this expansion must have occurred as humans lived in bigger and bigger social groups. Judging from the size of fossil skulls, Dunbar estimates that the earliest hominids, such as *Australopithecus afarensis* 3 million years ago, formed groups of around 55. Early species of *Homo* living 2 million years ago would have hung together in bands of 80 individuals. By a million years ago, *Homo erectus* groups had cracked 100, and by 100,000 years ago, when human brains had reached our own neocortex size, they were congregating in gaggles of 150.

The average size of the human neocortex hasn't changed since then, and Dunbar sees a lot of evidence that our biggest significant social groups have remained at 150 people. Clans in hunter-gatherer tribes in New Guinea average 150 people. The Hutterites, a group of fundamentalist Christians who live communally on farms, limit the size of their farming communities to 150, forming new ones if the group gets too big. Around the world, the average size of an army company is 150. "I think on average there are 150 people that each of us knows well and knows warmly," Dunbar claims. "We understand how they tick. We know about their history and how they relate to us."

As hominid bands expanded, their complexity grew as well. And once they crossed a certain threshold, Dunbar argues, the old ways in which primates interacted no longer worked. One of the most important ways that primate allies show their affection to each other is by grooming. Grooming not only gets rid of lice and other skin parasites, but it also is soothing. Primates turn grooming into a social currency that they can use to buy the favor of other primates. But grooming takes a lot of time, and the larger the group size, the more time primates spend grooming one another. Gelada baboons, for example, live on the savannas of Ethiopia in groups that average 110, and they have to spend 20 percent of their day grooming one another.

The size of hominid brains suggests that their group size reached 150 by 100,000 years ago, and at that point grooming became an impractical tool.

"You simply cannot fit enough grooming into the working day," says Dunbar. "If we had to bond our groups of 150 the way primates do, by grooming alone, we would have to spend about 40 or 45 percent of our total daytime in grooming. It would be just wonderful, because grooming makes you feel very warm and friendly toward the world, but it's impractical. If you have to get out there and find food on the savannas, you don't have that amount of time available."

Hominids needed a better way to bond. Dunbar thinks that better way was language.

Working out the origin of language remains one of the biggest challenges in evolutionary biology. Speech cannot turn to stone, so it leaves no direct record of its existence. Before the 1960s, most linguists didn't even think that language was, strictly speaking, a product of evolution. They thought that it was just a cultural artifact that humans invented at some point in their history, just as they invented canoes or square dances.

One of the reasons for this notion was the way linguists thought the brain produced language. Assuming that the brain was a general-purpose information processor, they concluded that babies learned to speak simply by using their brains to find the meaning of the words they heard. But Noam Chomsky, a linguist at the Massachusetts Institute of Technology, took the opposite stance: babies were born with the underlying rules of grammar already hardwired into their brains. How else, Chomsky asked, could one explain the fact that all languages on Earth share certain grammatical patterns, such as subjects and verbs? How else could a baby master the complexities of language in only three years? Words are as arbitrary as dates in history, and yet no one would expect a 3-year-old to memorize a time line of the battles of the Peloponnesian War. Not only do children learn individual words, but they quickly use the words they hear to discover grammatical rules. Their brains, Chomsky argued, must already be primed for learning language.

Research since the 1960s suggests that the human brain has special language modules, just as it has modules for seeing edges or for social intelligence. The human brain uses them for storing rules of grammar, syntax, and semantics—all the ingredients that give language its sense and complexity. Linguists can see the language organ at work in the mistakes that children make as they learn how to speak. They may use the standard rules for making plurals or past-tense verbs to create words that don't exist, such as *tooths, mouses,* and *holded*. Young children have the capacity to set up rules of grammar in their brains, but they are still having trouble overriding them with rote memorization of irregular words.

More evidence comes from certain types of brain damage that rob people of the ability to use language, or even just components of it. Some people have trouble only with proper names or words for animals. A team of British scientists studied a man with a rich vocabulary of nouns, including words like *sextant, centaur,* and *King Canute,* but who could use only three verbs: *have, make,* and *be.* In each case, a particular language module has been damaged, leaving the rest of the brain intact.

Obviously, 3-year-old children do not automatically burst into Shakespearean verse. They need to be immersed in a sea of words as their brains develop so that their inner rules of grammar can wrap themselves around a particular language. But the "language instinct" (a phrase coined by linguist Steven Pinker of MIT) is so strong that children can create new languages on their own. In 1986 Judy Kegl, a linguist at the University of Southern Maine, had the chance to watch one come into existence.

That year Kegl had gone to Nicaragua to visit schools for deaf children. The Nicaraguan government had organized several schools in the early 1980s, but the students were struggling. The children arrived knowing only a few crude gestures that they had developed on their own with their parents. Their teachers didn't teach them full-blown sign language, but only tried to use "finger spelling," in which different shapes represent each letter in a word. The finger spelling was supposed to help the students make the transition to speaking the words, but since they had no idea what the teachers were trying to teach them, the entire project failed miserably.

The teachers noticed that even as the children struggled to communicate with them, they had no trouble communicating with one another. They were no longer using the paltry collection of gestures that they had brought from their homes, but a rich, new system that the teachers could not understand. Kegl was invited to come to the schools and help the teachers understand what was happening.

The teenagers at the secondary school, she discovered, used a pidgin cobbled together out of makeshift gestures that they all shared. But the younger children at the primary school were doing something far more sophisticated. Kegl was shocked to see them signing to one another rapid-fire, with the sort of rhythm and consistency that reflected a real sign language, complete with rules of grammar. The younger the children, the more fluent they were. "You could see just by the way the signs were orchestrated and structured that

A girl speaks Nicaraguan sign language, a unique language that was invented by deaf school-children in the 1980s.

something more was going on there," says Kegl. "It became clear that I was there in the early stages of the emergence of a language."

For the first few years, Kegl struggled to decode the language, sometimes eliciting signs or sentences from the children, and sometimes simply watching them tell long narratives. In 1990 she and the children began to watch cartoons, and she would have them tell her what was happening. The cartoons became her Rosetta stone.

Kegl discovered that the children's signs were elegant, clever, and evocative. In the teenage pidgin, the word for *speak* was opening and closing four fingers and a thumb in front of the mouth. The children had taken this mimicry and enhanced it: they opened their fingers at the position of the speaker and closed them at the position of the person being spoken to. They had also invented a way to use prepositions like verbs. Where an English speaker would say, "The cup is on the table," a Nicaraguan signer would say something like, "Table cup ons." While this may seem weird to an English speaker, other languages such as Navajo make regular use of it.

In the years since she first came to Nicaragua, Kegl has worked with the deaf community to put together a dictionary of their language, which now contains more than 1,600 words. In the meantime she has also put together a theory for its origin. Children came to the schools with nothing more than their simple gestures. They pooled them together into a common set and then crafted that into the pidgin that the teenagers used. Younger children then came to the school with brains primed for learning language, seized on the gestures of the older children, and endowed them with grammar. These young children abruptly produced a language that from the beginning was as complex and complete as any spoken language. And once this true language had emerged, new experiences led to the creation of new words.

"What happens," Kegl says, "is that these gestures become gradually richer and richer and more varied. But we can't see the leap between them and the first signs of language because the grammar is inside the child."

If the grammar is indeed inside the child—if, in other words, its rules are hardwired into our brains—then evolution must have had a hand in its wiring. But that raises a difficult question: How could natural selection shape language in all its complexity? Scientists can't go back in time to watch language come into being, but recently they've discovered some tantalizing clues to its evolution by modeling it on a computer. They've found that just as legs or eyes could have evolved incrementally, language may have built up its complexity step by step.

Martin Nowak and his colleagues at the Institute for Advanced Study in Princeton designed a mathematical model of language evolution based on a few reasonable assumptions. One is that mutations that let an animal communicate more clearly may raise its reproductive fitness. Vervet monkeys, for example, have a set of distinct sounds that they use to alarm their fellow vervets about birds and snakes and other threats. Being able to tell the difference between those calls can make the difference between life and death. If a vervet mistakes a snake call for a bird call, it may rush to the ground, only to be devoured by a waiting python. Another assumption Nowak makes is that a bigger vocabulary—if it can be properly communicated—also brings an evolutionary advantage. A vervet that can understand both a bird alarm and a snake alarm will have better odds of survival than a vervet that has room in its brain for only one.

In Nowak's model, individuals were endowed with a simple, vervet-like communication system. Their vocabulary consisted of a collection of sounds, each of which corresponded to some particular thing out in the real world. As the individuals reproduced, mutations cropped up in their offspring that changed the way they spoke. Some of these mutations let the individuals handle a bigger vocabulary than their ancestors; in Nowak's model, these individuals were awarded more reproductive success.

Nowak found that his model consistently converged on the same results. Initially, the individuals communicated with one another with a few distinct calls. Their language gradually became more complicated as new calls were added. But as their vocabulary grew, it became harder to distinguish new calls from the old ones. The closer the sounds became, the easier it became to confuse them. (Think of the long *a* in *bake* and the short *a* in *back,* for example.)

While a bigger collection of calls may bring an evolutionary edge, the confusion that comes with it can cancel out its benefits. In trial after trial, Nowak found that his simulated vocabulary expanded to a certain size and then stopped growing. His results may explain why most animals aside from humans can communicate only with a small number of signals: they have no way of overcoming the inevitable confusion that comes with a big repertoire of sounds.

But what if our ancestors evolved a way out of this trap? To explore this possibility, Nowak changed his model. He allowed some of the individuals to start stringing together their simple sounds into sequences—to combine sounds into words. Now Nowak pitted a few of these word-speaking individuals against the original sound-speakers. He found that if the individuals had only a few messages that they needed to convey to one another, they could get by with a system of sounds. But if their environment was more complex and they needed to use more messages, word-speaking eventually won out. By combining a small number of sounds in a vast number of unique words, Nowak's individuals could avoid the confusion that similar sounds creates.

But Nowak discovered that word-speaking has its limits, too. In order for a word to survive in a language, people have to use it. If they forget it, the word sinks into oblivion. These days books and videotapes can help keep old words in circulation, but among our hominid ancestors, there were only spoken words, which had to be stored in hominid brains. Since brains don't have an infinite memory capacity, they limited the size of the vocabulary hominids could use. Hominids might invent new words, but only if old words were forgotten.

Nowak created a second twist on his language model to study this limit. Instead of simply using a single word for any given concept, some individuals could now combine words together to describe events. Some words could represent actions, others the people or things involved in that action, and others still would represent their relationship. In other words, Nowak gave these individuals syntax. Syntax allows a person to give a few hundred words millions of different meanings, depending on how the words are arranged. But syntax can create a confusion of its own if speakers aren't careful. Even though the same words are required for the headlines "Dewey defeats Truman" and "Truman defeats Dewey," their meanings are quite different.

When Nowak and his colleagues pitted syntax against simple word-concept communication, they discovered that syntax is not always best. A syntax-free language beats out syntax when there are only a few events that have to be described. But above a certain threshold of complexity, syntax became more successful. When a lot of things are happening, and a lot of people or animals are involved, speaking in sentences wins.

While Nowak's models are simple, they capture some of the crucial aspects of how language could have gradually evolved from a simple set of signals. The children who invented Nicaraguan Sign Language may have recapitulated its evolution from signs to words to syntax. Nowak's results also suggest how our ancestors got out of the communication trap that most other animals are stuck in. Something about the life of our ancestors became complex and created a demand for a complex way in which they could express themselves.

A strong candidate for that complexity, as Dunbar and others have shown, was the evolving social life of hominids. But even if hominids a million years ago had something to say, they might not have had the anatomy for saying it. We modern humans use a very peculiar sort of anatomy in order to speak, an anatomy unlike any other living mammal. Other mammals—including chimpanzees—have a voice box that rides high in their throats. This arrangement lets them breathe while they drink or eat, because the air passageway and the esophagus are divided. But it also creates a very small vocal tract between the voice box and the mouth. With so little room, the tongue cannot move around enough to make complex noises.

At some point in hominid evolution, the larynx must have dropped down to the low position that it takes in the human throat. This sort of anatomy comes with risks, because food or water can slip into our windpipes more easily than in other mammals and make us choke. But it also created enough room for our tongues to flick around and create the repertoire of sounds that a spoken language demands.

That's not to say that language couldn't have gotten its start before the voice box was in place. Hominids might have made signs with their hands—which were already capable of fine movements, judging from the tools they were making 2.5 million years ago. They might have combined these signs with simple sounds and movements to create a protolanguage. With such a system in place, evolution might have favored a bigger brain to handle more complex symbol processing and a more human-like throat to make more sophisticated speech possible.

No one knows the exact chronology of this evolution, because language leaves precious few traces on the human skeleton. The voice box is a flimsy piece of cartilage that rots away. It is suspended from a slender C-shaped bone called a hyoid, but the ravages of time usually destroy the hyoid too. Many researchers have turned instead to less direct clues that are left on hominid skeletons. They have looked at the angle at the base of the skull, in the hopes of calculating the length of the vocal tract. They have measured the width of the hole where a tongue-controlling nerve enters the skull. They have looked at the impression of the brain on its case, searching for language-related regions. In each case researchers have claimed to have found a clue to the first signs of speech. But skeptics have shown that none of these clues is actually a reliable guide to the presence of speech.

With all of this debate over a few shreds of hard evidence, it's not surprising that experts are divided over when language reached its modern form. Leslie Aiello of University College London, for example, maintains that the acceleration of brain size that started 500,000 years ago must have brought speech with it. Robin Dunbar, on the other hand, has proposed that language started only 150,000 years ago. He argues that only then were our ancestors living in groups that were too large for grooming to work as a social tool. People would have had to have substituted language for grooming and other primitive ways of interacting in order for hominid society to hold together.

With language, for example, you can keep tabs on what other people are doing and on what they're saying about you. You can manipulate other people with words as well and hold on to your place in a large society. Even today language still functions mainly as gossip. Dunbar has eavesdropped on people on trains and in cafeterias, and he consistently finds that two-thirds of their conversations are about other people. Language is, Dunbar argues, grooming by other means.

Yet other researchers think that even Dunbar's figure of 150,000 years is too old for the origin of language. They are convinced that full-blown language may have appeared only as recently as 50,000 years ago. It is only then that the human fossil record documents a spectacular mental explosion, in which people understood themselves and the world around them in ways that their ancestors never could have imagined. It was then that the modern mind was born, and language could well have been a crucial ingredient in its birth.

Modern Life, 50,000 B.C.
THE DAWN OF US

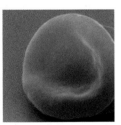

Jean-Marie Chauvet had no reason to be surprised on the afternoon of December 18, 1994. He was walking along a limestone gorge in the Ardèche region of southeast France with two friends, looking for caves. The Ardèche region is shot through with caves, and Chauvet, who had grown up there, had been spelunking since he was 12. In 1988 he began a systematic survey of the region with Christian Hillaire and Eliette Brunel Deschamps, expert cavers themselves. In the six years that followed, they found a number of new caves, 12 of which were decorated with ancient wall paintings. December 18 was a cold day, so Chauvet's team decided to explore a sunny area at the entrance to the gorge. It was not a particularly remote place; shepherds brought their flocks there, and presumably other spelunkers had been through many times. If anything spectacular was to be found, it should have been discovered long ago.

With the arrival of modern humans in Europe, simple all-purpose tools were replaced by complicated kits of scrapers, arrowheads, and other specialized implements.

The images in the Chauvet cave, dating back 32,000 years, are among the oldest known paintings on Earth.

Chauvet's team followed a mule path through the oaks and boxtrees until they reached a cliff, and there they found a hole. The hole was barely big enough for them to stoop their way inside, and they soon found themselves in a downward-sloping passageway a few yards long. It might well have been a dead end, but among the rubble at the end of the passageway they felt a slight draft.

The three of them took turns pulling the rocks away from the passageway, lying on their stomachs, heads downward. Finally they cleared a way through, and Deschamps, the smallest of the three, wriggled her way forward 10 feet. She found that the passageway opened at its end. When she cast her flashlight ahead, the beam soared out into a giant gallery, its floor 30 feet below.

The cavers ran a ladder from the passageway down to the gallery floor and descended into the darkness. Stalactites and stalagmites glittered like fangs in their flashlights. Columns of calcite were coated with jellyfish-like tendrils. They moved deeper into the cave. A mammoth suddenly lurched into the light. Then a rhinoceros, then a trio of lions. The animals were painted across the cave walls, some alone, some in giant stampedes—horses, owls, ibexes, bears, reindeer, bison—interspersed with the outlines of hands and mysterious rows of red dots. The spelunkers were familiar with cave paintings, but they had never seen anything on such a scale. They were confronted by a menagerie of at least 400 animal images.

The cave, which has since been named after Chauvet, is profoundly important, and not just for the paintings themselves. Archaeologists have measured the carbon 14 in the charcoal in the paintings and used it to estimate their age. People were painting animals on the walls of the Chauvet cave at least 32,000 years ago. That makes them the oldest paintings in the world.

The history of life is marked by dates like this one. They stand in a long row, like poles on a slalom course; any theory for the evolution of life must be agile enough to weave its way past them. The rocks of southwest Greenland show that life existed on Earth 3.85 billion years ago; those of the Karroo Desert in South Africa show that 250 million years ago almost all of life died out. The Chauvet paintings mark an event just as remarkable in the history of life: the moment our ancestors leapt into a world of art, symbols, complex tools, and culture—the things that make us most unique, most human.

The Chauvet cave and other archaeological sites that date from around the same time hint that this leap was sudden and long. The lineage that led to humans split from our closest living relatives, the chimpanzees, about 5 million years ago. It evolved in fits and starts, producing many branches that are long extinct and only one that has survived. Judging from the shapes of bones and sequences of genes, several teams of scientists have estimated that biologically modern humans evolved in Africa between 200,000 and 100,000 years ago. For tens of thousands of years, they left little mark on Earth except for the stone tools they used for butchering meat. Only around 50,000 years ago did they sweep out of Africa, and in a matter of a few thousand years they replaced all other species of humans across the Old World. These new Africans did not just look like us; now they acted like us. They invented tools far more sophisticated than those of their ancestors—hafted spears and spear-throwers, needles for making clothes, awls and nets—which they made from new materials like ivory, shells, and bones. They built themselves houses and adorned themselves with jewelry and carved sculptures and painted caves and cliff walls.

Most of the great transformations in evolution, such as the origin of life or the Cambrian explosion, took place hundreds of millions, or even billions, of years ago. In comparison, this human transformation happened only yesterday. But it is just as significant. Modern humans have become the world's dominant species, able to live just about anywhere on the planet. Our success is so staggering, in fact, that it threatens to destroy many other species. And while we may threaten the evolution of other species, we have also created a new form of evolution: the evolution of culture.

THE FIRST MODERNS

Only in the last 20 years has the possibility of such a revolution become clear. Earlier generations of scientists had a very different idea of how modern humans arose. In the old view, modern human evolution began a million years ago. At

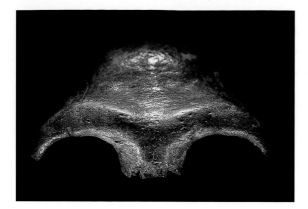

This skull cap was among the first bones of a Neanderthal found in 1856. Just over a century later, scientists discovered that these fossils still contained DNA, which turned out to be unlike that of any living human.

that point, a single species of hominid, *Homo erectus,* lived in Africa, Asia, and Australia. Although *H. erectus* was spread out across thousands of miles, each population continued to keep at least some contact with its neighbors. As men and women from these scattered bands mated, the genes of *H. erectus* flowed across its entire range. No group became isolated enough to fragment into a new species of its own. Some populations may have taken on a distinctive look here and there as they adapted to local conditions. In Europe, for example, where humans had to cope with harsh ice ages, the people we now call Neanderthals had stocky bodies and thick, low-browed skulls. In the tropics of Asia, on the other hand, populations of *H. erectus* became tall and slender. But together, the thinking went, all of these humans evolved into their modern form.

Neanderthal fossils can be found in Europe and the Near East from about 200,000 years ago to 30,000 years ago. Scientists initially believed that Neanderthals then evolved into modern Europeans, and along the way their tool kit evolved as well. Instead of the old stone tools and spears of the Neanderthals, the earliest Europeans (known as Cro-Magnons) used exquisitely crafted tools made of many different materials, such as fishhooks made of antlers and bones or spear-throwers with detachable foreshafts. Cro-Magnons buried their dead in elaborate rituals and wore necklaces and other ornaments. The fossil record of human evolution was spottier in Asia and Africa, but researchers assumed that *H. erectus* also evolved into modern humans in those regions and developed the new technology at the same time.

But in the 1970s a few paleoanthropologists began to contemplate a radically different vision of human evolution. They proposed that Neanderthals and *Homo erectus* in Asia were actually two distinct species, and that neither was the ancestor of *Homo sapiens.*

At the Natural History Museum in London, for example, paleoanthropologist Christopher Stringer found that fossils of Cro-Magnons looked less like Neanderthals than they looked like slightly older Africans. Instead of evolving from Neanderthals, Stringer proposed that modern Europeans descended from African immigrants. The Neanderthals who were alive 30,000 years ago did not evolve into modern Europeans, Stringer declared. They became extinct.

As Stringer stared at fossil skulls, a geneticist at the University of California at Berkeley named Allan Wilson was trying to reconstruct human history with biochemistry. He set out to analyze the DNA in human mitochondria—those energy-generating factories of the cell that carry their own DNA. He chose mitochondrial DNA rather than any of the genes in the nucleus because it passes from one generation to the next relatively unchanged. Unlike most genes, which are shuffled between the chromosomes we inherit from both our parents, mitochondrial genes come only from our mother. (This is because sperm cannot inject their mitochondria into an egg.) Any differences between a mother's mitochondrial DNA and her child's can arise only when the genes spontaneously mutate. As different mutations build up through the generations, it becomes possible to use mitochondrial DNA to distinguish different lineages.

Wilson's team analyzed samples of mitochondria from people throughout the world, sequencing the genes and grouping them together based on their similarity to one another. In the process they created an evolutionary tree of living humans. The branches of living Africans, Wilson discovered, all reached deepest into the tree of humanity. The tree suggested that Africa is the source of the common ancestor of living humans.

Most paleoanthropologists at the time might have gone along with this idea if Wilson had been talking about an early species of *Homo* that lived in Africa 2 million years ago, before any hominids had left the continent. But Wilson found that human genes were saying something very different. Once his team had constructed their evolutionary tree, they calculated how long ago the common ancestor of present-day humans lived. They estimated the rate at which the mitochondrial DNA mutates, and then they compared the variations in the genes to judge how much mutation had occurred in the different lineages. Their molecular clock then gave them an estimate for the age of the first modern human: somewhere in the neighborhood of 200,000 years.

"Mitochondrial Eve"—as this common ancestor came to be known—was certainly ancient, but to scientists who championed a multiregional origin of *Homo sapiens,* she was far too young. None of the older Neanderthals in Europe or *Homo erectus* in Asia could have contributed their genes to living humans. But for Stringer, Wilson's tree offered stunning support.

Stringer, Wilson, and other scientists began formulating a scenario for modern humans they called "Out of Africa." As *Homo* spread out of Africa, they proposed, it evolved into several distinct species that did not breed with one other. *Homo erectus* settled across much of Asia, while Neanderthals (also known as *Homo neanderthalensis,* a species in its own right) established themselves in Europe and the Near East. During this time, *Homo sapiens* was evolving from older hominids back in Africa. At some point, *Homo sapiens* migrated to Asia and Europe. The cave paintings of Chauvet, as well as the jewelry, weapons, clothing, and other artifacts that turn up in the fossil record after 50,000 years ago were all made by *Homo sapiens,* who left them behind as they explored the world. And when *Homo sapiens* arrived on the territory of *Homo erectus* or Neanderthals, these other humans disappeared.

Although the Out of Africa hypothesis drew heavy fire when it was first proposed, it has been drawing strength from recent fossil discoveries in Asia,

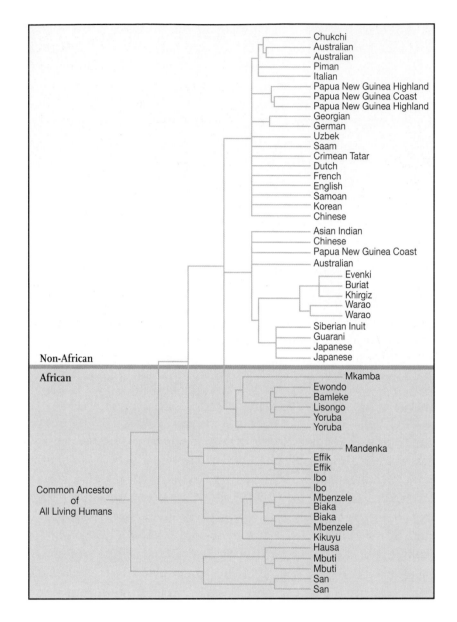

In 2000 researchers compared the DNA of 53 humans from around the world and created this evolutionary tree. Africans descend from the oldest lineages, suggesting that all living humans can trace their ancestry back to that continent. The variations in DNA suggests that this common African ancestor lived 170,000 years ago.

Europe, and Africa. Paleoanthropologists have found fossils of Neanderthals in Israel that lived alongside anatomically modern humans for 30,000 years without any sign of mixing before the Neanderthals disappeared. In Asia, *Homo erectus* survived long after *Homo sapiens*'s first fossils turned up. Some evidence suggests that *Homo erectus* was still alive as recently as 30,000 years ago on Java.

Meanwhile, discoveries in Africa have offered support of their own. "If we look at Europe 100,000 years ago," says Richard Klein of Stanford University, "they're exclusively Neanderthals. And then you look to Africa, and the people who were living in Africa are physically very modern in appearance."

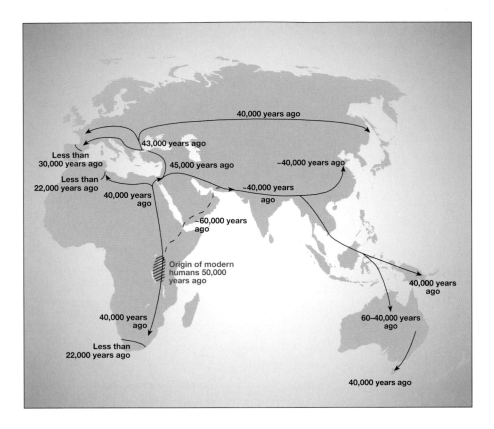

Fossil and genetic evidence suggest that modern humans emerged from Africa around 50,000 years ago and spread over the Old World.

Geneticists who have followed up on Wilson's initial work have generally confirmed the Out of Africa model. No matter which genes they analyze to construct the tree of human evolution, Africans consistently occupy branches that reach down closest to the base of the tree. With more sequences available to compare, mitochondrial genes now point to an origin 170,000 years ago. In the late 1990s, a team of geneticists compared the Y chromosomes of humans (the chromosome that determines whether a person is male). Those studies indicate that modern humans date back only 50,000 years. Of course, all the studies have a margin of error of tens of thousands of years, so they do not necessarily contradict one another.

These researchers will have to sort out who's precisely right and who's precisely wrong, but one thing is clear from all their work: we are a very young species.

NEANDERTHAL DNA

If more proof was needed of our collective youth, the genes of Neanderthals can now provide it. In 1995 the German government asked Svante Pääbo, an expert on fossil DNA at the University of Munich, to see whether the original

Neanderthal fossil found in 1856 still preserved any DNA. Pääbo was skeptical, since genes are so fragile, but he agreed to look. He and his graduate student Matthias Krings took a small sample of bone from the Neanderthal fossil's upper arm and analyzed it for amino acids, the building blocks of proteins. They were surprised to find some, and they embarked on a search for genes. Where amino acids might still survive, they reasoned, DNA might survive as well.

The search was a difficult one, because even a speck of dust could contaminate the fossil with the DNA of living human beings. To eliminate the risk, Krings bleached the outside of the bone and then set up his equipment in a sterilized room. Only when he was sure he was free of contamination did he grind up the Neanderthal bone and apply chemicals that would copy any fragments of DNA it held.

Krings's spine tingled as he watched the results come up on his computer: a sequence of 379 bases of DNA that was similar—but not identical—to human DNA. The champagne stayed corked, though, until the entire procedure was performed by researchers at the lab of Mark Stoneking at Penn State University on a second set of samples from the Neanderthal fossils. Stoneking independently discovered the same sequence.

Pääbo's team then built an evolutionary tree by comparing the Neanderthal DNA to almost 1,000 sequences of human DNA, as well as the DNA of chimpanzees. On their tree, Europeans and Africans cluster together on one branch, while the Neanderthal belong to a completely separate branch. And judging by the number of differences that have accumulated between the genes of Neanderthals and *Homo sapiens,* Krings and his colleagues estimated that their common ancestor may have lived as long as 600,000 years ago. This common ancestor presumably lived in Africa; one branch of its descendants migrated into Europe and became Neanderthals. The one that stayed behind evolved into us.

When Pääbo's team published their report in 1997, skeptics wondered if such a small fragment of Neanderthal DNA had enough information in it to put Neanderthals in their proper evolutionary place. But by 2000 two more pieces of Neanderthal DNA had come to light. Pääbo's team discovered one of them in a set of 42,000-year-old bones from Croatia, while a separate team found genes in 29,000-year-old Neanderthal fossils from the Caucasus Mountains. In both cases, the scientists isolated the same stretch of DNA that Pääbo had discovered in the original Neanderthal material. The three sequences of DNA resembled one another much more than any of them resembled the genes of any living human. These genetic fragments came from Neanderthals separated from one another by hundreds of miles and thousands of years. The chances that they should all just happen to have such similar sequences is close to impossible.

The new evidence from Neanderthal DNA supports the hypothesis that Neanderthals went extinct. And yet the fossil record shows that Neanderthals were not a delicate species waiting to be nudged into oblivion. They were tough, resourceful humans, rugged enough to survive the ice ages of Europe. They made spears as elegantly balanced as Olympic javelins, which they used to kill horses and other big mammals. They were so good at hunting that their diet consisted almost entirely of meat. Neanderthals took care of their sick, as demonstrated by a skeleton from the Shanidar Cave in Iraq: it belongs to a man whose head and body had been smashed, yet who had lived for several more years.

Likewise, *Homo erectus* was not some hothouse flower of a hominid. Its range reached from the bleak northern edges of China to the sticky jungles of Indonesia, and in this domain it survived for well over a million years. And yet *Homo sapiens* is still here, while *H. neanderthalensis* and *H. erectus* are now gone. What was the difference that allowed us to survive?

A NEW KIND OF MIND

The most obvious difference that paleoanthropologists can make out between modern humans 50,000 years ago and *Homo erectus* and Neanderthals is in the things that they made and left behind. *Homo erectus* in Asia never seems to have gotten beyond a hand-axe level of technology. Neanderthals could make spears and a collection of stone blades, but little more.

Modern humans, on the other hand, invented new tools that demanded great skill to make, and invented them at an astonishing rate. Modern humans made spears tipped with antler—a lightweight but strong material that had to be soaked for hours and sanded down in order to make a point. They invented spear-throwers that allowed them to flick their spears over their shoulders and hurl them much farther. Compared to Neanderthals, who charged their prey with wooden bayonets, modern humans could kill more game and put themselves at less risk in the process.

Not all of the inventions of modern humans were designed for practical purposes like hunting. In caves in Turkey, for example, scientists have found necklaces of snail shells and bird claws dating back at least 43,000 years. From the start, modern humans were wearing jewelry. It's possible that these ornaments were a sort of tribal identification or a way people marked their rank in their band.

"People were investing thousands of hours of labor in the production of body ornaments," says Randall White of New York University. "It was a priority in their lives, marking status and roles. When people put something on their bodies, it communicates immediately to other people who they are socially."

About 50,000 years ago modern humans began wearing jewelry and other ornaments, which may have served to identify their status or tribe. They also acquired a unique capacity for making art, such as this female figurine.

The artifacts that humans left behind speak to a profound shift in the way humans saw themselves and the world. And that shift may have given them a competitive edge. "Something happened about 50,000 years ago," explains Klein. "It happened in Africa. These people who already looked quite modern became behaviorally modern. They developed new kinds of artifacts, new ways of hunting and gathering, that allowed them to support much larger populations."

Researchers can only speculate for now about what brought the shift about. Some have proposed that the creative revolution was purely a matter of culture. Anatomically modern humans in Africa experienced some change—perhaps a population boom—that forced their society to cross some kind of threshold. Under these new conditions, people invented modern tools and art. "Cro-Magnons were perfectly capable of going to the moon neurologically, but they didn't because they weren't in a social context where the conditions were right," says White. "There was no challenge to provoke that kind of invention."

But Richard Klein, a paleoanthropologist at Stanford University, has grave doubts about such an explanation. If humans had the potential to paint the caves of Chauvet or build superior spears for hundreds of thousands of years, why was there such a long delay? If the revolution was purely cultural, then why didn't the Neanderthals living side by side with modern humans for thousands of years adopt the new tools and art and make them their own, in the way today's cultures borrow from one another?

Klein also points out that the ancestors of modern humans were probably not expanding their population when they suddenly changed their behavior. Geneticists can use the variability in the DNA of living humans to estimate how large their founding population was, and none of their estimates is very big. It now appears that all humans on the planet descend from just a few thousand Africans. "This emergence of fully modern humans appears to occur at a time when African populations were relatively sparse," says Klein.

A small group may not be a good place for cultural changes to take place, but biologists have long known it can be good for evolutionary changes. Mutations can sweep through their ranks quickly, rapidly altering them in the process. And with that fact in mind, Klein has proposed that the dawn of modern humanity was brought about by biology. New mutations to the genes that shape the human brain cropped up in Africa 50,000 years ago and gave it the capacity to make art and technology—a capacity that no other earlier humans possessed. "My own view," says Klein, "is that there was a brain change."

That brain change might have allowed humans to escape the rigid mental constraints that had trapped their ancestors. Instead of viewing animals only as food, modern humans could also recognize that their bone and antlers could be used as tools. Instead of using the same weapons for all sorts of game, modern humans began to invent ones that were specialized for different kinds of game, be it fish or ibex or red deer. This new way of thinking—what Stephen Mithen, a University of Reading archaeologist, calls "fluid intelligence"—even allowed people to think abstractly about nature and themselves and create symbolic representations in the form of paintings and sculptures.

Language, at least in its fullest flower, might have also been part of this late-blooming capacity. "It may be that what happened 50,000 years ago was the ability to produce speech rapidly, understandably, that other people could parse and make sense of; and then allow the use of that speech to spread information about new ways of doing things that people could not have spread in the same way before," says Klein.

The complexity of the new technology was too great to learn simply by example. In Russia, people boiled mammoth tusks and then buried them with their dead. Such a tradition would be impossible for an inexpressive Neanderthal to carry on. Modern humans could describe their new inventions to other people, so that new ideas could spread quickly. Modern humans began making tools out of stone, ivory, and other materials that

Neanderthals (left skull) lived alongside modern humans (right) for thousands of years in Europe and the Near East before going extinct.

were transported hundreds of miles from their sources: language could have made it possible for bands to communicate with one another about the goods they wanted to trade. With language, modern humans could give particular meanings to jewelry and art, whether that meaning was social or sacred.

Researchers do not yet know exactly what happened when modern humans, equipped with a new culture and perhaps with new brains, emerged from Africa and began encountering *Homo erectus* and Neanderthals. Did they wage all-out war? Did they bring devastating diseases to Europe and Asia, in the same way the Spanish brought smallpox to the Aztecs? Or perhaps, as many researchers suspect, the new brains of modern humans simply gave them a competitive edge. "They replaced Neanderthals in Europe principally because they were just behaving in a far more sophisticated way. And particularly, they were much more effective hunter-gatherers," says Klein.

Modern humans could trade with one another for supplies; they could use sophisticated language to settle disputes instead of spiraling into deadly battle. They could invent weapons and other tools to get more food and make clothes for themselves, and they could survive droughts and cruel winters that might have claimed other humans. The fossil evidence certainly suggests that they managed to live in higher densities than Neanderthals. The Neanderthals may have retreated to mountainous enclaves, where inbreeding and disasters eventually snuffed them out.

Not all modern humans headed into Europe, of course. The ones that spread into Asia may have hugged the coasts initially. Artifacts discovered along the Red Sea show that Africans had settled on its shores, living on shell-fish, as early as 120,000 years ago. It's possible that their descendants depended on this way of life as they colonized the coasts around the Arabian Peninsula and India and continued spreading toward Indonesia. When these troublesome newcomers arrived on the territory of *Homo erectus,* the resident hominids may have pulled back into the inland jungles to find refuge. Eventually they became so isolated that they winked out of existence 30,000 years ago. While some modern humans headed upriver into the heart of Asia, others set out to sea, using boats to get to New Guinea and Australia, where no hominid had ever set foot before. By 12,000 years ago humans had traveled from Asia to the New World, racing all the way to the southern tip of Chile. In an evolutionary flash, every major continent except for Antarctica was home to *Homo sapiens.* What had once been a minor subspecies of chimp, an exile from the forests, had taken over the world.

UNNATURAL SELECTION

Evolution made the cultural revolution possible 50,000 years ago, but human culture became so powerful that it turned the tables: now culture could influence the course of biological evolution. The fitness of genes determines the course of natural selection, but human inventions can change their fitness. Scientists can even see the imprint of our cultures in our DNA.

The coevolution of culture and genes has made it possible, for instance, for some people to drink milk. Among mammals, this is a bizarre gift. One of the main ingredients in milk is a sugar called lactose, and in order to absorb it, mammal intestines produce an enzyme called lactase that cuts it into digestible pieces. As a rule, mammals only make lactase as nursing infants; as they grow, they stop producing it. For a typical adult mammal, making lactase is a pointless exercise, since it has no need to digest milk.

Like other mammals, the majority of humans stop making lactase by the time they grow up. For them fresh milk is disagreeable—instead of being broken down and digested, lactose builds up and becomes food for a thriving colony of bacteria. The wastes that the bacteria excrete cause gas and diarrhea. (Cheese and yogurt have less lactose, so they're easier to digest.)

Yet some people (including half of all Americans) have mutations in their DNA that have disabled the lactase switch. They go on making the enzyme as adults, and they have no trouble drinking milk. This ability became widespread only after cattle were domesticated 10,000 years ago. Among groups of people who came to depend on cattle herding for their survival—such as tribes in northern Europe or the southern edge of the Sahara—being able to drink milk beyond infancy represented an evolutionary advantage. Mutations that let adults continue making lactase spread through the tribes. But among people who never relied on cattle, such as Australian Aborigines and Native Americans, genes that favored drinking milk conferred no advantage and never became common.

Over the course of history people have not only taken on new diets but have had to face different kinds of diseases. People who have settled in tropical regions had to struggle against malaria, because the mosquitoes that carry the disease thrive there. In malaria-prone regions, blood diseases that are rare or nonexistent elsewhere become common. People of African and Mediterranean descent, for example, suffer from sickle-cell anemia. Their red blood cells contain a defective form of hemoglobin, the molecule that the cell uses to carry oxygen from the lungs. When these cells are deprived of oxygen, the hemoglobin collapses and the cell shrinks from a plump bag to a scrawny tube. In this form red blood cells can get stuck in small blood vessels, forming dangerous clots or even tearing the vessels apart. When they pass through the spleen, white blood cells there can sense that they are defective and destroy them. The clots and the loss of red blood cells can make bones rot and retinas detach; ultimately sickle-cell anemia can kill.

About 150,000 babies in sub-Saharan Africa are born with the disease annually, and very few survive to childhood. To get sickle-cell anemia a person has to inherit two copies of a mutant hemoglobin gene, one from each parent. Far more people carry a single copy of the sickle-cell gene; with a single copy, a person's hemoglobin is only slightly defective. But given the harm that sickle-cell anemia causes, it should be far rarer than it is today. The death of people with two copies of the gene should have drained it almost completely out of the human gene pool.

The sickle-cell gene survives because it can give life as well as take it away. The parasite that causes the nastiest form of malaria, *Plasmodium falciparum,* invades red blood cells and eats their hemoglobin. It causes intense fevers, and the blood cells it invades can clump up, quickly forming lethal clots. As the parasite devours the hemoglobin in a cell, the cell loses its oxygen. If an infected cell has a defective hemoglobin gene, the loss of oxygen will make it collapse into a sickle. Now it can no longer clump together with other cells, sparing a person the dangerous blood clots. And sickle cells are so clearly deformed that they are quickly destroyed in the spleen, along with the parasites they carry.

People with a single copy of the sickle-cell gene thus can survive a bout of malaria that might kill someone who lacked it. Natural selection destroys copies of the gene in people who carry two copies, but it spreads them by allowing people with single copies to have children.

The sickle-cell gene marks the spread of agriculture. Before people farmed, malaria was probably not as important a scourge as it is today. Five thousand years ago, for example, much of sub-Saharan Africa was covered in forests. The floors of African forests are relatively mosquito-free, with most species sucking the blood of birds, monkeys, snakes, and other residents of the canopy. But then waves of farmers spread through sub-Saharan Africa and began turning the forests into fields. Puddles formed in the eroded, exposed earth, providing the perfect breeding spots for *Anopheles gambiae,* a formerly rare species of mosquito with a taste for humans. Farmers working in fields and sleeping in villages were easy targets for these mosquitoes, and they carried *Plasmodium* from one person to another. As malaria became more common, defenses against it began to evolve. Sickle-cell anemia, in other words, is one of the prices the world pays for farming.

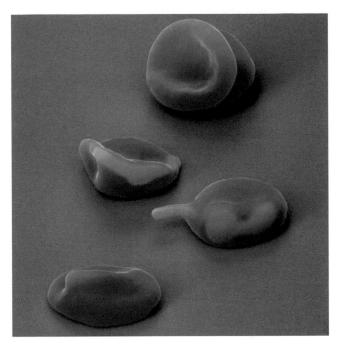

These red blood cells are infected with *Plasmodium,* the parasite that causes malaria. Over the last few thousand years, humans in many parts of the world have adapted to this devastating disease by evolving blood cells that are harder to invade.

While culture can sometimes drive natural selection in new directions, on the whole it may be slowing human evolution down to a crawl. Genes that might once have lowered our reproductive fitness no longer pose such a danger. One in 10,000 babies in the United States, for example, is born with a genetic disorder known as phenylketonuria, which prevents them from breaking down an amino acid known as phenylalinine. As these children eat food, levels of phenylalinine in their bloodstream build up, until it damages their developing brain and causes mental retardation. Once the mutations that cause phenylketonuria would have lowered a person's fitness. But now parents of children with phenylketonuria have the medical knowledge to save their children. A diet low in phenylalinine will allow their children to grow up with healthy brains. Thanks to medicine and other inventions, we have blunted the stark differences between successful and failing genes, making it harder for natural selection to create much change.

In the future, cultural evolution may slow biological evolution down even more. Natural selection works fastest when there's a big difference in the reproductive success of individuals—some individuals have no offspring at all, while others have big families. Today people around the world are enjoying better levels of food, health, and income, and as a result, they are raising smaller families. As the differences in human reproductive success dwindle, natural selection becomes weaker.

Another blow against future evolution is the human genome itself. All humans on Earth descend from a small group of people who lived in Africa somewhere between 60,000 and 170,000 years ago. That small founding population had relatively little genetic diversity to speak of, and little evolutionary time has passed for new diversity to emerge. There is more genetic diversity in the chimpanzees that live in the Tai forest of the Ivory Coast than in the world's entire human population. A few mutations have cropped up in different populations of humans from time to time, and natural selection has been able to nurture these new genes in some places—places where drinking milk or fighting off malaria is important, for example. But genes also have a way of spreading and mixing, as people come into contact with one another. Human history is, above all, a story of mingling.

In Italy, Luigi Cavalli-Sforza of Stanford University can see the imprint of genes brought to the country by Greek colonists who settled in the boot and in eastern Sicily, of Phoenicians and Carthaginians in western Sicily, of Celts in the north. There are traces of the mysterious Ligurians, subjugated by the Romans, near Genoa; near Tuscany the genes of Etruscans still survive 2,500 years after their civilization vanished. Around Ancona a cluster of genes survives from a civilization that existed 3,000 years ago, called the Osco-Umbro-Sabellian civilization. These genes were able to mingle together through Italy during an age when humans could travel only by masted ships and horseback. As transportation has improved, the mingling has only accelerated. Europeans came to the New World, bringing with them African slaves, and the genes of three continents began to mix together. Today airborne immigrants are accelerating the flow of genes around the planet even more. In such a swirling sea of DNA, there is not enough isolation anymore for any new species of human to emerge. For the foreseeable future, the great hominid tree of 15 species or more will remain pruned down to our own. And even within our own species, it may be very hard for natural selection to produce much change.

MAN-MADE EVOLUTION

Regardless of whether human biological evolution rolls on into the future or creeps to a halt, a new sort of evolution has come into being. Culture itself evolves. Languages evolve, airplanes evolve, music evolves, mathematics evolves, cooking evolves, even fashions in hats evolve. And the ways human creations change with time mirror biological evolution in uncanny ways. Languages branch into new ones much as animal species do: the longer a dialect is isolated from its mother tongue, the more distinct it becomes. Scientists can even chart the history of languages by drawing evolutionary trees, using the same methods they would use to chart the evolution of hominids.

One common feature of biological evolution is exaptation, in which an old structure gets borrowed to perform a new function—such as legs on a fish used to walk on land. Stephen Jay Gould points out that the same thing happens as technology evolves. In the markets of Nairobi, for example, you can buy sandals made out of automobile tires. "Tires make very good sandals," Gould writes, "but one would never argue that Goodrich (or whoever) built the tires to provide footwear in Third World nations. Durability for sandals is a latent potential of auto tires, and the production of such sandals defines a quirky functional shift."

Biological evolution and cultural evolution are just similar enough that scientists wonder if some of the same principles are at work in both of them. In his 1976 book *The Selfish Gene,* Richard Dawkins noted that our thoughts have surprisingly genelike properties. A song, for example, is a piece of information that is encoded in our brains, and when we sing it to others, it becomes encoded in theirs as well. It is as if we had sneezed and passed on a cold virus to them. Dawkins dubbed these virus-like chunks of information "memes." The same laws that govern the rise and fall of genes may apply to memes as well. Some genes are better at getting themselves replicated than others. A gene that makes a defective photoreceptor in the eye won't spread as easily as one that lets its owner see. Likewise, some memes spread more easily than others, and a mutation to a meme may give it a competitive edge it lacked before.

In order to succeed, a meme doesn't have to be superior in any intellectual sense to other memes. All it needs to do is get itself copied. One of Dawkins' personal favorites is the Saint Jude letter, which promises good luck to its recipients if they send it to their friends. Saint Jude letters don't make people win lotteries or get cured of cancer; they only manipulate people's hopes to get themselves carried around the world. The Internet now surges with bad jokes and nude pictures of celebrities, passed on from one computer to another, thanks to the desires or boredom of Web surfers.

But there are many ways in which culture doesn't evolve like viruses. People do not cause little mutations to an idea and then send these malformed notions out into the world to thrive or fail. They think things through hard; they jumble them up with other ideas into new fusions. Memes do not jump directly from one brain to another as DNA gets copied letter for letter from one generation to the next. People observe the actions of other people and try to imitate them, sometimes successfully, sometimes not.

In some ways cultural evolution is more Lamarckian than Darwinian. Lamarck thought that if a giraffe's neck became longer during its life, it could pass on that long neck to its offspring. A father teaches his son how to make a sword, but in the son's experience he may find a better way to apply the clay or fold the steel; he can then pass on a changed recipe to his own son.

Genes, it's now becoming clear, do not always wait patiently for one generation to copy them into a new one. They have traveled across the tree of life many times, as viruses carry them to new species or as one organism devours

Ideas, trends, beliefs, and fads (such as the hula-hoop craze of the 1950s) can spread like viruses through human cultures.

another. Likewise, a piece of one culture can leap into another one and combine into something new. Witness Marco Polo bringing gunpowder to Europe, or African slaves bringing syncopated rhythm to the United States. The English in which I write these words is not a word-for-word descendant of the English that Chaucer spoke. It has been invaded by words from all over the world.

We live in symbiosis with our culture. The idea of a plow can survive only if there is a person thinking of it. And without plows, most humans would starve. For the most part, these technological inventions have served as extensions of our bodies. A spear lets a hunter kill an elk without claws or fangs. A windmill can grind seeds and grains far more effectively than our own teeth. Books are extensions of our brains, giving us a collective memory far more powerful than any one person can have.

In the 1940s, however, humans stumbled across a new form of culture: the computer. Computers are even more profound brain extensions than books. They can store information far more densely (compare the bound version of the Oxford English Dictionary filling a shelf to a single compact disk holding the same information). What's more, computers are the first tools that can process information in the same basic ways a brain can, perceiving, analyzing, planning. Of course, the computer itself cannot do anything; it must be commandeered by programming code, much like cells are commandeered by their DNA.

In the 1950s only a fraction of a single megabyte of random-access memory existed on the entire planet. Today a single cheap home computer may contain 50 megabytes, and it is no longer a lonely island of silicon: since the 1970s, the world's computers have begun joining together as the World Wide Web has spread like a mesh of fungus threads. The Web is encircling Earth, subsuming not only computers but cars and cash registers and televisions. We have surrounded ourselves with a global brain, which taps into our own brains, an intellectual forest dependent on a hidden fungal network.

Computers can dutifully do the things we tell them to do. They can control the path of a spacecraft as it slingshots around the rings of Saturn. They can track the rise and fall of insulin in a diabetic's body. But it's possible that once their global network becomes complex enough, it will spontaneously take on something like our own intelligence, perhaps even our own consciousness. Research on artificial life and evolutionary computing has already suggested that computers can evolve an intelligence that doesn't resemble our own. If a computer is allowed to come up with its own ways to solve problems, it evolves solutions that may make no sense to our brains. There is no telling what the global web of computers may evolve into. In time, our culture may become an intimate stranger to us, a symbiotic brain. The lions on the walls of Chauvet will begin to dance.

What about God?

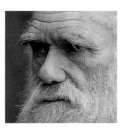

ELIZABETH STEUBING is a senior at Wheaton
College in Illinois, a school whose motto is *Christo
et Regno Ejus* ("For Christ and His Kingdom").
Steubing is a devout Christian herself and finds
in her faith a deep reservoir of strength. "I think your worldview
really does matter in where you're going in life," she says. "I want
to become a doctor and help people. It's not like you can't do that
without my worldview. But it just makes the picture so much bigger
if I want to serve God and do what He wants with my life. It's comforting too that I'm not just in it by myself."

Steubing is also a Christian who asks a lot of questions, about
troubling subjects like the existence of evil and the suffering of
innocent people. "It's very hard to explain away people's pain, and
I've thought a lot about that," she says. "I have professors that I go
to a lot and ask really tough questions, and they certainly don't

Garden of Eden,
by Jan Brueghel
(1568–1625).

have all the answers either. But I love listening to what they have to say about it, because they've been on this road a lot longer than I have."

Steubing grew up in Zambia before coming to the United States for high school. "We lived outside of town in the forest, and we just kept all sorts of pets like chameleons, bush babies, snakes, you name it. So we were always surrounded by nature." At Wheaton she takes classes in biology, and her professors are teaching her about evolution. "I wouldn't swear by it," she says, "but I think it's what fits the evidence better than any other theory that may be out there."

Steubing knows that some of her fellow students at Wheaton object to being taught evolution, and that some parents won't even send their children to Wheaton because of it. "I just can't relate," she says. "If it is an intellectual enterprise, I don't want to hide from it just because some parts of it might pinch my theology."

But Steubing is asking questions. What does evolution mean for her faith? On the one hand, she doesn't feel that she has to deny evolution in order to be a Christian. "I think evolution as a process is fine. There's a lot of evidence for it. But to say that it excludes God is pretty presumptuous. If God wants to work through His natural creation, I mean, what's wrong with that? We would like to say that God works through material processes today in how He deals with us. Why can't He have done that in the past?"

Steubing knows that some people have argued that there's a conflict between the two. "If you extrapolate the evolutionary model to say that there is no other influence in how we came to be who we are, and it's just strict natural selection, that does pose a problem for what it means to be human," she says.

Ultimately, her questions about evolution boil down to one: "Where is God's place if everything does have a natural cause?" she asks. "I've wrestled with that quite a bit."

She is not alone.

Ever since the publication of *Origin of Species,* people have been pondering the significance of evolution for the meaning of their lives, and of life in general. Are we a biological accident or a cosmic imperative? For some people, the only way to deal with Steubing's question has been to deny the evidence for evolution altogether. That was what Bishop Wilberforce did in 1860. And today a vocal opposition to Darwin remains in force, primarily in a country in which Darwin never set foot: the United States.

AMERICA MEETS DARWIN

Although Darwin never visited the United States, his theory was escorted into the country by his friend the botanist Asa Gray. In Gray's review of *Origin of*

Species in the *Atlantic Monthly,* he declared that Americans should no longer rely on a literal reading of Genesis for answers to biology's questions. After all, physicists were not content with simply being told that God created the heavens and Earth; they were beginning to work out how the solar system had evolved from a dusty cloud. "The mind of such an age cannot be expected to let the old belief about species pass unquestioned," Gray wrote.

Like their European counterparts, most American scientists accepted evolution by the end of the 1800s, although some of them had their doubts about the particulars of natural selection. As far as historians can tell, reading Darwin caused none of them to give up their religion. Gray himself was a devout Christian, and he argued that divine guidance might somehow channel the course of natural selection.

While some American Protestant leaders were hostile to Darwinism when it first came to this country, they didn't speak out against it. As long as scientists were expressing doubts about the theory, they were content to stay on the sidelines. But by the end of the 1800s, as those scientific doubts faded away, Protestant leaders began to speak out. They saw Darwinism as not just wrong but dangerous. Only if God had directly created mankind in His image was there any basis for morality. Darwinism seemed to render mankind nothing more than an animal. If we were the product of nothing more than natural selection, how could we be God's special creation? And if we were not, then why should people adhere to the Bible?

Their moral objections overshadowed any scientific ones. Most American Protestants at the turn of the century considered the Bible to be, for the most part, literal truth, but few actually believed that the planet was created in six days a few thousand years ago. Some believed that the first line of Genesis— "In the beginning God created the heavens and the Earth"—referred to a vast length of time, in which matter, life, and even fossils were created. God then destroyed that pre-Adamic creation and constructed an Edenic restoration in 4004 B.C., the age calculated by Bishop Ussher. In this restoration, God created a new set of animals, and plants, and the rest of life in six actual days. Such a cosmology could accommodate an old universe and an old Earth, complete with its extinct life and ancient geology, by fitting all of it into a capacious "beginning."

Others argued that the word *day* in Genesis was a poetic expression that didn't specifically mean 24 hours. They interpreted the six days of creation as six vast gulfs of time during which God brought the world and life into existence. An ancient Earth with a long biological history was no threat to them. No matter how old the world was, what mattered was that God—not evolution—had created life, and humans in particular.

Even at the end of the nineteenth century, Darwinism was only a distant threat to conservative American Christians, the preoccupation of scientists in a few American universities. But by the 1920s the controversy over Darwin had turned from a slow simmer to a rapid boil.

DRAWING BATTLE LINES

One major reason for this change was the rise of public schools. In 1890, only 200,000 American children attended public schools, but by 1920, 2 million were enrolled. The textbooks that the students were given in their biology classes embraced evolution, describing it as a process of genetic mutations and natural selection. For opponents of evolution, the textbooks came to represent a godless intrusion into their children's lives.

Another reason for the conflict was evolution's role in some disturbing cultural changes in both the United States and Europe. World War I was an unprecedented butchery, and some Germans tried to justify the carnage with evolution. They drew their inspiration from the influential German biologist Ernst Haeckel, who portrayed humans as the pinnacle of all evolution and claimed that we were evolving to still loftier heights. In his book *History of Creation,* Haeckel wrote, "We are proud of having so immensely outstripped our lower animal ancestors and derive from it the consoling assurance that in the future also, mankind as a whole will follow the glorious career of progressive development and attain a still higher degree of mental perfection."

For Haeckel, some humans were more progressive than others. He divided them into 12 different species and ranked them from lowest to highest. At the lowest were various species of Africans and New Guineans, and at the very summit were Europeans—"*Homo mediterraneus.*" And within *H. mediterraneus,* Haeckel's fellow Germans stood at the very summit. "It is the Germanic race in North-western Europe and in North America, which above all others, is in the present age spreading the network of its civilization over the whole globe, and laying the foundations for a new era of higher mental culture," he wrote. Sooner or later most of the other races would "completely succumb in the struggle for existence to the superiority of the Mediterranean races."

Haeckel's vision of human destiny became the foundation of a new, biologically based religion that he called Monism. A group of his devotees, called the Monist League, declared that the next stage of evolution required Germany to become a world power and urged their countrymen to enter World War I not for political ends but to realize their evolutionary destiny.

In England and the United States, meanwhile, evolution was being misused in another way, to justify laissez-faire capitalism. The British philosopher Herbert Spencer, promoting a jumble of Darwinian natural selection and Lamarckian evolution, claimed that a free-market struggle would make humans evolve greater intelligence. He granted that there would have to be suffering along the way—Spencer thought that the Irish famines were an example of people taking "the high road to extinction"—but this suffering would prove worthwhile, because humanity would be elevated to moral perfection.

Spencer won admirers from around the world, particularly among the new robber barons of the Industrial Revolution. Andrew Carnegie called Spencer "Master Teacher." Spencer's admirers codified his ideas into a school of thought known as social Darwinism, which held that the incredible disparity of rich and poor at the end of the 1800s was not an injustice but simply biology. "The millionaires are a product of natural selection, acting on the whole body of men to pick out those who can meet the requirement of certain work to be done," said Yale sociologist William Graham Sumner, a leading social Darwinist.

Social Darwinism was as scientifically baseless as Monism. It took natural selection out of its proper place in biology and put it in a social environment, while also muddling Darwin's theory with Lamarck's discredited one. But despite being unscientific, social Darwinism lent authority to government efforts to control the evolution of the human race. In the early 1900s, the United States and other countries sterilized retarded people and others they judged to be degenerates so that they wouldn't contaminate the evolution of their nations.

Some of the textbooks used in American public high schools celebrated this controlled breeding. In *A Civic Biology* (1914), the author wrote of families in which criminality and other vices were thought (wrongly) to be a hereditary curse. "If such people were lower animals, we would probably kill them off to prevent them from spreading. Humanity will not allow this, but we do have the remedy of separating the sexes in asylums or other places and in various ways preventing intermarriage and the possibilities of perpetuating such a low and degenerate race."

Out of the turmoil of the early twentieth century emerged fundamentalism. Fundamentalists were determined to return Protestantism to its traditional roots, and that required putting a stop to teaching evolution in public schools. Leading the fundamentalists in this fight was the statesman William Jennings Bryan. Bryan's political celebrity—he had run as the Democratic nominee for president three times and served as secretary of state under Woodrow Wilson—earned the anti-evolution crusade national attention.

Bryan's hostility to evolution was not based on any particular scientific point of view. Like many other fundamentalists of his day, Bryan believed that

Opponents of evolution attacked Darwin for one of the most disturbing aspects of his theory: the way it linked humans to other animals. If we are nothing more than animals, then why should we act like anything more than animals?

the six days of creation were metaphorical, rather than a strict 144 hours. Bryan wasn't even opposed to the idea that animals and plants might have evolved from older species. What bothered him about evolution was the effect he thought Darwinism had on the soul. "I object to the Darwinian theory," Bryan declared, "because I fear we shall lose the consciousness of God's presence in our daily life, if we must accept the theory that through all the ages no spiritual force has touched the life of man and shaped the destiny of nations. But there is another objection. The Darwinian theory represents man as reaching his present perfection by the operation of the law of hate—the merciless law by which the strong crowd out and kill off the weak."

Bryan would have been quite right to be concerned if he was referring to social Darwinism or Monism, which people used to justify brutality, poverty, and racism. But both philosophies were based on misreadings of *Origin of Species,* mixed with a healthy portion of Lamarckian pseudoscience. Bryan did not recognize this difference, and instead he made Darwin his sworn enemy. In the process, he helped sow a confusion that still thrives in the United States.

A SHOW TRIAL FOR EVOLUTION: THE SCOPES CASE

In 1922 Bryan learned that Kentucky's Baptist State Board of Missions had passed a resolution calling for a state law against the teaching of evolution in public schools. He embraced the cause, traveling around the state to speak in its favor. When the ban came up for consideration at the Kentucky House of Representatives, it lost by a single vote. By then, Bryan and his fellow creationists had carried the crusade to other southern states, and in 1925 Tennessee was the first state to heed the call with legislation.

The American Civil Liberties Union was opposed to the law, on the grounds that it robbed schoolteachers of their freedom of speech. They set out to overthrow it by announcing they would defend any Tennessee teacher who would break it. Their plan was to attract a lot of attention to a trial they knew they'd lose and then appeal the decision on the grounds that the law was unconstitutional. If they won the appeal on those grounds, the law itself would be wiped out.

In the sleepy town of Dayton, Tennessee, a few town leaders heard of the ACLU's offer. They had no particular feelings one way or the other about the law, but they all agreed that a show trial would be a good way to earn some publicity. They met at a local drugstore with a young teacher and football coach named John Scopes. Scopes, who normally taught physics, told them that as a substitute teacher he had taught chapters on human evolution from *A Civic Biology.* The town leaders asked him if he would stand in court for a

Clarence Darrow (left) and William Jennings Bryan (right) pose for a photograph in a rare show of camaraderie during the Scopes Monkey Trial.

test case, and after some thought, Scopes said he would be glad to. Once a warrant had been written out for his arrest, he left the drugstore to play some tennis.

The ACLU expected Scopes to be found guilty and promptly fined, whereupon they could get down to the real business of an appeal. But that wasn't what happened. Bryan, who happened to be in Tennessee at the time, announced that he would come to Dayton to "assist" the prosecution, making an already prominent trial into a sensation. The ACLU then got an offer it couldn't refuse, from the defense lawyer Clarence Darrow. Darrow had just made national headlines in 1924 for his defense of Nathan Leopold and Richard Loeb, two university students who had killed a boy for sport. Darrow had convinced the judge that although they were guilty, they should not be executed. The murder was not for any cold-blooded motive but simply the product of two "diseased brains."

It was just this sort of materialistic view of human nature that Bryan despised. And the outspoken agnostic Darrow had no great love for Bryan either, considering him little more than a huckster. Darrow sent a telegram to the ACLU volunteering his services, sending a copy to the press as well. With reporters waiting to hear its response, the ACLU had no choice but to accept Darrow's offer. And at that moment they lost all control of their cause.

With two such famous, fiery men opposed to each other, the trial drew national attention, and with radios in many American homes for the first time, it even drew a national audience. The Scopes Monkey Trial, as it came to be known, certainly proved to be high drama (a play, *Inherit the Wind,* was written about it and was later made into a movie). Darrow went so far as to call Bryan himself to the witness stand, where he grilled his opponent on the contradictions of creationism. Darrow forced Bryan to admit that he did not believe in a literal six days of creation, demonstrating that the Bible was open to many different interpretations—some of which might accommodate evolution. But the following day the judge threw out Bryan's testimony. Darrow promptly demanded that Scopes be found guilty. He was already thinking about an appeal.

Darrow is usually remembered as the winner of the Scopes Monkey Trial, but the truth is that the teaching of evolution actually suffered from his grandstanding. The jury found Scopes guilty, and the judge fined him $100. Darrow appealed the case to the Tennessee Supreme Court, and it was overturned a year later—but not on the constitutional grounds the ACLU had hoped for. The judge in the original case had set the fine for Scopes, but Tennessee law allowed only juries to set fines over $50. On this technicality, the Tennessee Supreme Court dismissed the case. "Nothing is to be gained by prolonging the life of this bizarre case," the judges declared.

Darrow was apparently more interested in making great speeches than in paying close attention to his case; because he had failed to object to the judge's fine in the original trial, the ACLU had lost its chance to challenge the ban on teaching evolution. In fact, Tennessee's anti-evolution law would stay on the books for more than 40 years. And by the end of the 1920s Bryan's allies had passed similar laws in Mississippi, Arkansas, Florida, and Oklahoma.

THE RISE OF "CREATION SCIENCE"

In the 1940s and 1950s, the United States became a hotbed of research in evolutionary biology. It was home to leaders of the modern synthesis, such as Theodosius Dobzhansky, Ernst Mayr, and George Simpson; American paleontology, genetics, and zoology were becoming the envy of the world. But little of that new knowledge filtered out of museums and biology departments to the public at large, because creationists pressured textbook publishers to drop any mention of evolution. Afraid to lose business, publishers bowed to their demands.

The tide began to turn in the 1960s, in part because of the Soviet launch of the Sputnik satellite in 1957. That triumph of Soviet science created a national panic over the state of American science education—including the teaching of evolution. Textbooks began surveying evolution again, and by 1967 even the Tennessee legislature had repealed the law that had gotten Scopes arrested.

Meanwhile, a young Arkansas biology teacher named Susan Epperson was playing the part of Scopes in a new court case. Epperson challenged Arkansas's anti-evolution law, arguing that it established religion in public schools. The Arkansas Supreme Court refused to hear the case, saying simply that the law was a valid exercise of a state's rights. Finally the ACLU saw a chance to make the case that had eluded it 42 years before. They took Epperson's appeal to the U.S. Supreme Court in 1968, and the Court struck down the Arkansas law. Justice Abe Fortas wrote that the law was "an attempt to blot out a particular theory from public education" in order to advance a religious agenda.

Ironically, Fortas may have helped steer creationists toward a new strategy, one that they still use today. Rather than simply claim that Darwinism was immoral, a growing number of anti-evolutionists instead claimed that creationism was actually a viable alternative scientific account of the rise of life. Anyone who would keep this "creation science" out of the classrooms, they claimed, was guilty of trying to blot out a particular theory from public education. Instead, the creationists demanded, public schools should offer both theories and let students decide for themselves.

The first manifesto of "creation science" appeared in 1961, a book called *Genesis Flood,* written by a hydraulic engineer named Henry Morris. Morris revived an old strain of creationism based on an extremely strict

Some creationists claim that Noah's Flood was responsible for all of the fossils and geological formations on Earth. Geologists abandoned this idea 300 years ago.

reading of the Bible. He argued that Earth was created in six literal days, that it was only a few thousand years old, and that all the geological formations and fossils on Earth had been set in their current place during Noah's Flood. Morris claimed that all the new evidence of an ancient Earth, such as geological clocks based on radioactive decay, were flawed, tainted, or otherwise unreliable. And he offered what he claimed were scientific explanations for the events in the Bible. Adam really could have named all the animals in a single day, for example, because he was much more intelligent in his Edenic perfection than living humans. In 1972 Morris founded the Institute for Creation Research, which has published books, magazines, videos, and Web sites ever since.

Morris inspired other creationists to coat their own opposition to evolution in a scientific lacquer. Some, who go by the name Old-Earth creationists, accept a 13-billion-year-old universe and a 4.5-billion-year-old Earth but claim that humans were created uniquely and only recently by God. Not surprisingly, each camp scorns the other. Old-Earth creationists attack Flood Geology creationists for ignoring the facts of geology and astronomy. Flood Geology creationists accuse Old-Earth creationists of abandoning the word of God and inching their way down the slippery slope toward Darwinism and atheism. Creationists may stand opposed to Darwin, but they certainly don't stand together.

THE TEST OF SCIENCE

The creation science movement convinced the state of Arkansas to pass a statute requiring equal time for both creation science and evolution in science classes. But the courts have not been impressed with creationism's new packaging. When Arkansas's statute was challenged in district court in 1982, Judge William Overton threw it out on the grounds that it violated the First Amendment. Overton ruled that creation science was no science at all, but rather an attempt to promote religion in public schools.

Science is the search for natural explanations for what we observe in the world around us. Central to science is the construction of theories. In ordinary speech, *theory* may sometimes connote a rough guess or a hunch, but in science it has a specific meaning: a theory is an overarching set of proposals about some aspect of the universe. The germ theory claims that certain diseases are caused by living organisms. Newton's theory of gravity holds that every object in the universe exerts an attractive force on every other object.

Theories cannot be directly proven right or wrong. Instead, they spawn hypotheses—testable predictions about specific applications of the theories. If the hypotheses tend to hold up, if they guide scientists to new discoveries about how the world works, and if different scientific approaches offer interlocking support, a theory is confirmed. Yet scientific knowledge is always subject to more testing and to uncovering deeper levels of reality. Newton's theory of gravity works astonishingly well and can be used to send spacecraft to other planets. But it turns out to be embedded in an even broader theory, Einstein's theory of relativity. Gravity, Einstein proposed, actually curves the fabric of space. One hypothesis that emerged from his theory was that light from distant stars would bend around the sun—a prediction that was confirmed during an eclipse in 1919. Physicists are now trying to establish an even broader theory of how the universe works, in which they can embed relativity and other theories, such as quantum physics.

The predictive power of science lets us perceive things in nature far beyond our senses. No one has seen the center of the Earth, but geologists know that it is an iron core by measuring the magnetic field it creates and the ways it alters seismic waves of earthquakes that pass through it. It is impossible to actually see a black hole, but the theory of relativity not only predicts that black holes will form from some collapsing stars, but that they will leave certain signatures—such as beams of x-rays shot out by matter being sucked into them. And astronomers are now documenting the existence of those predicted beams.

Much of evolution's work is likewise hidden from view, but it is not beyond our mental grasp. No one was alive 200 million years ago, but we can deduce from the geological record that the animals alive at the time inhabited a planet in many ways like Earth today. Living animals are subject to evolution, and so those ancient animals must have been as well. We don't have a moment-by-moment record of life's history, but we can get some understanding of what has actually happened over the past 4 billion years from the evidence that has been left behind.

As Judge Overton recognized in his 1982 decision, creationism cannot live up to science's demands. Scientists come up with explanations for how nature works, and those explanations have consequences that they can test. It's possible to test some of the consequences of creationist explanations, and they fail. Henry Morris, for example, has claimed that *Homo sapiens* must be only a few thousand years old, based on the current rate of population growth. He made this claim by extrapolating back from today's population to a time when there were only two people on Earth (presumably Adam and Eve). "The most probable date for man's origin, based on the known data from population statistics, is about 6,300 years ago," Morris declared.

Morris's claim has consequences. For one thing, it can be used to calculate the human population at any point in history. Historical records clearly show, for example, that the Egyptian pyramids were built around 4,500 years ago. If you use Morris's own timetable to calculate how many people were on the planet at the time, you get a grand total of 600. Since there are historical records from people who lived outside of Egypt 4,500 years ago, those 600 people could not all have lived in Egypt. Given that Egypt constitutes only 1 percent of the Earth's landmass, it might have contained only six people. In order to believe Morris's claim that humanity is only 6,300 years old, we have to believe that a handful of people could have built the pyramids.

In order to make a case for creationism, its advocates try marshaling evidence from scientists, but they do so selectively, ignoring the full picture. For example, creationists sometimes point to the Cambrian explosion as proof of God's work. "Significant from our standpoint is that at a certain time in the supposed geological calendar, popularly called the Cambrian era, are found a host of fossils which are virtually absent from older layers of rock," write Wayne Frair and Percival Davis in *A Case for Creation*. "From a scientific standpoint alone it is evident that a spectacular event must have occurred at this time. It seems reasonable that the abrupt change at the period designated as Cambrian is a result of God's creative activity."

These creationists leave out a few key facts. The fossil record runs back well over 3 billion years before the Cambrian explosion, and paleontologists have found fossils of multicellular animals as old as 575 million years, some of which are clearly relatives of the groups that appeared during the Cambrian explosion 40 million years later. The Cambrian explosion certainly happened quickly, as geological history goes, but it still took around 10 million years. And judging from the way different animal embryos are built, a few relatively small genetic changes may have triggered many of the dramatic transformations to animal bodies seen during that time.

Most arguments for creationism aren't really arguments for it at all, but are just quibbles with evolutionary biology. Many creationists, for example, will grant that evolution can happen on a small scale within a species: bacteria can evolve resistance and the length of beaks on Darwin's finches can change size. But this sort of microevolution, creationists say, cannot produce macroevolution, such as the creation of totally new body plans. For evidence, they claim that no one has ever seen such transformations happen, and that there are no fossils of the intermediates involved.

Darwin addressed this issue in *Origin of Species* when he pointed out how rarely fossils form and how sparse their record must be. Despite the scarcity of fossils, paleontologists have found many intermediate forms that creationists have claimed could not exist. Creationists used to get enormous pleasure out of the lack of walking whales, for example. That was before paleontologists starting digging up whale feet.

The many whales with legs that paleontologists have now uncovered are probably ancient cousins of today's whales rather than direct ancestors. Nevertheless, their position on the evolutionary tree shows how early whales moved from land to the sea. A whale may look radically different from its closest living relatives such cows and hippos—in fact it may be hard to imagine how one could evolve into the other—but the transitional fossils of whales show how the transformation could have occurred in small steps. *Ambulocetus,* for example, is an alligator-like whale that could have swum like an otter by kicking its short legs. *Protocetus,* on the other hand, had smaller hind legs and looser hips, which allowed it to get more power out of its tail.

Like other transformations documented in the fossil record, the origin of swimming whales seems to have taken place over millions of years—a rate of evolution far slower than the sort that has been documented in wild animals, such as the guppies of Trinidad. It's not possible to predict macroevolutionary patterns from such short-term change, but the fact remains that if you accept microevolution, you get macroevolution for free.

PALEY RIDES AGAIN

In the wake of Judge Overton's 1982 rejection of "creation science," creationists headed back to the drawing board. They searched for new ways to get their ideas back into the public schools. Today their preferred tactic is simply to strip their claims free of all mention of God and the Bible.

This new strain of creationism goes by the name of Intelligent Design. Life, the argument goes, is so complex that it could not have evolved. Instead, it must have been created by a designer. Exactly who was this designer, you may ask. Arguments for Intelligent Design coyly leave that question open. But traditional creationists see Intelligent Design as a sharp wedge they may be able to use to crack the public school shell. *Of Pandas and People,* a 1989 book about Intelligent Design intended for children, got this rave review from Answers in Genesis, a Flood Geology organization: "Intended for textbook use in public schools, this superbly written book has no Biblical content, yet contains creationists' interpretations for classic evidences usually found in standard textbooks supporting evolution."

Advocates of Intelligent Design claim that they're at the scientific forefront. Instead of challenging evolution with the old attacks on missing links or the age of the planet, they draw on biochemistry and genetics. They claim that there is an irreducible complexity to the molecular working of life. In order to clot a wound, for instance, a long cascade of chemical reactions creates the clotting molecules. Take out one of the elements in the cascade, and a person bleeds to death. How, then, could evolution have built up this system from simpler parts?

Intelligent Design should sound familiar. It is a new spin on William Paley's watch lying on the heath over 200 years ago. Paley declared that if you encounter something that is complex, made of parts that each are essential for the whole thing to work, you have found something that must have been designed. The trouble with Paley is that a complex design does not necessarily require a designer.

Lungs, for example, appear to have evolved in fish long before any air-breathing land vertebrate existed. There are still some primitive, air-breathing fish alive today, such as the bichir of Africa. Lungs are helpful to the bichir, but not absolutely essential, because it can get oxygen through its gills. But by breathing through its lungs from time to time, a bichir can boost its swimming stamina with an extra supply of oxygen to the heart. Around 360 million years ago, one lineage of air-breathing fish began spending some time on dry land. As they increased their time out of the water, they adapted their limblike fins to support their weight as they walked. Eventually their gills disappeared altogether. Over the course of millions of years, these early tetrapods became completely dependent on their lungs—a process documented with fossils.

Here we have a complex system (the tetrapod body) that collapses if you take away one part (the lungs). And yet a study of the fossil record and living animals shows that this system is not irreducibly complex. Evolution can add something like a lung because it is advantageous; in time, that added piece of anatomy may become essential, and thus impossible to remove.

Evolution can produce complex biochemistry from simpler precursors in much the same way as it can produce complex anatomy. In recent years, scientists have been able to put together strong hypotheses about two cases: how Antarctic fish keep from freezing to death and how our blood clots.

First the unfrozen fish: a family of species called the notothenioids survive in subfreezing waters thanks to a natural antifreeze in their bloodstream.

Lungs are absolutely essential to us but not to air-breathing fish, such as this lungfish.

Their liver creates a sugar-studded protein that bonds to the surface of microscopic ice crystals and stops them from growing. Antifreeze has allowed notothenioids to thrive in the Antarctic Ocean: 94 species are known so far, with new ones discovered every year.

Making antifreeze is a complex process, and without it the notothenioids would die. But just because it is complex doesn't mean that evolution could not have produced it. The biochemist Chi-Hing C. Cheng and her fellow researchers at the University of Illinois have discovered clues as to how the antifreeze gene evolved. They found that it has some remarkable resemblances to another gene expressed not in the liver but in the pancreas, where it produces a digestive enzyme that is injected into the intestines. Cheng found that the instructions for making an antifreeze molecule are contained in a nine-base sequence that is repeated dozens of times in a single gene. (The repetition allows a single gene to produce a lot of antifreeze.) Cheng discovered that the same nine-base sequence is embedded in the digestive-enzyme gene. The only reason that the enzyme gene doesn't make antifreeze is that the sequence is located in a section of "junk DNA" that gets edited out of the gene before its sequence is used to build a protein.

Cheng discovered other similarities between the antifreeze gene and the digestive-enzyme gene. At the front of each gene is a sequence that acts like a shipping label, telling cells to secrete their proteins rather than keep them within the cell, and the labels match almost perfectly. And at the end of each gene is a command that tells a cell to stop translating it into RNA; the sequences of the two commands are almost identical.

With these discoveries in hand, Cheng proposed how the antifreeze gene came into being. At some point in the distant past, the digestive-enzyme gene was accidentally duplicated. The original version continued making its enzyme, while the extra copy went through a series of mutations. The nine-base sequence was moved to a new part of the gene where it was no longer edited out as junk DNA, but produced a protein: antifreeze. Later, mutations duplicated the nine bases many times over so that the gene could make even more antifreeze. And as the antifreeze portion of the gene grew, the original part that made the digestive enzyme was eliminated. Eventually all that remained of the old gene was the shipping label at the front and the translation-stopping signal at the end.

Because the original digestive enzyme is made in the pancreas, Cheng suggests that the earliest versions of the antifreeze protein were made there

Scientists catch fish that live in polar waters to study their natural antifreeze. The genes responsible for this remarkable trait still retain traces of their evolution from other genes.

as well. The pancreas produces digestive enzymes that it injects into the intestines to help break down food. Because the fish swallows cold water, the intestines are a prime place for ice to form. As a result, a primitive antifreeze protein in its gut would have allowed a fish to survive in cold waters where it would otherwise freeze to death.

Eventually the signals that control when and where the antifreeze genes become active evolved as well. Instead of switching on in the cells of the pancreas, the antifreeze genes started becoming active in the liver. While the pancreas sends its enzymes to the intestines, the liver can inject enzymes into the bloodstream. By filling the fish's blood with antifreeze, it could protect the fish's entire body against ice and let it withstand even colder temperatures.

In 1999 Cheng's group found a remarkable confirmation of their hypothesis: a chimera gene. In the DNA of an Antarctic fish they discovered a gene that contained the instructions both for making antifreeze *and* for making the digestive enzyme. This was exactly the sort of intermediate on the path from a digestive gene to a true antifreeze gene that the scientists had predicted.

THE HISTORY OF A CLOT

As remarkable as the antifreeze in notothenioids may seem—as tempted as some might be to call it Intelligently Designed—the evidence strongly suggests that gene duplications and less drastic mutations gradually created it. With these sorts of mutations, evolution can do even more: it can create entire systems of molecules that we depend on for our survival.

Consider the molecules that create blood clots. When we are healthy these molecules (called clotting factors) course through our blood, doing nothing. But if you should cut yourself with a knife and the blood in the ruptured vessels mingles with the surrounding tissues, that's no longer the case. Some of the proteins in the tissue will react with one type of clotting factor and activate it. Now the clotting factor can start a chain reaction: it grabs a second type of clotting factor and activates it; it in turn activates a third type, and so on through a series of reactions. A final clotting factor slices apart a molecule called fibrinogen, turning it into a sticky substance that forms a clot. The complexity of the clotting system is its strength: a single original clotting factor can activate several factors in the next step, and these in turn can switch on many more factors in the third step. From a tiny trigger, millions of fibrinogen molecules can be activated.

It's a remarkable system for stopping wounds, no doubt. And it depends on all its parts—if people are born without one type of clotting factor they become hemophiliacs, for whom a scratch may mean death. But that doesn't mean that it had to have been Intelligently Designed.

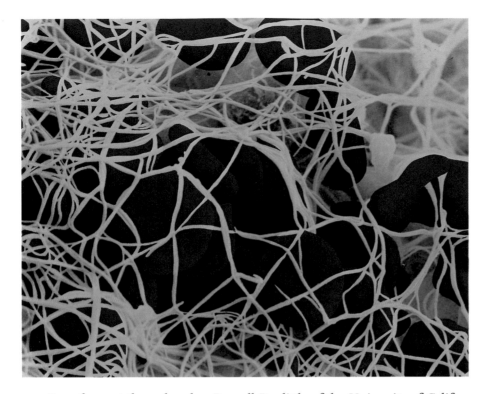

A blood clot forms around a wound. Scientists are working out the evolution of the biochemistry behind this process.

Over the past three decades, Russell Doolittle of the University of California at San Diego has been testing a hypothesis for how blood clotting in vertebrates evolved. The fact that clotting factors can activate other clotting factors is nothing special; all animals have enzymes that activate proteins to make them ready to carry out many different kinds of jobs. One of these enzymes might be the ancestor of all the clotting factors.

Imagine an early vertebrate that lacked any clotting factor whatsoever. That actually isn't so hard to picture, given that animals such as earthworms and starfish don't have any either. They don't bleed to death, because they have cells in their bloodstream that can become sticky and form crude clots. Now imagine that the gene for a slicing enzyme was duplicated. The extra copy evolved into a simple clotting factor made only in the bloodstream. It would be activated in a wound and slice apart proteins in the blood, some of which would turn out to be sticky. A clot would form, one that was superior to the old kind. If this initial clotting factor was duplicated, the chain reaction would double in length and become more sensitive. Add another factor, and it gets more sensitive still. Gradually the entire clotting process could have evolved this way.

Doolittle has put this hypothesis to the test and found support wherever he has looked. The clotting factors turn out to be very similar to one another, and Doolittle discovered they are all closely related to a digestive enzyme. Doolittle predicted that fibrinogen—the protein that clotting factors turn into a sticky, clot-forming substance—descended from a protein in our invertebrate ancestors that did some other job. Doolittle looked for cousins of fibrinogen in our close invertebrate relatives and found one in the sea cucumber. Even though sea cucumbers can't carry out a clotting cascade, they have a fibrinogen-like protein in their bodies.

Testing these hypotheses hasn't been easy. The tale of the antifreeze gene has required scientists to trawl for fish in iceberg-choked oceans. The blood-clotting story has required 30 years of lab work. They do not account for the evolution of cholesterol or of collagen, or of the hundreds of thousands of other molecules manufactured by life on Earth. Advocates of Intelligent Design make a great fuss over how little evolutionary biologists know about biochemical evolution. They take this ignorance as evidence that these molecules are too complex to submit to evolutionary explanations, and that Intelligent Design must be right. But all it demonstrates is that 50 years after the discovery of DNA, scientists still have plenty to learn about the history of life.

PUSHING BACK THE TEST

Once Intelligent Design is shorn of its distracting attacks on evolution, there's very little real science left to consider. How does Intelligent Design account for all of the evidence in favor of evolution, from the fossil record to mutation rates to the similarities and differences between species? At what exact point did the designer intervene in the evolution of the horse, or bird flight, or the Cambrian explosion? And what did the designer do? How can we test these claims? What predictions has Intelligent Design made that have resulted in important new discoveries? If you look for answers to these questions, you end up only with contradictions, untestable claims, or, most often, silence.

In 1996 Michael Behe attempted to make the case for Intelligent Design in *Darwin's Black Box*. Behe, a biochemist at Lehigh University, presented some examples of complex biochemistry and declared that they could not have evolved. Yet at the same time, he granted that "on a small scale, Darwin's theory has triumphed." In other words, in the world of Intelligent Design, things do evolve: finches change the size of their beaks; HIV adapts to new hosts; birds introduced into the United States have diversified into new groups. But these sorts of small changes can't produce the complexity of life.

The problem for Intelligent Design is that small changes add up to big effects. As time passes, mutations build up in the DNA of populations of animals and other organisms. Once enough small changes have amassed, populations can evolve into distinct species. Scientists can use the genetic differences to work out the relationships of these species. If Behe accepts microevolution, he has no choice but to accept the tree of life as well. (According to this tree, incidentally, humans are close cousins to chimpanzees. Creationists who don't savor the idea that their ancestors were apes must recognize that Intelligent Design surrenders this point.) And since Behe offers no objection to the fossil record and isotopic dating, he apparently accepts that the tree of life has branched out over the course of the last 4 billion years.

So where does evolution stop and "design" begin? Hard to say. Did an Intelligent Designer step in 500 million years ago and install the clotting cascade in the earliest vertebrates? Did the designer step in 150 million years ago, when mammals evolved a complex set of molecules that allowed a placenta to implant itself into a mother's uterus and prevent her immune system from rejecting a fetus? Or every time a species of milkweed has invented a new kind of poison to fight off insects? Behe never tells us.

Making matters even murkier, Behe even grants that some molecules don't look designed. Hemoglobin, the molecule that carries oxygen in red blood cells, has a structure remarkably like myoglobin, an oxygen-storing molecule in our muscles. Behe therefore says that hemoglobin is not a good example of Intelligent Design. "The behavior of hemoglobin can be achieved by a rather simple modification of the behavior of myoglobin," he writes. But myoglobin—now, that *has* to be irreducibly complex, according to Behe, because he cannot conceive how it could have evolved.

By muddling its claims with evolution, Intelligent Design cannot create hypotheses that can be tested. If I propose that a molecule is irreducibly complex and evidence emerges that it could have evolved by gene duplication or some other process, I can write it off as a product of evolution and move the whole argument back to the earlier molecule. Behe himself toys with pushing Intelligent Design all the way back to the very beginning of life's history. He speculates that the first cell might have been designed with all the complex networks of genes that were used later in different organisms. Different kinds of organisms continued to use certain genes, while others were silenced.

"This notion leaves so much of molecular evolution unexplained that it's hard to know where to start," says H. Allen Orr, a biologist at the University of Rochester. It is true that some genes do get silenced over time. A gene may get duplicated, for example, and one of the copies mutates until it can't make a protein. These useless genes are known as pseudogenes. But the pseudogenes we carry, Orr points out, resemble our own active genes. If Behe were correct, you'd expect to find pseudogenes in our DNA that looked like the active genes in all sorts of other species. Why don't we carry pseudogenes for making rattlesnake venom or flower petals? Why do we instead share so many pseudogenes with chimpanzees?

Evolution offers a straightforward explanation: only after our own ancestors diverged from those of flowers and rattlesnakes did these pseudo-genes evolve. Intelligent Design, on the other hand, can claim only that it just so happened that silenced genes were silenced the way they were. As with previous versions of creationism, Intelligent Design leaves us with a designer who goes to enormous lengths to trick us into thinking that life evolved.

Intelligent Design fails because it abandons the central quest of science. "If you're allowed just to postulate something complicated enough to design a universe intelligently, then you've sold the past," says Richard Dawkins. "You've simply allowed yourself to assume the existence of exactly the thing which we're trying to explain. The beauty of the theory of evolution by natural selection is that it starts with simple things, and it builds up slowly and gradually to complex things, including things complex enough to design things—brains, in other words. If you allow yourself to use the idea of design right from the start, then you are simply giving up the beginning. You're simply not providing any kind of explanation at all."

THE LIMITS OF SCIENCE

Unable to mount an effective scientific argument, champions of Intelligent Design and older forms of creationism resort to rhetoric. They claim, for instance, that evolution is actually ideology, spawned from a cult of naturalism, which claims that God has no role in the universe and events have only natural causes. Darwinists "have adhered to the myth out of self-interest and a zealous desire to put down God," writes Phillip Johnson, a law professor and outspoken creationist. Johnson claims that evolutionary biologists refuse to consider the possibility that supernatural intervention has influenced the universe and are blind to the weaknesses of evolution. In a fair hearing—in which divine intervention could be considered as a possible explanation for life's history—Johnson claims that creationism would win.

Yet science, whether it takes the form of chemistry, physics, or evolutionary biology, can explain only the lawlike regularities of the world. If God were to change the mass of protons every morning, it would be impossible for physicists to make any predictions about how atoms work. The scientific method does not claim that events can have only natural causes but that the only causes that we can understand scientifically are natural ones. As powerful as the scientific method may be, it must be mute about things beyond its scope. Supernatural forces are, by definition, above the laws of nature, and thus beyond the scope of science.

Johnson and other creationists direct their fury at evolutionary biology, but in effect they are attacking every branch of science. When microbiologists study an outbreak of resistant tuberculosis, they do not research the possibility that it is an act of God. When astrophysicists try to figure out the sequence of events by which a primordial cloud condensed into our solar system, they do not simply draw a big box between the hazy cloud and the well-formed planets and write inside it, "Here a miracle happened." When meteorologists fail to predict the path of a hurricane, they do not claim that God's will pushed it off course.

Science cannot simply cede the unknown in nature to the divine. If it did, there would be no science at all. As University of Chicago geneticist Jerry Coyne puts it, "If the history of science shows us anything, it is that we get nowhere by labeling our ignorance 'God.'"

"Creation science" in any form doesn't influence the way practicing scientists study the history of life. Paleontologists continue to discover fossils crucial to our understanding of how humans, whales, and other animals came into being. Developmental biologists continue to listen to the symphony of embryo-building genes to understand how the Cambrian explosion took place. Geochemists continue to uncover isotopic clues about when life first appeared on Earth. Virologists continue to discover the strategies that viruses such as HIV evolve in order to outwit their hosts. For all of them, evolutionary biology, not creationism, remains the foundation of their work.

And yet, despite this failure as a science, creationists are trying as hard as ever to gain control of the way American public schools teach science. For the

Pope John Paul II greets the faithful gathered in St. Peter's Square on May 27, 1998, after delivering an address on evolution. He declared that evolution was compatible with Christianity—although any claims that the spirit could emerge from living matter "are incompatible with the truth about man."

most part, their work has gone unnoticed by the public at large, but in 1999 a scandal in Kansas brought creationism back to national headlines.

CREATIONISM IN KANSAS

Kansas high school students take a statewide exam based on a set of standards approved by the Kansas State Board of Education. In 1998 the board asked a committee of scientists and science teachers to revise the standards. The committee based their work on standards that had been created by the National Research Council in 1995, and along the way they consulted the American Association for the Advancement of Science and other major science organizations. In May 1999, the committee presented the board with their proposed standards, based on the latest consensus among scientists on everything from astronomy to ecology. Among other things, the standards would have called for students to understand the fundamentals of evolution—how lineages adapt to their environment and how biologists use evolutionary theory to explain the nature of life's diversity.

But when the committee presented its standards, something peculiar happened. One of the board members came forward with a different set of standards, which, it was later discovered, had been written by a creationist organization based in Missouri. The writing team refused to accept them but agreed to try to address the concerns of the conservative members of the board in their own standards. The board demanded that the standards include a statement about tolerating different points of view, which the writing team inserted. The board demanded that the writing team define microevolution and macroevolution, so the writing team explained how generation-by-generation changes (microevolution) produce large-scale patterns and processes that fit under the label of macroevolution, such as the origin of new body plans and changing extinction rates. But the board then tried to get the writing team to drop any further discussion of macroevolution, which the writing team refused to do.

When the board met in August, the writing committee decided to take a stand: they demanded that the board vote to either accept or reject their standards. But the board abruptly switched the committee's standards with yet another version of its own. At first glance, these new standards looked like the ones the writing committee had presented, but a closer look revealed that most references to evolution had been deleted. In the few passages that survived, the standards now declared that natural selection should be taught as a process that "does not add new information to the existing genetic code." The state exams would not test students on evolution, or even on continental drift, the age of Earth, or the Big Bang. The board approved the new standards by a vote of 6 to 4.

By deleting original passages and adding false statements, the board would have made science classes perfectly amenable to strict creationists. Consider how students would be taught that "natural selection does not add new information to the existing genetic code." In fact mutations such as gene duplications, combined with natural selection, create new kinds of genetic information all the time. By inserting this falsehood into the standards, the board was supporting the "microevolution yes, macroevolution no" claims of creationism.

The geology requirements also fell in line with creationism. The board dropped the requirement that students understand continental drift, the foundation of all modern studies of our planet. Instead, they suggested that students learn that "at least some stratified rocks may have been laid down quickly, such as Mount Etna in Italy or Mount St. Helens in Washington State." Here they raised a specious argument favored by Flood Geology creationists to try to explain how geological formations could have formed in a few thousand years.

In making the classroom safe for creationism, the state board was destroying the committee's attempts to teach students the fundamental nature of science. A theory was no longer "a well-substantiated explanation" but merely "an explanation"—in other words, a guess. Science was no longer "the human activity of seeking natural explanations for what we observe in the world around us" but one of seeking "logical explanations." With that wording, the board implied that scientists could discover supernatural forces.

Journalists quickly got wind of the board's decision, and the state board of education was suddenly yanked out of obscurity. Governor Bill Graves announced he was disgusted by the board's actions, and the presidents and chancellors of every Kansas state university condemned the vote. The board members who had voted in favor of the creationist standards suddenly found themselves besieged by the national press and claimed that they had acted only in the interests of good science. But in the process they ended up revealing more of their ignorance. "Where is the evidence for that canine-looking creature that somehow has turned into a porpoise-looking creature, or that cow that has somehow turned into a whale?" board of education chairwoman Linda Holloway asked a reporter for NBC, apparently unaware of the fossil record of whales with legs.

A grassroots opposition to the denatured standards sprang up in Kansas. It gathered strength in the months that followed, and in the next round of school board elections in 2000 the creationist-leaning bloc suffered heavy losses. Two members (including Holloway) were defeated by moderate Republicans in the primary, and a third member resigned and was replaced with another moderate Republican. In February 2001 the board finally approved the original standards, with the teaching of evolution intact.

Creationists may have lost this round in Kansas, but they continue their political fight throughout the United States. In May 2000, Intelligent Design proponents were welcomed to Capitol Hill by conservative congressmen to describe their ideas. The Oklahoma legislature has passed a law declaring that biology textbooks must inform students that the universe was created by God. In Alabama, textbooks are pasted with warnings that evolution is a controversial theory, not a fact. In the spring of 2001 a bill was introduced in Louisiana preventing the state government from distributing false information—such as radiometric dating.

These sorts of laws are not the only way to stop teachers from teaching evolution—intimidation works as well. In order to avoid controversy and stand-offs with some parents, high school biology teachers often shy away from Darwin. "I talk to teachers at science-teacher conferences and they tell me that their principal has told them just to skip evolution this year because it's an election year," says Eugenie Scott, executive director of the National Center for Science Education. "There is a new school board coming on and they don't want any problems. That is crazy. I mean, that is just not the way to run a coherent curriculum."

PAYING THE PRICE

The result of these conflicts is not a new generation of creationists, but a generation of students who don't understand evolution. This is a bad state of affairs, and not simply because the theory of evolution stands as one of the greatest scientific accomplishments of the past 200 years. Many careers that students might want to pursue actually depend on a solid understanding of evolution.

To search for oil and minerals, for example, you have to understand the history of life on Earth. For 4 billion years, species have evolved, given rise to new species, and become extinct. Their fossils can act as markers for rocks that were formed while they were alive. If geologists find some distinctive fossil plankton in a formation of rock that's rich in oil, they know they may find oil somewhere else if they can find that same plankton again.

Evolution is even more important to biotechnology, because when researchers tinker with life itself, they have to deal with the fact that it evolves. The resistance that bacteria have to many antibiotics didn't just happen: it unfolded according to the principles of natural selection, as the bacteria with the best genes for fighting the drugs prospered. Without understanding evolution, a researcher has little hope of figuring out how to create new drugs and determine how they should be administered.

The same goes for vaccines. As microbes evolve, they become isolated into genetically distinct populations, which create new branches on the evolutionary tree. A vaccine may work against one strain of a disease like AIDS, but fail against more common ones because they're only distantly related. The evolutionary tree also tells scientists where diseases come from (in the case of AIDS, most likely chimps). That in turn can guide them to possible cures.

Evolution on its grandest scale can be just as crucial to business. Some of the biggest efforts in biotechnology these days are going into genome sequencing—decoding the complete sequence of our genetic code, as well as that of other life-forms such as bacteria, protozoa, insects, and worms. The money's going in because big profits may come out. Scientists are studying the genes of fruit flies because humans have very similar genes. Experiments on the flies may someday lead to medical miracles such as extending the human life span. But scientists will first need to learn how that similarity evolved. Medicine, in other words, has its roots in the Cambrian explosion.

The same kinds of applications may also come from understanding how different species have fused together over time. Take malaria. This disease, which kills around 2 million people every year, defies the best efforts of modern medicine. Recently scientists have discovered that the parasite that causes malaria carries genes that come from algae. Perhaps a billion years ago, the ancestor of this parasite engulfed a species of algae. Instead of digesting it, it turned the algae into a symbiotic partner, and today some of the algae's genes remain. This discovery may reveal a way to attack malaria. If the parasite has some algae-like qualities, the poisons that are known to kill plants might be able to kill it as well. Without an evolutionary framework, scientists would probably never have thought to try to destroy malaria with weed killers.

Biotechnology will keep speeding ahead, and it will keep relying on evolution as its central organizing principle. And it won't wait for people who don't understand how life evolves because someone else decided they didn't need to.

IN THE COUNTRY OF ASA GRAY

When William Jennings Bryan waged his battle against evolution in the 1920s, he was motivated not so much by any real scientific objection, but by his disgust at the thought of a world given over to Darwin. For Bryan, evolution threatened the notion of a moral universe created by God, with humanity created specifically in his image. All that was left was a brutal struggle for supremacy with no purpose whatsoever.

Bryan may have confused evolutionary biology with some of the social movements of his day, but he raised a fundamentally important question, one that no amount of evidence in favor of evolution can ever obscure: Is there a place for God in a world where evolution operates, where natural selection plays the role once given to the designer?

God and evolution are not mutually exclusive. Evolution is a scientific phenomenon, one that scientists can study because it is observable and predictable. But digging up fossils does not disprove the existence of God or a higher purpose for the universe. That is beyond science's power. Asa Gray put it best when he said that claiming that Darwinism is a religious belief "seems much the same as saying my belief is Botany."

The United States has been home to a long line of religious evolutionists ever since Gray, an evangelical Christian, introduced the country to *Origin of Species*. Gray once commented that Darwin's theory "can be held theistically or atheistically. Of course, I think the latter wrong & absurd." When the Kansas State Board of Education tried to eliminate evolution from high schools in 1999, one of the leading critics of the decision was another evangelical Christian, Kansas State University geologist Keith Miller. "God is the creator of all things, and nothing would exist without God's continually willing it to be," Miller declares, but he nevertheless accepts the evidence for evolution. "If God used and providentially controlled evolutionary mechanisms in the creation of plants and animals, I see no reason to reject an evolutionary origin for humankind."

For Kenneth Miller, a biochemist at Brown (and no relation to Keith Miller), evolution affords plenty of room for the God of his Catholic faith. In his 1999 book, *Finding Darwin's God*, he points out that the mutations that make evolution possible take place on a quantum level, and as a result we can never know with perfect certainty whether a particular mutation will take place. When a cosmic ray whistles into the inner sanctum of a cell's nucleus and collides with DNA, it may or may not transform one of the bases. "Evolutionary history can turn on a very very small dime—the quantum state of a single subatomic particle," says Miller. And thanks to the uncertainty of quantum physics, if God influences evolution by mucking with mutations, His effects will be scientifically undetectable.

But even if God does influence mutations, that doesn't mean that He controls life like a micromanager. Miller points out that many Christians have long accepted that human history is influenced by chance and contingency, although it may indeed have an overall purpose we cannot fully comprehend. Nature, he argues, is no different. And thanks to that chance and contingency, life itself can evolve. "A God who presides over an evolutionary process is not an impotent, passive observer," says Miller. "Rather, He is one whose genius fashioned a fruitful world in which the process of continuing creation is woven into the fabric of matter itself."

Miller suspects that evolution has a destiny embedded within it, and we are part of it. "Sooner or later it would have given the Creator exactly what He was looking for—a creature who, like us, could know Him and love Him, could perceive the heavens and dream of the stars, a creature who would eventually discover the extraordinary process of evolution that filled His earth with so much life."

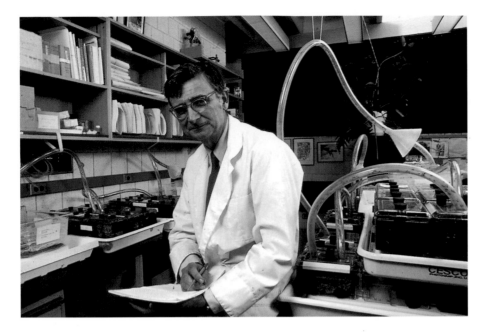

Rejecting what he calls a "biological God," biologist Edward O. Wilson allows for the possibility of Deism: a belief in a God who created the universe and allowed it to unfold according to natural laws.

Because God constructed the universe to follow certain natural laws, He made it possible for us to comprehend His creation, but thanks to chance and contingency, we have the freedom that Christianity demands. "God stands back from His creation not to abandon His creatures but to allow His people true freedom," Miller writes. "He used evolution as the tool to set us free."

Edward O. Wilson, the champion of sociobiology, has offered a very different vision of God in his writings. He grew up in a Southern Baptist family, and at age 14 he chose to be baptized. In a Pensacola, Florida, church a pastor dipped him into a tank of water, "like a ballroom dancer," he later recalled, "backward and downward, until my entire body and head dipped beneath the surface."

The baptism affected Wilson deeply, but in a physical way rather than in the spiritual way he had expected. He wondered if everything—the world itself—was only physical. "And something small somewhere cracked. I had been holding an exquisite, perfect spherical jewel in my hand, and now, turning it over in a certain light, I discovered a ruinous fracture."

Wilson abandoned what he calls "a biological God, one who directs organic evolution and intervenes in human affairs." Instead, he now leans toward Deism, the belief that God set in motion the universe and did not need to tinker with it afterward. But Wilson is not disheartened by the thought of living in such a universe:

> The true evolutionary epic, retold as poetry, is as intrinsically ennobling as any religious epic. Material reality discovered by science already possesses more content and grandeur than all religious cosmologies combined.

The continuity of the human line has been traced through a period of deep history a thousand times as old as that conceived by the Western religions. Its study has brought new revelations of great moral importance. It has made us realize that *Homo sapiens* is far more than an assortment of tribes and races. We are a single gene pool from which individuals are drawn in each generation and into which they are dissolved the next generation, forever united as a species by heritage and a common future. Such are the conceptions, based on fact, from which new intimations of immortality can be drawn and a new mythos evolved.

These three scientists, one an evangelical Christian, one a Catholic, one a Deist, cannot speak for all scientists, let alone all people. Science is the business of finding theories that explain the natural world and generating hypotheses that can be tested with evidence from our senses. It is up to all of us—nonscientists and scientists, Christians and Jews, Muslims and Buddhists, believers and agnostics and atheists—to ponder what the world actually means.

DARWIN'S SILENCE

Some readers may not be happy to be plunged into such a cacophony of opinions at the end of this book. It might be more comforting to be guided to some single gleaming truth. But this ending is probably as Darwin would have wanted it.

Darwin wrestled with his spirituality for most of his adult life, but he kept his struggles private. When he boarded the *Beagle* at age 22 and began his voyage around the world, he was a devout Anglican. As he read Lyell and saw the slow work of geology in South America, he began to doubt a literal reading of Genesis. And as he matured as a scientist on the journey, he grew skeptical of miracles. Nevertheless, Darwin still attended Captain FitzRoy's weekly services on the ship, and on shore he sought churches whenever he could find them. While in South Africa, he and FitzRoy wrote a letter together in which they praised the role of Christian missions in the Pacific. When Darwin returned to England, he was no longer a parson in the making but was certainly no atheist.

In the notebooks Darwin began keeping on his return, he explored every implication of evolution by natural selection, no matter how heretical. If eyes and wings could evolve without help from a designer, then why couldn't behavior? And wasn't religion just another type of behavior? All societies had some type of religion, and their similarities were often striking. Perhaps religion had evolved in our ancestors. As a definition of religion, Darwin jotted down, "Belief allied to instinct."

Yet these were little more than thought experiments, speculations that distracted Darwin every now and then from his main work: discovering how evolution could produce the natural world. Darwin did experience an intense spiritual crisis during those years, but science was not the cause.

At age 39, Darwin watched his father Robert slowly die over the course of months. He thought about his father's private doubts about religion, and he wondered what those doubts would mean to Robert in the afterlife. At the time Darwin happened to be reading a book by Coleridge called *Aids to Reflection,* about the nature of Christianity. Nonbelievers, Coleridge declared, should be left to suffer the wrath of God.

Robert Darwin died in November 1848. Throughout Charles's life, his father had shown him unfailing love, financial support, and practical advice. Was Darwin now supposed to believe that his doubting father was going to be cast into eternal suffering in hell? If that were so, then many other nonbelievers, including Darwin's brother Erasmus and many of his best friends, would follow him as well. If that was the essence of Christianity, Darwin wondered why anyone would want such a cruel doctrine to be true.

Shortly after his father's death, Darwin's health turned for the worse. He vomited frequently and his bowels filled with gas. He turned to hydropathy, a Victorian medical fashion in which a patient was given cold showers and steam baths and wrapped in wet sheets. He would be scrubbed until he looked "very like a lobster," he wrote to Emma. His health improved, and his spirits rose even more when Emma discovered that she was pregnant again. In November 1850 she gave birth to their eighth child, Leonard. But within a few months death would return to Down House.

In 1849 three of the Darwin girls—Henrietta, Elizabeth, and Anne—suffered bouts of scarlet fever. While Henrietta and Elizabeth recovered, 9-year-old Anne remained weak. She was Darwin's favorite, always throwing her arms around his neck and kissing him. Through 1850 Anne's health still did not rebound. She would vomit sometimes, making Darwin worry that "she inherits, I fear with grief, my wretched digestion." The heredity that Darwin saw shaping all of nature was now claiming his own daughter.

In the spring of 1851 Anne came down with the flu, and Darwin decided to take her to Malvern, the town where he had gotten his own water cure. He left her there with the family nurse and his doctor. But soon after, she developed a fever and Darwin rushed back to Malvern alone. Emma could not come because she was pregnant again and only weeks away from giving birth.

When Darwin arrived in Anne's room in Malvern, he collapsed on a couch. The sight of his ill daughter was awful enough, but the camphor and ammonia in the air reminded him of his nightmarish medical school days in Edinburgh, when he had seen children operated on without anesthesia. For a week—Easter week, no less—he watched her fail, vomiting green fluids. He wrote agonizing letters to Emma. "Sometimes Dr. G. exclaims she will get through the struggle; then, I see, he doubts.—Oh my own it is very bitter indeed."

Anne died on April 23, 1851. "God bless her," Charles wrote to Emma. "We must be more & more to each other my dear wife."

When his father had died, Darwin had felt a numb absence. Now, when he came back to Down House, he mourned in a different way: with a bitter, rageful, Job-like grief. "We have lost the joy of our household, and the solace of our old age," he wrote. He called Anne a "little angel," but the word gave him no comfort. He could no longer believe that Anne's soul was in heaven, that her soul had survived her unjustifiable death.

It was then, 13 years after Darwin discovered natural selection, that he gave up Christianity. Many years later, when he put together an autobiographical essay for his grandchildren, he wrote, "I think generally (and more and more as I grow older), but not always, that an agnostic would be the most correct description of my state of mind."

Darwin did not trumpet his agnosticism. Only by poring over his private autobiography and his letters have scholars been able to piece together the nature of his faith after Anne's death. Darwin wrote a letter of endorsement, for example, to an American magazine called *The Index,* which championed what it called "free religion," a humanistic spirituality in which the magazine claimed "lies the only hope of the spiritual perfection of the individual and the spiritual unity of the race."

Yet when *The Index* asked Darwin to write a paper for them, he declined. "I do not feel that I have thought deeply enough [about religion] to justify any publicity," he wrote to them. He knew that he was no longer a traditional Christian, but he had not sorted out his spiritual views. In an 1860 letter to Asa Gray, he wrote, "I am inclined to look at everything as resulting from designed laws, with the details, whether good or bad, left to the working out of what we may call chance. Not that this notion at all satisfies me. I feel most deeply that the whole subject is too profound for human intellect. A dog might as well speculate on the mind of Newton."

While Haeckel and others tried to use evolution to overturn conventional religion, Darwin stayed quiet. In private he complained about the way social Darwinism twisted his own work. Once, in a letter to Lyell, he wrote sarcastically, "I have received in a Manchester newspaper rather a good squib, showing that I have proved 'might is right' and therefore that Napoleon is right, and every cheating tradesman is also right." But Darwin decided not to write his own spiritual manifesto. He was too private a man for that.

Despite his silence, Darwin was often pestered in his later years for his thoughts on religion. "Half the fools throughout Europe write to ask me the stupidest questions," he groused. The inquiring letters reached deep into his most private anguish. To strangers, his responses were much briefer than the one he had sent to Gray. To one correspondent, he simply said that when he had written *Origin of Species,* his own beliefs were as strong as a prelate's. To another, he wrote that a person could undoubtedly be "an ardent theist and an evo-lutionist," and pointed to Asa Gray as an example.

Yet to the end of his life, Darwin never published any-thing about religion. Other scientists might declare that evo-lution and Christianity were perfectly in harmony, and others such as Huxley might taunt bishops with agnosticism, but Darwin would not be drawn out. What he actually believed or didn't, he said, was of "no consequence to any one but myself."

Darwin (shown here in an 1878 photograph) kept his spiritual views far from public view. His struggles with faith had less to do with evolution than with the death of his daughter.

Darwin and Emma rarely spoke about his faith after Anne's death, but he came to rely on her more with every passing year, both to nurse him through his illnesses and to keep his spirits up. At age 71, he looked over the letter she had written to him just after they married, urging him to remember what Jesus had done for him. On the bottom he wrote, "When I am dead, know that many times, I have kissed & cryed over this."

Two years later Emma caught him in her arms when he collapsed at Down House. For the next six weeks she cared for him as he cried out to God and coughed up blood and slipped into unconsciousness. On April 19, 1882, he was dead.

Emma planned to have her husband buried in the local churchyard, but Huxley and other scientists thought the nation should honor him instead. When Darwin had started as a scientist, the word *scientist* did not even exist yet. Natural history was butterfly collecting in the service of piety. Fifty years later scientists were society's leaders, looking deeper every year into the work-ings of life itself. Westminster Abbey was not only for kings and clergy—the explorer David Livingstone was buried there, as was James Watt, the inventor of the steam engine. Colonies and industry had made England great, and so too, it was agreed, had Darwin.

A few days later Westminster Abbey filled with mourners, and Darwin's coffin was brought to the center of the transept. A choir sang a hymn adapted from the Book of Proverbs.

Happy is the man that findeth wisdom, and the man that getteth understanding.

She is more precious than rubies: and all the things thou canst desire
   are not to be compared unto her.
Length of days is in her right hand; and in her left hand riches and
   honour.
Her ways are ways of pleasantness, and all her paths are peace.

Darwin was lowered into the floor of the abbey, close to Newton. Now he would be silent forever about his faith. He had left us behind, in the natural world he had unveiled. It is an ancient world, in which we are an infant species; a braided river of genes flows around us and through us, its course altered by asteroids and glaciers, by rising mountains and spreading seas. When Darwin wrote *Origin of Species,* he promised his readers "a grandeur in this view of life," and life now displays far more grandeur than even Darwin appreciated. He began the exploration of this remarkable world, and he left us to walk deeper into it, without him.

FURTHER READING

CHAPTER 1: DARWIN AND THE *BEAGLE*

Appel TA. *Cuvier-Geoffroy debate: French biology in the decades before Darwin.* New York: Oxford University Press, 1987.

Browne EJ. *Charles Darwin: A biography.* New York: Knopf, 1995.

Coleman WR. *Georges Cuvier, zoologist: A study in the history of evolution theory.* Cambridge, Mass.: Harvard University Press, 1964.

Darwin C. *Journal of researches into the natural history and geology of the countries visited during the voyage of H.M.S. 'Beagle' round the world, under the command of Capt. Fitz Roy, R.N.* London: Henry Colborn, 1839.

Desmond AJ, Moore JR. *Darwin.* London: Michael Joseph, 1991.

Koerner L. *Linnaeus: Nature and nation.* Cambridge, Mass.: Harvard University Press, 1999.

Lovejoy AO. *The great chain of being: A study of the history of an idea.* Cambridge, Mass.: Harvard University Press, 1936.

Lyell C. *Principles of geology; or, The modern changes of the earth and its inhabitants, considered as illustrative of geology.* London: John Murray, 1850.

Mayr E. *The growth of biological thought: Diversity, evolution, and inheritance.* Cambridge, Mass.: Belknap Press, 1982.

Paley W. *Natural theology; or, Evidences of the existence and attributes of the Deity.* London: Wilks and Taylor, 1802.

Rudwick MJS. *The great Devonian controversy: The shaping of scientific knowledge among gentlemanly specialists.* Chicago: University of Chicago Press, 1985.

———. *The meaning of fossils: Episodes in the history of palaeontology.* Chicago: University of Chicago Press, 1985.

Thomson KS. *HMS Beagle: The story of Darwin's ship.* New York: Norton, 1995.

CHAPTER 2: "LIKE CONFESSING A MURDER": THE ORIGIN OF *ORIGIN OF SPECIES*

Bowler PJ. *The eclipse of Darwinism: Anti-Darwinian evolution theories in the decades around 1900.* Baltimore: Johns Hopkins University Press, 1983.

Browne EJ. *Charles Darwin: A biography.* New York: Knopf, 1995.

Chambers R. *Vestiges of the natural history of creation.* New York: Wiley/Putnam, 1845.

Darwin C. *Autobiography and selected letters.* New York: Dover, 1958.

———. *A monograph on the sub-class Cirripedia, with figures of all the species.* London: Ray Society, 1851.

———. *A monograph on the fossil Balanidae and Verrucidae of Great Britain.* London: Printed for the Palaeontographical Society, 1854.

———. *On the origin of species by means of natural selection, or The preservation of favored races in the struggle for life.* London: John Murray, 1859.

Desmond AJ. *Huxley: The devil's disciple.* New York: Penguin, 1994.

———, Moore JR. *Darwin.* London: Michael Joseph, 1991.

Jones S. *Darwin's ghost.* New York: HarperCollins, 2000.

Malthus TR. *An essay on the principle of population as it affects the future improvement of society.* London: J. Johnson, 1798.

Ospovat D. *The development of Darwin's theory: Natural history, natural theology, and natural selection, 1838–1859.* Cambridge: Cambridge University Press, 1981.

Rupke NA. *Richard Owen: Victorian naturalist.* New Haven, Conn.: Yale University Press, 1994.

CHAPTER 3: DEEP TIME DISCOVERED: PUTTING DATES TO THE HISTORY OF LIFE

Bowring SA, Housh T. "The Earth's early evolution." *Science,* 1995, 269: 1535–1540.

Briggs DEG, Erwin DH, Collier FJ. *The fossils of the Burgess Shale.* Washington: Smithsonian Institution Press, 1994.

Brocks JJ, Logan GA, Buick R, Summons RE. "Archean molecular fossils and the early rise of eukaryotes." *Science,* 1999, 285: 1033–1036.

Budd GE, Jensen S. "A critical reappraisal of the fossil record of the bilaterian phyla." *Biological Reviews of the Cambrian Philosophical Society,* 2000, 75: 253–295.

Burchfield JD. *Lord Kelvin and the age of the Earth.* New York: Science History Publications, 1975.

Carroll RL. *Vertebrate paleontology and evolution.* New York: Freeman, 1988.

Dalrymple GB. *The age of the Earth.* Stanford, Calif.: Stanford University Press, 1991.

Fortey RA. *Life: A natural history of the first four billion years of life on Earth.* New York: Knopf, 1998.

Gould SJ. *Wonderful life: The Burgess Shale and the nature of history.* New York: Norton, 1989.

Knoll AH. "A new molecular window on early life." *Science,* 1999, 285: 1025–1026.

Lunine JI. *Earth: Evolution of a habitable world.* Cambridge: Cambridge University Press, 1999.

McPhee JA. *Annals of the former world.* New York: Farrar Straus & Giroux, 1998.

Prothero DR, Prothero DA. *Bringing fossils to life: An introduction to paleobiology.* New York: McGraw-Hill, 1997.

Schopf JW. *Cradle of life: The discovery of Earth's earliest fossils.* Princeton, N.J.: Princeton University Press, 1999.

———. "Solution to Darwin's dilemma: Discovery of the missing Precambrian record of life." *Proceedings of the National Academy of Sciences,* 2000, 97: 6947–6953.

Xiao S, Zhang Y, Knoll AH. "Three-dimensional preservation of algae and animal embryos in a Neoproterozoic phosphorite." *Nature,* 1998, 391: 553–558.

Zimmer C. "Ancient continent opens window on the early earth." *Science,* 1999, 286: 2254–2256.

CHAPTER 4: WITNESSING CHANGE: GENES, NATURAL SELECTION, AND EVOLUTION IN ACTION

Adami C. *Introduction to artificial life.* New York: Springer, 1998.

———, Ofria C, Collier TC. "Evolution of biological complexity." *Proceedings of the National Academy of Sciences,* 2000, 97: 4463–4468.

Adams MB. *The evolution of Theodosius Dobzhansky: Essays on his life and thought in Russia and America.* Princeton, NJ.: Princeton University Press, 1994.

Albertson RC, Markert JA, Danley PD, Kocher TD. "Phylogeny of a rapidly evolving clade: The cichlid fishes of Lake Malawi, East Africa." *Proceedings of the National Academy of Sciences,* 1999, 96: 5107–5110.

Bowler, PJ. *The Norton history of the environmental sciences.* New York: Norton, 1993.

Cook LM, Bishop JA. *Genetic consequences of man-made change.* London, England & Toronto: Academic Press, 1981.

Coyne JA, Orr HA. "The evolutionary genetics of speciation." *Philosophical Transactions of the Royal Society of London Series B: Biological Sciences,* 1998, 353: 287–305.

Dawkins R. *The blind watchmaker: Why evidence of evolution reveals a universe without design.* New York: Norton, 1996.

———. *The selfish gene.* New York: Oxford University Press, 1976.

Depew DJ, Weber BH. *Darwinism evolving: Systems dynamics and the genealogy of natural selection.* Cambridge, Mass.: MIT Press, 1995.

Dieckmann U, Doebeli M. "On the origin of species by sympatric speciation." *Nature,* 1999, 400: 354–357.

Dobzhansky TG. *Genetics and the origin of species.* New York: Columbia University Press, 1937.

Goldschmidt T. *Darwin's dreampond: Drama in Lake Victoria.* Cambridge, Mass.: MIT Press, 1996.

Grant PR, Grant BR. "Non-random fitness variation in two populations of Darwin's finches." *Proceedings of the Royal Society of London Series B,* 2000, 267: 131–138.

———, Royal Society (Great Britain). Discussion Meeting. *Evolution on islands.* Oxford & New York: Oxford University Press, 1998.

Harris RS, Kong Q, Maizels N. "Somatic hypermutation and the three R's: repair, replication and recombination." *Mutation Research,* 1999, 436: 157–178.

Henig RM. *The monk in the garden: How Gregor Mendel and his pea plants solved the mystery of inheritance.* Boston: Houghton Mifflin, 2000.

Howard DJ, Berlocher SH. *Endless forms: Species and speciation.* New York: Oxford University Press, 1998.

Huxley J. *Evolution, the modern synthesis.* New York & London: Harper & Brothers, 1942.

Janeway C. *Immunobiology: The immune system in health and disease.* New York: Garland, 1999.

Johnson TC, Kelts K, Odada E. "The Holocene history of Lake Victoria." *Ambio,* 2000, 29: 2–11.

Kondrashov AS, Kondrashov FA. "Interactions among quantitative traits in the course of sympatric speciation." *Nature,* 1999, 400: 351–354.

Koza JR. *Genetic programming III: Darwinian invention and problem solving.* San Francisco: Morgan Kaufmann, 1999.

Lenski RE, Ofria C, Collier TC, Adami C. "Genome complexity, robustness and genetic interactions in digital organisms." *Nature,* 1999, 400: 661–664.

Mayr E. *Systematics and the origin of species, from the viewpoint of a zoologist.* New York: Columbia University Press, 1942.

———. *The growth of biological thought: Diversity, evolution, and inheritance.* Cambridge, Mass.: Belknap Press, 1982.

Provine WB. *The origins of theoretical population genetics.* Chicago: University of Chicago Press, 1971.

Rensberger B. *Life itself: Exploring the realm of the living cell.* New York: Oxford University Press, 1996.

Reznick DN, Shaw FH, Rodd FH, Shaw RG. "Evaluation of the rate of evolution in natural populations of guppies (*Poecilia reticulata*)." *Science,* 1997, 275: 1934–1937.

Ridley M. *Evolution.* Cambridge, Mass.: Blackwell Science, 1998.

Sato A, Ohigin C, Figueroa F, Grant PR, Grant BR, Tichy H, et al. "Phylogeny of Darwin's finches as revealed by mtDNA sequences." *Proceedings of the National Academy of Sciences,* 1999, 96: 5101–5106.

Simpson GG. *Tempo and mode in evolution.* New York: Columbia University Press, 1984.

Stiassny M, Meyer A. "Cichlids of rift lakes." *Scientific American,* 1999, 280: 64–69.

Taubes G. "Evolving a conscious machine." *Discover,* June 1998, 73–79.

Wade MJ, Goodnight CJ. "Wright's shifting balance theory: An experimental study." *Science,* 1991, 253: 1015–1018.

Weiner J. *The beak of the finch: A story of evolution in our time.* New York: Knopf, 1994.

———. *Time, love, memory: A great biologist and his quest for the origins of behavior.* New York: Knopf, 1999.

CHAPTER 5: ROOTING THE TREE OF LIFE: FROM LIFE'S DAWN TO THE AGE OF MICROBES

Andersson SG, Zomorodipour A, Andersson JO, Sicheritz-Ponten T, Alsmark UC, Podowski RM, et al. "The genome

sequence of *Rickettsia prowazekii* and the origin of mitochondria." *Nature*, 1998; 396: 133–140.

Barns SM, Fundyga RE, Jeffries MW, Pace NR. "Remarkable
archaeal diversity detected in a Yellowstone National Park
hot spring environment." *Proceedings of the National Academy of Sciences*, 1994, 91: 1609–1613.

Cech TR. "The ribosome is a ribozyme." *Science*, 2000, 289:
878–879.

Doolittle WF. "Uprooting the tree of life." *Scientific American*,
February 2000, 90–95.

Freeland SJ, Knight RD, Landweber LF. "Do proteins predate
DNA?" *Science*, 1999, 286: 690–692.

Ganfornina MD, Sanchez D. "Generation of evolutionary
novelty by functional shift." *Bioessays*, 1999, 21: 432–439.

Gee H. *In search of deep time: Beyond the fossil record to a new
history of life*. New York: Free Press, 1999.

Gesteland RF, Cech T, Atkins JF. *The RNA world: The nature
of modern RNA suggests a prebiotic RNA*. Cold Spring Harbor, N.Y.: Cold Spring Harbor Laboratory Press, 1999.

Holland PW. "Gene duplication: Past, present and future."
Seminars in Cellular and Developmental Biology, 1999, 10:
541–547.

Kerr RA. "Early life thrived despite earthly travails." *Science*,
1999, 284: 2111–2113.

Landweber LF. "Testing ancient RNA-protein interactions."
Proceedings of the National Academy of Sciences, 1999, 96:
11067–11078.

Lazcano A, Miller SL. "The origin and early evolution of life:
prebiotic chemistry, the pre-RNA world, and time." *Cell*,
1996, 85: 793–798.

Levin BR, Bergstrom CT. "Bacteria are different: Observations, interpretations, speculations, and opinions about
the mechanisms of adaptive evolution in prokaryotes."
Proceedings of the National Academy of Sciences, 2000, 97:
6981–6985.

Margulis L. *Symbiotic planet: A new look at evolution*. New
York: Basic Books, 1998.

Maynard Smith J, Szathmáry E. *The origins of life: From the
birth of life to the origin of language*. Oxford & New York:
Oxford University Press, 1999.

Mojzsis SJ, Arrhenius G, McKeegan KD, Harrison TM, Nut-
man AP, Friend CR. "Evidence for life on Earth before
3,800 million years ago." *Nature*, 1996, 384: 55–59.

Muller M, Martin W. "The genome of *Rickettsia prowazekii*
and some thoughts on the origin of mitochondria and
hydrogenosomes." *Bioessays*, 1999, 21: 377–381.

Pace NR. "A molecular view of microbial diversity and the
biosphere." *Science*, 1997, 276: 734–740.

Sapp J. *Evolution by association: A history of symbiosis*. New
York: Oxford University Press, 1994.

Wills C, Bada J. *The spark of life: Darwin and the primeval soup*.
Cambridge, Mass.: Perseus Publishing, 2000.

Woese C. "The universal ancestor." *Proceedings of the National
Academy of Sciences*, 1998, 95: 6854–6859.

CHAPTER 6: THE ACCIDENTAL TOOL KIT: CHANCE AND CONSTRAINTS IN ANIMAL EVOLUTION

Allman JM. *Evolving brains*. New York: Scientific American
Library/Freeman, 1999.

Arthur W. *The origin of animal body plans: A study in evolutionary developmental biology*. Cambridge, England & New
York: Cambridge University Press, 1997.

Averof M, Patel NH. "Crustacean appendage evolution asso-
ciated with changes in Hox gene expression." *Nature*,
1997, 388: 682–686.

Bateson W. *Materials for the study of variation treated with espe-
cial regard to discontinuity in the origin of species*. London:
Macmillan, 1894.

Briggs DEG, Erwin DH, Collier FJ. *The fossils of the Burgess
Shale*. Washington: Smithsonian Institution Press, 1994.

Budd GE, Jensen S. "A critical reappraisal of the fossil record
of the bilaterian phyla." *Biological Reviews of the Cambrian
Philosophical Society*, 2000, 75: 253–295.

Butterfield NJ. "Plankton ecology and the Proterozoic to
Phanerozoic transition." *Paleobiology*, 1997, 23: 247–262.

Carroll RL. *Vertebrate paleontology and evolution*. New York:
Freeman, 1988.

Conway Morris S. "The Cambrian 'explosion': Slow fuse or
megatonnage?" *Proceedings of the National Academy of Sciences*, 2000, 97: 4426–4429.

————. *The crucible of creation: The Burgess Shale and the rise of
animals*. Oxford & New York: Oxford University Press, 1998.

————. "Showdown on the Burgess Shale: The challenge."
Natural History, December 1998: 48–55.

Eldredge N. *Time frames: The rethinking of Darwinian evolution and the theory of punctuated equilitria*. London: Heine-
mann, 1986.

Fortey RA. *Life: A natural history of the first four billion years of
life on Earth*. New York: Knopf, 1998.

Ganfornina MD, Sanchez D. "Generation of evolutionary
novelty by functional shift." *Bioessays*, 1999, 21: 432–439.

Gee H. *Before the backbone: Views on the origin of the vertebrates*. London; New York: Chapman & Hall, 1996.

————. *Shaking the tree: Readings from nature in the history of
life*. Chicago: University of Chicago Press, 2000.

Gehring WJ. *Master control genes in development and evolution:
The homeobox story*. New Haven, Conn.: Yale University
Press, 1998.

Gerhart J, Kirschner M. *Cells, embryos, and evolution: Toward
a cellular and developmental understanding of phenotypic
variation and evolutionary adaptability*. Malden, Mass.:
Blackwell Science, 1997.

Gould SJ. *Ontogeny and phylogeny*. Cambridge, Mass.: Belknap Press, 1977.

————. *Wonderful life: The Burgess Shale and the nature of history*. New York: Norton, 1989.

Hoffman PF, Kaufman AJ, Halverson GP, Schrag DP. "A neoproterozoic snowball earth." *Science*, 1998, 281: 1342–1346.

Holland LZ, Holland ND. "Chordate origins of the vertebrate
central nervous system." *Current Opinion in Neurobiology*,
1999, 9: 596–602.

Keys DN, Lewis DL, Selegue JE, Pearson BJ, Goodrich LV,
Johnson RL, et al. "Recruitment of a hedgehog regulatory
circuit in butterfly eyespot evolution." *Science*, 1999, 283:
532–534.

Knoll AH, Carroll SB. "Early animal evolution: Emerging
views from comparative biology and geology." *Science*,
1999, 284: 2129–2137.

Lundin LG. "Gene duplications in early metazoan evolution." *Seminars in Cellular and Developmental Biology*, 1999, 10: 523–530.

McNamara KJ. *Shapes of time: The evolution of growth and development*. Baltimore: Johns Hopkins University Press, 1997.

O'Leary M, Uhen M. "The time of origin of whales and the role of behavioral changes in the terrestrial-aquatic transition." *Paleobiology*, 1999, 25: 534–556.

Prothero DR, Prothero DA. *Bringing fossils to life: An introduction to paleobiology*. New York: McGraw-Hill, 1997.

Shankland M, Seaver EC. "Evolution of the bilaterian body plan: What have we learned from annelids?" *Proceedings of the National Academy of Sciences*, 2000, 97: 4434–4437.

Shubin N, Tabin C, Carroll S. "Fossils, genes and the evolution of animal limbs." *Nature*, 1997, 388: 639–648.

Sumida SS, Martin KLM. *Amniote origins: Completing the transition to land*. San Diego: Academic Press, 1997.

Williams GC. *The pony fish's glow: And other clues to plan and purpose in nature*. New York: Basic Books, 1997.

Zimmer C. *At the water's edge: Macroevolution and the transformation of life*. New York: Free Press, 1998.

———. "In search of vertebrate origins: Beyond brain and bone." *Science*, 2000, 287: 1576–1579.

CHAPTER 7: EXTINCTION: HOW LIFE ENDS AND BEGINS AGAIN

Bowring SA, Erwin DH, Isozaki Y. "The tempo of mass extinction and recovery: The end-Permian example." *Proceedings of the National Academy of Sciences*, 1999, 96: 8827–8828.

Burney DA. "Holocene lake sediments in the Maha'ulepu Caves of Kaua'i: Evidence for a diverse biotic assemblage from the Hawaiian lowlands and its transformation since human arrival." *Ecological Monographs*, in press.

Chapin FSr, Zavaleta ES, Eviner VT, Naylor RL, Vitousek PM, Reynolds HL, et al. "Consequences of changing biodiversity." *Nature*, 2000, 405: 234–242.

Cohen AN, Carlton JT. "Accelerating invasion rate in a highly invaded estuary." *Science*, 1998, 279: 555–558.

Daszak P, Cunningham AA, Hyatt AD. "Emerging infectious diseases of wildlife: Threats to biodiversity and human health." *Science*, 2000, 287: 443–449.

Dobson AP. *Conservation and biodiversity*. New York: Scientific American Library, 1996.

Drake F. *Global warming: The science of climate change*. New York: Oxford University Press, 2000.

Erwin DH. "After the end: Recovery from extinction." *Science*, 1998, 279: 1324–1325.

———. *The great Paleozoic crisis: Life and death in the Permian*. New York: Columbia University Press, 1993.

———. "Life's downs and ups." *Nature*, 2000, 404: 129–130.

Flannery TF. *The future eaters: An ecological history of the Australasian lands and people*. New York: George Braziller, 1995.

Gaston KJ. "Global patterns in biodiversity." *Nature*, 2000, 405: 220–227.

Holdaway RN, Jacomb C. "Rapid extinction of the moas (Aves: Dinornithiformes): Model, test, and implications." *Science*, 2000, 287: 2250–2254.

Inouye DW, Barr B, Armitage KB, Inouye BD. "Climate change is affecting altitudinal migrants and hibernating species." *Proceedings of the National Academy of Sciences*, 2000, 97: 1630–1633.

Kaiser J. "Does biodiversity help fend off invaders?" *Science*, 2000, 288: 785–786.

Kasting JF. "Long-term effects of fossil fuel burning." *Consequences*, 1998, 4: 15–27.

Kemp TS. *Mammal-like reptiles and the origin of mammals*. London & New York: Academic Press, 1982.

Kirchner JW, Weil A. "Delayed biological recovery from extinctions throughout the fossil record." *Nature*, 2000, 404: 177–179.

Kyte F. "A meteorite from the Cretaceous/Tertiary boundary." *Nature*, 1998, 396: 237–239.

Lawton JH, May RM. *Extinction rates*. Oxford & New York: Oxford University Press, 1995.

MacPhee RDE. *Extinctions in near time: Causes, contexts, and consequences*. New York: Kluwer Academic/Plenum Publishers, 1999.

McCann KS. "The diversity-stability debate." *Nature*, 2000, 405: 228–233.

Miller GH, Magee JW, Johnson BJ, Fogel ML, Spooner NA, McCulloch MT, et al. "Pleistocene extinction of *Genyornis newtoni*: Human impact on Australian megafauna." *Science*, 1999, 283: 205–208.

Mooney HA, Hobbs RJ, eds. *Invasive species in a changing world*. Washington, D.C.: Island Press, 2000.

Myers N, Mittermeier RA, Mittermeier CG, da Fonseca GA, Kent J. "Biodiversity hotspots for conservation priorities." *Nature*, 2000, 403: 853–858.

National Research Council (U.S.). Board on Biology. *Nature and human society: The quest for a sustainable world*. Washington, D.C.: National Academy Press, 2000.

Norris RD, Firth J, Blusztajn JS, Ravizza G. "Mass failure of the North Atlantic margin triggered by the Cretaceous-Paleogene bolide impact." *Geology*, 2000, 28: 1119–1122.

Pimm S, Askins R. "Forest losses predict bird extinctions in eastern North America." *Proceedings of the National Academy of Sciences*, 1995, 92: 10871–10875.

———, Raven P. "Biodiversity: Extinction by numbers." *Nature*, 2000, 403: 843–845.

Powell JL. *Night comes to the Cretaceous: Dinosaur extinction and the transformation of modern geology*. New York: Freeman, 1998.

Prothero DR, Prothero DA. *Bringing fossils to life: An introduction to paleobiology*. New York: McGraw-Hill, 1997.

Purvis A, Hector A. "Getting the measure of biodiversity." *Nature*, 2000, 405: 212–219.

Quammen D. "Planet of weeds." *Harper's*, 1990, 297: 57–69.

———. *The song of the dodo: Island biogeography in an age of extinctions*. New York: Scribner, 1996.

Ricciardi A, MacIsaac HJ. "Recent mass invasion of the North American Great Lakes by Ponto-Caspian species." *Trends in Ecology and Evolution*, 2000, 15: 62–65.

Rosenzweig ML. *Species diversity in space and time*. Cambridge: Cambridge University Press, 1995.

Sala OE, Chapin FS, 3rd, Armesto JJ, Berlow E, Bloomfield J, Dirzo R, et al. "Global biodiversity scenarios for the year 2100." *Science*, 2000, 287: 1770–1774.

Shackleton, NJ. "The 100,000-year ice-age cycle identified and found to lag temperature, carbon dioxide, and orbital eccentricity." *Science*, 2000, 289: 1897–1902.

Sheehan PM, David E. Fastovsky DE, Barreto C, Hoffmann RG. "Dinosaur abundance was not declining in a '3 m gap' at the top of the Hell Creek Formation, Montana and North Dakota." *Geology*, 2000, 28: 523–526.

Simberloff D, Schmitz DC, Brown TC. *Strangers in paradise: Impact and management of nonindigenous species in Florida.* Washington, D.C.: Island Press, 1997.

Smil V. *Cycles of life: Civilization and the biosphere.* New York: Scientific American Library/Freeman, 1997.

Stott PA, Tett SFB, Jones GS, Allen MR, Mitchell JFB, Jenkins GJ. "External control of 20th century temperature by natural and anthropogenic forcings." *Science*, 2000, 290: 2133–2137.

Thornton IWB. *Krakatau: The destruction and reassembly of an island ecosystem.* Cambridge, Mass.: Harvard University Press, 1996.

Van Driesche J, Van Driesche R. *Nature out of place: Biological invasions in the global age.* Washington, D.C.: Island Press, 2000

Vitousek PM, D'Antonio CM, Loope LL, Westbrooks R. "Biological invasions as global environmental change." *American Scientist*, 1996, 84: 468–479.

Ward PD. *The call of distant mammoths: Why the ice age mammals disappeared.* New York: Copernicus, 1997.

———. *The end of evolution: On mass extinctions and the preservation of biodiversity.* New York: Bantam Books, 1994.

———, Montgomery DR, Smith R. "Altered river morphology in South Africa related to the Permian-Triassic extinction." *Science*, 2000, 289: 1740–1743.

Wilson EO. *The diversity of life.* Cambridge, Mass.: Belknap Press, 1992.

CHAPTER 8: COEVOLUTION: WEAVING THE WEB OF LIFE

Barlow C. *The ghosts of evolution: Nonsensical fruits, missing partners, and other evolutionary anachronisms.* New York: Basic Books, 2001.

Brodie ED, III, Brodie ED, Jr. "Predator-prey arms races and dangerous prey." *Bioscience*, 1999, 49: 557–568.

Buchmann SL, Nabhan GP. *The forgotten pollinators.* Washington, D.C.: Island Press, 1996.

Currie CR, Mueller UG, Malloch D. "The agricultural pathology of ant fungus gardens." *Proceedings of the National Academy of Sciences*, 1999, 96: 7998–8002.

Darwin C. *On the various contrivances by which British and foreign orchids are fertilised by insects, and on the good effects of intercrossing.* London: John Murray, 1862.

DeMoraes CM, Lewis WJ, Pare PW, Alborn HT, Tumlinson JH. "Herbivore-infested plants selectively attract parasitoids." *Nature*, 1998, 393: 570–573.

Diamond J. "Ants, crops, and history." *Science*, 1998, 281: 1974–1975.

Ehrlich PR, Raven PH. "Butterflies and plants: A study in coevolution." *Evolution*, 1964, 18: 586–608.

Evans EP. *The criminal prosecution and capital punishment of animals.* London: Faber & Faber, 1987.

Farrell BD. " 'Inordinate fondness' explained: Why are there so many beetles?" *Science*, 1998, 281: 555–559.

Georghiou GP, Lagunes-Tejeda A. *The occurrence of resistance to pesticides in arthropods.* Rome: UN Food and Agriculture Organization, 1991.

Melander AL. "Can insects become resistant to sprays?" *Journal of Economic Entomology*, 1914, 7: 167–173.

Mueller UG, Rehner SA, Schultz TR. "The evolution of agriculture in ants." *Science*, 1998, 281: 2034–2038.

Murray DR. *Seed dispersal.* Sydney & Orlando: Academic Press, 1986.

National Research Council (U.S.). Committee on Strategies for the Management of Pesticide Resistant Pest Populations. *Pesticide resistance: Strategies and tactics for management.* Washington, D.C.: National Academy Press, 1986.

Pimentel D. *Techniques for reducing pesticide use: Economic and environmental benefits.* Chichester, England & New York: Wiley, 1997.

———, Lach L, Zuniga R, Morrison D. "Environmental and economic costs of nonindigenous species in the United States." *Bioscience*, 2000, 50: 53–65.

———, Lehman H. *The pesticide question: Environment, economics, and ethics.* New York: Chapman & Hall, 1993.

Sheets TJ, Pimentel D. *Pesticides: Contemporary roles in agriculture, health, and environment.* Clifton, N.J.: Humana Press, 1979.

Thompson JN. *The coevolutionary process.* Chicago: University of Chicago Press, 1994.

von Helversen D, von Helversen O. "Acoustic guide in bat-pollinated flower." *Nature*, 1999, 398: 759–760.

Winston ML. *Nature wars: People vs. pests.* Cambridge, Mass.: Harvard University Press, 1997.

CHAPTER 9: DOCTOR DARWIN: DISEASE IN THE AGE OF EVOLUTIONARY MEDICINE

Carrington M, Kissner T, Gerrard B, Ivanov S, O'Brien SJ, Dean M. "Novel alleles of the chemokine-receptor gene CCR5." *American Journal of Human Genetics*, 1997, 61: 1261–1267.

Chadwick D, Goode J. *Antibiotic resistance: Origins, evolution, selection, and spread.* New York: Wiley, 1997.

Cohen J. "The hunt for the origin of AIDS." *Atlantic Monthly*, 2000, 286: 88–103.

Ewald P. *The evolution of infectious diseases.* Oxford: Oxford University Press, 1997.

Farmer P. "Social inequalities and emerging infectious diseases." *Emerging Infectious Diseases*, 1996, 2: 259–269.

Gao F, Bailes E, Robertson DL, Chen Y, Rodenburg CM, Michael SF, et al. "Origin of HIV-1 in the chimpanzee *Pan troglodytes troglodytes*." *Nature*, 1999, 397: 436–441.

Garrett L. *Betrayal of trust: The global collapse of public health.* New York: Hyperion, 2000.

Hahn BH, Shaw GM, De Cock KM, Sharp PM. "AIDS as a zoonosis: Scientific and public health implications." *Science*, 2000, 287: 607–614.

Lawrence JG, Ochman H. "Molecular archaeology of the *Escherichia coli* genome." *Proceedings of the National Academy of Sciences*, 1998, 95: 9413–9417.

Levy SB. *The antibiotic paradox: How miracle drugs are destroying the miracle.* New York: Plenum Press, 1992.
———. "The challenge of antibiotic resistance." *Scientific American,* 1998, 278: 46–53.
Nesse RM, Williams GC. *Why we get sick: The new science of Darwinian medicine.* New York: Times Books, 1994.
Ploegh HL. "Viral strategies of immune evasion." *Science,* 1998, 280: 248–253.
Stearns SC. *Evolution in health and disease.* Oxford & New York: Oxford University Press, 1999.
Weiss RA, Wrangham RW. "From *Pan* to pandemic." *Nature,* 1999, 397: 385–386.
Witte W. "Medical consequences of antibiotic use in agriculture." *Science,* 1998, 279: 996–997.
Zimmer C. *Parasite rex: Inside the bizarre world of nature's most dangerous creatures.* New York: Free Press, 2000.

CHAPTER 10: PASSION'S LOGIC: THE EVOLUTION OF SEX

Alcock J. *Animal behavior: An evolutionary approach.* Sunderland, Mass.: Sinauer Associates, 1998.
Andersson MB. *Sexual selection.* Princeton, N.J.: Princeton University Press, 1994.
Andrade MCB. "Sexual selection for male sacrifice in the Australian redback spider." *Science,* 1996, 271: 70–72.
Birkhead TR. *Promiscuity: An evolutionary history of sperm competition.* Cambridge: Harvard University Press, 2000.
———, Møller AP. *Sperm competition and sexual selection.* San Diego & London: Academic Press, 1998.
Boesch C, Boesch-Achermann H. *The chimpanzees of the Taï Forest: Behavioural ecology and evolution.* Oxford & New York: Oxford University Press, 2000.
Cronin H. *The ant and the peacock: Altruism and sexual selection from Darwin to today.* New York: Cambridge University Press, 1991.
Dawkins R. *The selfish gene.* New York: Oxford University Press, 1976.
Dusenbery DB. "Selection for high gamete encounter rates explains the success of male and female mating types." *Journal of Theoretical Biology,* 2000, 202: 1–10.
Gould JL, Gould CG. *Sexual selection.* New York: Scientific American Library/Freeman, 1989.
Haig D. "Genetic conflicts in human pregnancy." *Quarterly Review of Biology,* 1993, 68: 495–532.
Hausfater G, Hrdy SB. *Infanticide: Comparative and evolutionary perspectives.* New York: Aldine, 1984.
Holland B, Rice WR. "Experimental removal of sexual selection reverses intersexual antagonistic coevolution and removes a reproductive load." *Proceedings of the National Academy of Sciences,* 1999, 96: 5083–5088.
Hrdy SB. *Mother nature: A history of mothers, infants, and natural selection.* New York: Pantheon Books, 1999.
Møller AP, Swaddle JP. *Asymmetry, developmental stability, and evolution.* Oxford: Oxford University Press, 1997.
Petrie M. "Improved growth and survival of offspring of peacocks with more elaborate trains." *Nature,* 1994, 371: 598–599.
———, Krupa A, Burke T. "Peacocks lek with relatives even in the absence of social and environmental cues." *Nature,* 1999, 401: 155–157.

Pizzari T, Birkhead TR. "Female feral fowl eject sperm of subdominant males." *Nature,* 2000, 405: 787–789.
Rice WR. "Sexually antagonistic male adaptation triggered by experimental arrest of female evolution." *Nature,* 1996, 381: 232–234.
Ridley M. *The Red Queen: Sex and the evolution of human nature.* New York: Penguin Books, 1995.
Waal FBMd. *Chimpanzee politics: Power and sex among apes.* Baltimore, Md. & London: Johns Hopkins University Press, 1998.
———. *Good natured: The origins of right and wrong in humans and other animals.* Cambridge, Mass.: Harvard University Press, 1996.
———, Lanting F. *Bonobo: The forgotten ape.* Berkeley, Calif.: University of California Press, 1997.
Welch AM, Semlitsch RD, Gerhardt HC. "Call duration as an indicator of genetic quality in male gray tree frogs." *Science,* 1998, 280: 1928–1930.
Wrangham RW. "Subtle, secret female chimpanzees." *Science,* 1997, 277: 774–775.
———, Peterson D. *Demonic males: Apes and the origins of human violence.* Boston: Houghton Mifflin, 1996.

CHAPTER 11: THE GOSSIPING APE: THE SOCIAL ROOTS OF HUMAN EVOLUTION

Allman JM. *Evolving brains.* New York: Scientifc American Library/Freeman, 1999.
Barkow JH, Cosmides L, Tooby J. *The adapted mind: Evolutionary psychology and the generation of culture.* New York: Oxford University Press, 1992.
Baron-Cohen S, Tager-Flusberg H, Cohen DJ. *Understanding other minds: Perspectives from developmental cognitive neuroscience.* Oxford & New York: Oxford University Press, 2000.
Buss DM. *The dangerous passion: Why jealousy is as necessary as love and sex.* New York: Free Press, 2000.
———. *Evolutionary psychology: The new science of the mind.* Boston & London: Allyn & Bacon, 1999.
———, Haselton MG, Shackelford TK, Bleske AL, Wakefield JC. "Adaptations, exaptations, and spandrels." *American Psychologist,* 1998, 53: 533–548.
Cosmides L. "The logic of social exchange: Has natural selection shaped how humans reason? Studies with the Wason selection task." *Cognition,* 1989, 31: 187–276.
Coyne JA, Berry A. "Rape as an adaptation." *Nature,* 2000, 404: 121–122.
Cummins DD, Allen C. *The evolution of mind.* New York & Oxford: Oxford University Press, 1998.
Darwin C. *The descent of man and selection in relation to sex.* London: John Murray, 1871.
Deacon TW. *The symbolic species: The co-evolution of language and the brain.* New York & London: Norton, 1997.
Dunbar RIM. *Grooming, gossip and the evolution of language.* London: Faber & Faber, 1996.
Emlen ST. "An evolutionary theory of the family." *Proceedings of the National Academy of Sciences,* 1995, 92: 8092–8099.
Fitch WT. "The evolution of speech: A comparative review." *Trends in Cognitive Sciences,* 2000, 4: 258–267.
Gagneux P, Wills C, Gerloff U, Tautz D, Morin PA, Boesch C, et al. "Mitochondrial sequences show diverse evolu-

tionary histories of African hominoids." *Proceedings of the National Academy of Sciences,* 1999, 96: 5077–5082.

Gould SJ. "The pleasures of pluralism." *New York Review of Books,* June 26, 1997, 46–52.

———, Lewontin RC. "The spandrels of San Marco and the Panglossian paradigm: A critique of the adaptationist programme." *Proceeding of the Royal Society of London Series B,* 1979, 205: 581–598.

Hare B, Call J, Agnetta B, Tomasello M. "Chimpanzees know what conspecifics do and do not see." *Animal Behaviour,* 2000, 59: 771–785.

Hauser M. *Wild minds: what animals really think.* New York: Henry Holt, 2000.

Holden C. "No last word on language origins." *Science,* 1998, 282: 1455–1459.

Hrdy SB. *Mother nature: A history of mothers, infants, and natural selection.* New York: Pantheon Books, 1999.

Johanson DC, Edgar B. *From Lucy to language.* New York: Simon & Schuster, 1996.

Kegl J. "The Nicaraguan sign language project: An overview." *Signpost,* 1994, 7: 24–31.

———, Senghas A, Coppola M. "Creation through contact: Sign language emergence and sign language change in Nicaragua." In: DeGraff M, ed., *Language contact and language change: The intersection of language acquisition, creole genesis, and diachronic syntax.* Cambridge, Mass.: MIT Press, 1999, pp. 179–237.

Klein RG. *The human career: Human biological and cultural origins.* Chicago: University of Chicago Press, 1999.

Miller GF. *The mating mind: How sexual choice shaped the evolution of human nature.* New York: Doubleday, 2000.

Nowak MA, Krakauer DC. "The evolution of language." *Proceedings of the National Academy of Sciences,* 1999, 96: 8028–8033.

———, Krakauer DC, Dress A. "An error limit for the evolution of language." *Proceedings of the Royal Society of London, Series B: Biological Sciences,* 1999, 266: 2131–2136.

O'Connell JF, Hawkes K, Blurton Jones NG. "Grandmothering and the evolution of *Homo erectus.*" *Journal of Human Evolution,* 1999, 36: 461–485.

Osborne L. "A linguistic big bang." *New York Times Magazine,* October 24, 1999: 84–89.

Penton-Voak IS, Perrett DI, Castles DL, Kobayashi T, Burt DM, Murray LK, et al. "Menstrual cycle alters face preference." *Nature,* 1999, 399: 741–742.

Perrett DI, Lee KJ, Penton-Voak I, Rowland D, Yoshikawa S, Burt DM, et al. "Effects of sexual dimorphism on facial attractiveness." *Nature,* 1998, 394: 884–887.

———, May KA, Yoshikawa S. "Facial shape and judgements of female attractiveness." *Nature,* 1994, 368: 239–242.

Pinker S. *How the mind works.* New York: Norton, 1997.

———. *The language instinct.* New York: Morrow, 1994.

Richmond BG, Strait DS. "Evidence that humans evolved from a knuckle-walking ancestor." *Nature,* 2000, 404: 382–386.

Sahlins MD. *The use and abuse of biology: An anthropological critique of sociobiology.* Ann Arbor: University of Michigan Press, 1976.

Scheib JE, Gangestad SW, Thornhill R. "Facial attractiveness, symmetry and cues of good genes." *Proceedings of the Royal Society of London Series B,* 1999, 266: 1913–1917.

Segerstråle UCO. *Defenders of the truth: The battle for science in the sociobiology debate and beyond.* New York: Oxford University Press, 2000.

Singh D. "Adaptive significance of female physical attractiveness: Role of waist-to-hip ratio." *Journal of Personality and Social Psychology,* 1993, 65: 293–307.

Tattersall I, Schwartz JH. *Extinct humans.* Boulder, Colo.: Westview Press, 2000.

Thornhill R, Gangestad SW. "Facial attractiveness." *Trends in Cognitive Sciences,* 1999, 3: 452–460.

———, Møller AP. "Developmental stability, disease and medicine." *Biological Reviews of the Cambridge Philosophical Society,* 1997, 72: 497–548.

———, Palmer C. *A natural history of rape: Biological bases of sexual coercion.* Cambridge, Mass.: MIT Press, 2000.

Waal FBMd. "The chimpanzee's service economy: Food for grooming." *Evolution and Human Behaviour,* 1997, 18: 375–386.

Wedekind C, Seebeck T, Bettens F, Paepke AJ. "MHC-dependent mate preferences in humans." *Proceedings of the Royal Society of London Series B,* 1995, 260: 245–249.

Whiten A. "Social complexity and social intelligence." In J Goode J, ed., *The nature of intelligence.* New York: Wiley, 2000, pp. 185–201.

———, Byrne RW. *Machiavellian intelligence II: Extensions and evaluations.* London & New York: Cambridge University Press, 1997.

———, Goodall J, McGrew WC, Nishida T, Reynolds V, Sugiyama Y, et al. "Cultures in chimpanzees." *Nature,* 1999, 399: 682–685.

Wilson EO. *Sociobiology: The new synthesis.* Cambridge, Mass.: Belknap Press, 1975.

Wright R. *The moral animal: Evolutionary psychology and everyday life.* New York: Pantheon Books, 1994.

CHAPTER 12: MODERN LIFE, 50,000 B.C.: THE DAWN OF US

Bertranpetit J. "Genome, diversity, and origins: The Y chromosome as a storyteller." *Proceedings of the National Academy of Sciences,* 2000, 97: 6927–6929.

Blackmore SJ. *The meme machine.* Oxford & New York: Oxford University Press, 1999.

Cavalli-Sforza LL. *Genes, peoples, and languages.* New York: North Point Press, 2000.

Chauvet J-M, Brunel Deschamps E, Hillaire C. *Dawn of art: The Chauvet Cave, the oldest known paintings in the world.* New York: Abrams, 1996.

Dyson G. *Darwin among the machines: The evolution of global intelligence.* Reading, Mass.: Addison-Wesley, 1997.

Ehrlich P. *Human natures.* Washington, D.C.: Island Press, 2000.

Holden C, Mace R. "Phylogenetic analysis of the evolution of lactose digestion in adults." *Human Biology,* 1997, 69: 605–628.

Klein RG. *The human career: Human biological and cultural origins.* Chicago: University of Chicago Press, 1999.

Krings M, Capellia C, Tschendtscher F, Geisert H, Meyer S, von Haeseler A, et al. "A view of Neandertal genetic diversity." *Nature Genetics*, 2000, 26: 144–146.

———, Geisert H, Schmitz RW, Krainitzki H, Pääbo S. "DNA sequence of the mitochondrial hypervariable region II from the neandertal type specimen." *Proceedings of the National Academy of Sciences*, 1999, 96: 5581–5585.

———, Stone A, Schmitz RW, Krainitzki H, Stoneking M, Pääbo S. "Neandertal DNA sequences and the origin of modern humans." *Cell*, 1997, 90: 19–30.

Lewontin RC. *Human diversity*. New York: Freeman, 1995.

Mithen SJ. *The prehistory of the mind: A search for the origins of art, religion and science*. London: Thames & Hudson, 1996.

Nemecek S. "Who were the first Americans?" *Scientific American*, 2000, 283: 80–87.

Ovchinnikov IV, Gotherstrom A, Romanova GP, Kharitonov VM, Liden K, Goodwin W. "Molecular analysis of Neanderthal DNA from the northern Caucasus." *Nature*, 2000, 404: 490–493.

Pääbo S. "Human evolution." *Trends in Cell Biology*, 1999, 9: 13–16.

———. "Neolithic genetic engineering." *Nature*, 1999, 398: 194–195.

Richards MP, Pettitt PB, Trinkaus E, Smith FH, Paunovic M, Karavanic I. "Neanderthal diet at Vindija and Neanderthal predation: The evidence from stable isotopes." *Proceedings of the National Academy of Sciences*, 2000, 97: 7663–7666.

Shen P, Wang F, Underhill PA, Franco C, Yang WH, Roxas A, et al. "Population genetic implications from sequence variation in four Y chromosome genes." *Proceedings of the National Academy of Sciences*, 2000, 97: 7354–7359.

Shreeve J. *The Neandertal enigma: Solving the mystery of modern human origins*. New York: Morrow, 1995.

Tattersall I. *Becoming human: Evolution and human uniqueness*. New York: Harcourt Brace, 1998.

Zimmer C. "After you, Eve." *Natural History*, March 2001, 32–35.

CHAPTER 13: WHAT ABOUT GOD?

Behe MJ. *Darwin's black box: The biochemical challenge to evolution*. New York: Free Press, 1996.

Cartmill, M. "Oppressed by evolution." *Discover*, March 1998, 78–83.

Chen L, DeVries AL, Cheng CH. "Evolution of antifreeze glycoprotein gene from a trypsinogen gene in Antarctic notothenioid fish." *Proceedings of the National Academy of Sciences*, 1997, 94: 3811–3816.

Cheng CH, Chen L. "Evolution of an antifreeze glycoprotein." *Nature*, 1999, 401: 443–444.

Darwin C. *Autobiography and selected letters*. New York: Dover Publications, 1958.

Dawkins R. *Unweaving the rainbow: Science, delusion, and the appetite for wonder*. Boston: Houghton Mifflin, 1998.

Dupree, A. H. *Asa Gray, 1810–1888*. Cambridge, Mass.: Harvard University Press, 1959.

Gasman, D. *The scientific origins of National Socialism: social Darwinism in Ernst Haeckel and the German Monist League*. New York: Elsevier, 1971.

Gould SJ. *Rocks of ages: Science and religion in the fullness of life*. New York: Ballantine, 1999.

———. "Non-overlapping Magisteria." *Skeptical Inquirer*, 1999, 23: 55–61.

Haeckel EHPA, Lankester ER. *The history of creation: or, The development of the earth and its inhabitants by the action of natural causes*. London: Henry S. King, 1876.

Larson EJ. *Summer for the gods: The Scopes trial and America's continuing debate over science and religion*. Cambridge, Mass.: Harvard University Press, 1998.

Miller KB. "Theological implications of an evolving creation." *Perspectives on Science and Christian Faith*, 1993, 45: 150–160.

Miller KR. *Finding Darwin's god: A scientist's search for common ground between God and evolution*. New York: Cliff Street Books, 1999.

Moore JR. "Of love and death: Why Darwin 'gave up Christianity.'" In JR Moore, ed., *History, humanity, and evolution: Essays for John C. Greene*. Cambridge, Mass.: Cambridge University Press, 1989, pp. 195–229.

National Academy of Sciences. *Science and creationism: A view from the National Academy of Sciences*. Washington, D.C.: National Academy Press, 1999.

Numbers RL. *The creationists*. New York: Knopf, 1992.

Orr HA. "Darwin v. intelligent design (again)." *Boston Review*, December 1996, 28–31.

Peel JDY. *Herbert Spencer: The evolution of a sociologist*. New York: Basic Books, 1971.

Pennock RT. *Tower of Babel: The evidence against the new creationism*. Cambridge, Mass.: MIT Press, 1999.

Wilson EO. "The biological basis of morality." *Atlantic Monthly*, 1998, 281: 53–70.

———. *Hardwired for God*. *Forbes ASAP*. Supplement, 1999: 132–134.

———. *Naturalist*. Washington, D.C.: Island Press, 1994.

———. "The two hypotheses of human meaning." *The Humanist*, 1999, 59: 30–31.

INDEX

Page numbers in *italics* refer to illustrations.

Acanthostega, 134
Acasta rocks, 57–58, *58*, 62–64
acritarchs, 127
Adami, Christoph, 94–95
Africa, 132, 170, 175–76
  AIDS epidemic in, 219–22, *219*, 224
  human evolution in, *71*, 260–68, 272–73, 275–76, 295, 296, 298, 300, 302, 307–8
  sickle-cell anemia in, 306
  speciation in, 89–92, *90*
agriculture, 175–76, 184
  fungus-growing ants and, 204–7, *206, 207*
  genetically altered crops and, 203–4
  malaria and, 306–7
  pesticide use in, 200–203
Agriculture Department, U.S., 183
AIDS (acquired immunodeficiency syndrome), xiii, 217–25, *219*, 336–37
  search for origin of, 218–22
Aids to Reflection (Coleridge), 341
Aiello, Leslie, 291
Alabama, textbooks in, 336
Alaska, glaciers in, 169
algae, 69, 154, 337
  acritarchs, 127
  blue-green, *see* cyanobacteria
  chloroplasts of, 112–13
  eaters of, 127–28
  photosynthetic, 23, 126
  red, 66
alleles, 81
altruism, 248–53
  reciprocal, 251–53, 273
Alvarez, Luis, 160–62, 164
Alvarez, Walter, 160–62, 164
Ambulocetus, x, 137, *138*, 139, *139*, 325
American Association for the Advancement of Science, 334

American Civil Liberties Union (ACLU), 318–21
amino acids, 76, 77, 104, 106, 300, 307
amniotes, 70–71, *71*
amphibians, 70, 71, 148
Andersson, Siv, 113
Andes Mountains, 19, 20–21
Andrade, Maydianne, 242
Andrewsarchus, 137–38
anecdotes and legends, ix, x, xii, 54
Angraecum sesquipedale, 189–90, 193
annelids, 67
Anopheles gambiae, 307
Answers in Genesis, 325
Antarctica, 166, 305
Antarctic Ocean, notothenioids in, 327
anteaters, 159
antelopes, 266
anthropologists, objections to sociobiology by, 274
antibiotic resistance, x, xiii, xvi, 212–17, *212*, 336
antibodies, 92–94
antifreeze, 326–28, *327*, 330
antigens, 92–94
ants:
  coevolution and, 204–7, *206, 207*
  sex of, 229, 230, 249, 273, 274
apatite, 65
Apatosaurus, 157
apes, 282
  human evolution and, 14, 51, *52*, 53
  humans compared with, 262–65
  social life of, 268–69
  see also specific apes
archaea, in tree of life, *102*, 103, 104, 107, 108, 109, 111
Archeoglobus fulgidus, 108
archetypes, Owen's views on, 40–41, 44, 50, 51
Archonata, 159
Ardèche region, 293–95
Ardipithecus ramidus, 263
Argentina, 18–19

Arkansas, 322
  anti-evolution law in, 320, 321
  Supreme Court of, 321
armadillos, 159, 167
art, 293–95, *294, 297*, 302–4, *302*
arthropods, 29, 120, 149
  nervous system of, 123, *123*
artificial life, 94–97, 311
Asia:
  AIDS in, 224
  humans in, 170, 296, 297–98, 301, 304–5
Asterales, 198
asteroids, 65–66, 144, 161–64, *163, 164*
atomic structure, 60
aurochs, 176
Australia, 30, 66, 109, 132, 144, 166
  cane toads in, 179
  Ediacaran fauna in, 67
  humans in, 169, 170, 171, 305, 306
Australopithecus afarensis, 264, *265*, 285
Australopithecus anamensis, 264
autistic children, 270
Avida, 94–96
avocados, 208
axes, hand, *260, 260*, 261, 262, 267, 285

baboons, 268, 270, 285
Bacillus thuringiensis, 203–4
bacteria, x, xi, xiii, *93*, 154, 209, 337
  antibiotic resistance developed by, x, xiii, 212–17, *212*, 336
  asexual reproduction of, 230
  cyanobacteria, 66, 70, *102, 107*, 109, 112, 152
  fossils of, 66, 69
  photosynthetic, 126
  Streptomyces, 206–7
  in stromatolites, *110*
  symbiotic, 111–14
  in tree of life, *102*, 103, 104, 107–8, *107*, 109, 111–14, *114*
Baer, Karl von, 31, 49–50

Balkan Wars (1912), 225, *225*, 226
Bangladesh, 166
barnacles, 40–43, *41*, 78, 121
   evolution of, xv, 41, 42, 135
barn swallows, 238, *239*, 240, 246
Baron-Cohen, Simon, 269–70
bases, DNA, 76, 94, 104
Basilosaurus, 140
Bateson, William, 119
bats, 41, *42*, 49, 50, 127, 159, 165, 173
   coevolution and, 194, 208
   vampire, 251–52
B cells, 92–94
Beagle, HMS, 3–4, 15–24, *15, 16, 18*, 32,
   36, 37, 40, 54, 340
   in Galápagos, 21–23, *22*
   mission of, 3, 4, 7
bears, 136, 137, 159, 169, 246
Beaufort, Francis, 7–8
Becquerel, Henri, 60, 61
bee-eaters, 250–51
   reproductive choices of, 275, *275*
bees, 248–49
   coevolution and, 190–91, *192*, 199,
   209
beetles, 179
   coevolution and, 197–200, *199*, 208
   dung, 246
   ladybird, 80–81
   on trial, 200
Behe, Michael, 330–31
Belize, face symmetry in, 277
Berry, Andrew, 282, 283
Bible, 36
   science vs., 9–12, 53, 54, 59, 315–16,
   320–25
   see also creationism
Bigland, John, 136
Binumarien people, 274, 283
biochemical warfare, 195–97
biodiversity, 89–92, *90*, 159, *168*, 169,
   173, 184
   coevolution and, 197–99
   hot spots of, 184, *185*
biological invasion, extinction and, 174,
   178–84, 186
biology, biologists, 27, 135, 185
   German, 28, 30, 39
   modern synthesis and, 74, 80–84
biotechnology, 336–37
bipedalism, 14, 264–65
birds, x, 49, 65, 71, 179, 330
   coevolution and, 194, 208
   difficulty in determining species of,
   82–84, *83*
   flightless, 28, 169, 170, 173, 175
   forest fragmentation and, 176–78
   of Galápagos, 21–22, *23*, 32–33, *32*,
   35, 87–89

"half-life" in extinction of, 177–78
   in Hawaii, 171, 173, 174, 178
   infidelity in, 240, 279
   vision of, 131
   wings of, 41
birth defects, 39
Bivelich, Alexander, 211–13
Black Death, 219, 222–24, *222, 223*
Black Sea, 180, 181
blind spots, 129, 131
blood, clotting of, 326, 328–30, *329*
blood cells:
   red, 306, 307
   white, *93*, 218, 223, 224, 306
Bmp-4, 123–24, *123*
Boag, Peter, 87
Bonobo (Waal), 255
bonobos (pygmy chimpanzees), 252,
   254–56, *254, 263*, 265–66
boobies, blue-footed, *23*
botanists, botany, 107, 182
Bottaccione Gorge, rocks from, 159–62
Bowring, Samuel, 149–50
brain, 125, 127, 130
   of apes, 51, *52*
   neocortex of, 156–57, 270, 285
brain, human, 51, *52*, 259, 261, 277,
   282, 303
   blind spots and, 129
   edge building by, 269, *269*
   increase in size of, x, 267–70, 285,
   291
   language and, 286–91
brain damage, 269–70, 286
Brazil, 17–18, 30, 132, 184, 193
breeding:
   as analogy for natural selection, 46,
   46, 48
   Darwin's studies of, x, 34, 43, 44, 46,
   48
Bridges, Calvin, 119–20
British Association for the Advancement of Science, 52–54
Brocks, Jochen, 66
Brodie, Edmund, Jr., 195–97, *196*
Brodie, Edmund, III, 195–97, *196*
Brooks, Thomas, 177–78
Brueghel, Jan, *313*
Bryan, William Jennings, 317–20, *319*,
   337
bryozoans, 127, 155
Bt, 203–4
bubonic plague, 219, 222–24, *222, 223*
Bufo marinus (cane toad), 179
bulls, musk ox, 235
burials, 303, 343–44
Burney, David, 171–75, *172*, 186
bush beard-grass, 180
Buss, David, 278, 279–80, 282–83

Calyceraceae, 198
Cambrian explosion, 68, 69, *69, 71*,
   118, 122–28, 156, 295, 324, 333,
   337
   gene duplication and the dawn of
   vertebrates in, 124–26, 132
   genes behind, 122–24
Cambrian period, 11, 64–65, 71, 146
Cambridge University, 6, 8, 9, 12
camels, 159, 166, 169, 208
Cameroon, chimpanzees in, 221
Campbell, George, 96
Campylobacter jejuni, 217
Canada, 84, 165, 201
   Acasta rocks in, 57–58, *58*, 62–64
Canary Islands, 6, 8, 16
cancer, 203
Cape Verde Islands, 16, 17
capitalism, laissez faire, 316–17
carbon, photosynthesis and, 194–95
carbon 12, 60, 65
carbon 13, 60, 65, 149
carbon 14, 60, 64, 294–95
carbon dioxide:
   in atmosphere, 126, 164, 165, 166,
   168, 183, 185–86
   in ocean, 149, 150–51
   in photosynthesis, 111–12, 194
Carnegie, Andrew, 317
Carnivora (carnivores), 159, 161, 169
Carroll, Lewis, 231
Case for Creation, A (Frair and Davis),
   324
Caspian Sea, 180, 181
Cassia grandis, 208
Catasetum saccatum, 191, *192*–93
catastrophic mass extinctions, xiii,
   143–44, 182
   Cretaceous-Tertiary, 159–65, *163,
   164*, 167
   Cuvier's views on, 10–11, 143
   Permian-Triassic, 146–51, 155, 156,
   157, 162
   rebirth after, 151–56
caterpillars, 65, 197, 213
cats, 46, 159, 165, 167, 220
Caucasus Mountains, 300
Cavalli-Sforza, Luigi, 308
cave paintings, 293–95, *294, 297*, 302
CCR5, *222*, 223–24
Cech, Thomas, 105–6
Cenozoic era, 145
Central America, cholera in, 226
Chad, 265
Chambers, Robert, 38–40, 50, 52–53
Charles Island (Santa María), 22
Chatham Island (San Cristobál), 21
Chauvet, Jean-Marie, 293–95
Chauvet cave, 293–95, *294, 297*, 302

Cheng, Chi-Hing C., 327–28
chickens, 174, 178
    antibiotics given to, 217
Chicxulub, 162–64
child abuse, 280–81
Chile, 31, 40, 226
    first humans in, 169, 305
chimpanzees, xi, 51, 52, *52*, 175, 213,
        254–56, *254*, 261–65, 282
    ancestors of HIVs in, 221–22
    brains of, *259*, 267, 270
    in evolutionary tree, 262, *263*
    humans compared with, 262–67,
        270–71, 290, 308
    human split from, *71*, 295
    pygmy, *see* bonobos
    sexual politics of, 251–53, *253*
    SIVs in, 221
    toolmaking of, 265–66, *266*
    violence of, 252–53, *253*, 254
China, 124, 175–76, 183, 201
    extinctions in, 149–50
chloroplasts, 111–13, *112*
cholera, *225*, 226, *226*
cholesterol, 66
Chomsky, Noam, 286
chordates, 68
chordin, *123*, 124
Christianity, *see* creationism; God
chromosomes, 77, 81, 232, 297
    Hox genes in, 120, 121
    Y, 299
cichlids, 127–28
    speciation of, 89–92, *90*
Civic Biology, A (textbook, 1914), 318
Clack, Jennifer, 133–34
Claraia, 154
Clarkston Valley, 201
climate, 150–51, 165–70, 183–84
    in Galápagos, 87–88
    human evolution and, 264, 266
clocks, testing of, 3, 4
cloning, 231–33
clover, red, 192
coelacanths, 132, 133
coevolution, 189–210
    ants and, 204–7, *206*, *207*
    biochemical warfare and, 195–97
    defined, 190
    disease and, 211–27
    extinction and, 207–9
    matrix of, 194–95
    and resistance to pesticides,
        200–203
    sexual go-betweens and, 190–93
colds, 219, 226
    natural selection in fighting of,
        92–94, *93*
Coleridge, Samuel Taylor, 341

comets, 104, *104*, 161
comparative anatomy, 28–29, *29*, 41,
    42, 49
competition, 92
    sexual, 230, 235
    between species, 48–49, 155, 156,
        157, 166, 167
    within species, 45, 48, 230, 235,
        239–42
computers, 310–11
    artificial life and, 94–97
Concepción, 20
conifers, 198
continents, plate tectonics and, 64
Cook, James, 174, 178
Copenhagen, 161
Copernicus, Nicolaus, xi, 41
coral reefs, 23–24, 33, 67, 134, 155, 181,
    184
Cosmides, Leda, 276
Costa Rica, 207–9
Cousin Island, 245
cows, 136, 137, 166, 174, 176, 220
    antibiotics given to, 217
    bacteria in, 111
Coyne, Jerry, 282, 283, 333
crabs, 29, 121, 173, 174, 175, 179
crater hypothesis, 23
creationism, 315–37
    and creation of Earth, 9, 12, 318
    evidence of evolution as viewed by, x,
        50, 52–54
    Flood Geology, 322, 325, 335
    Intelligent Design, 325–26, 330–32,
        336
    in Kansas, 334–36
    limits of science and, 332–34
    Old-Earth, 322
creation science, rise of, 320–22, 325
Cretaceous period, 71, 141, 159–65,
    167
    K-T boundary and, 162–65, 167
crickets, 237, 238
Croatia, 300
crocodiles, 148, 155
Cro-Magnons, 296, 302
Cruel Sister, The (Faed), 280
crustaceans, 40–42, 49, 69, 135
    eyes of, *128*
    parasitic, 213
cultural evolution, 305–11
Curie, Marie and Pierre, 60, *60*, 61
Currie, Cameron, 206–7
Cuvier, Georges, 10–11, *11*, 123, 143,
    155
    Lamarck attacked by, 14–15
cyanobacteria, 66, 70, 102, 107, 109,
    112, 152
cynodonts, 156

Daeschler, Ted, 134
daisies, 198, 200
damselflies, 241, *241*
dandelions, 198
Dangerous Passion, The (Buss), 282–83
Daphne Island, 87–88
Darrow, Clarence, 319–20, *319*
Darwin, Annie, xv, 42, 341–42, 343, *343*
Darwin, Charles, ix–xvii, 4–55, *5*, 86,
        134, 340–44
    and age of Earth, 58–60
    autobiography of, 15, 17, 342
    barnacle studies of, xv, 40–43, *41*, 78
    on *Beagle* voyage, 3–4, 6–9, 15–24,
        36, 37, 40, 54, 340
    biological invasions and, 43, 178
    bird collecting of, 21–22
    Cambrian period as viewed by,
        64–65, 69
    coevolution and, 190–94, *191*
    confrontation avoided by, 5, 25, 37
    coral reefs as viewed by, 23–24
    diary of, 18, 20
    Earth and life as viewed by contem-
        poraries of, 9–15
    in Edinburgh, 5–6, 14, 342
    education of, 5–6, 9, 12, 342
    Emma's correspondence with, 36,
        341, 342
    extinction as viewed by, 143–44, 146
    family background of, 5, 13–14
    finances of, 8, 25
    in Galápagos, 21–23
    health problems of, 33, 42, 46, 54,
        341
    heredity as problem for, 71, 73–74,
        101
    on human evolution, 260–62
    Huxley as ally of, 44, *44*, 50–54, 59
    independent scholarship as ambition
        of, 24–25
    Lyell's influence on, 16–17, 19, 20,
        24–25, 40, 59, 143
    marriage of, 35–36
    medical profession rejected by, 4,
        5–6
    on origin of whales, 135–36, 137
    private notebooks of, 27, *27*, 33, 340
    religious studies of, 4, 6, 8
    sexual selection and, 235–37
    in South America, 17–21
    spiritual views of, 340–44, *343*
    Wallace's correspondence with, 45,
        46
Darwin, Elizabeth, 341
Darwin, Emma Wedgwood, 35–36, *37*,
        38, 42, 43, 341–43
Darwin, Erasmus (Darwin's brother), 5,
        25, 27, 37, 341

Darwin, Erasmus (Darwin's grandfather), 13–14, *13*, 27
Darwin, Henrietta, 341
Darwin, Leonard, 341
Darwin, Robert, 5, 6, 8, 13, 25, 37, 341, 342
Darwin, Susan, 21, 25
Darwin, Susannah Wedgwood, 5, 35
Darwin's Black Box (Behe), 330–31
Darwin's finches, 22, 32–33, 87–89, *89*
　beak shape alterations in, x, 32, *32*, 87–88, 324, 330
Davis, Percival, 324
Dawkins, Richard, 309, 332
DDT, 202
deaf children, sign language of, 287–88, *287*
deer, 167, 175
Dennett, Daniel, xvi
Descent of Man and Selection in Relation to Sex, The (Darwin), 261
Deschamps, Eliette Brunel, 293–94
diet, of hominids, 264–65, 267
digestive enzymes, 327–30
digestive system, *123*, 166
dinosaurs, x, 141, 148, 149, 155, 156, 166
　emergence of, 71, *71*
　extinction of, 71, *157*, 159–65
　mammals compared with, 157–58, *157*, 160
diploblasts, 67, 122
disease, 211–27, 304
　AIDS, xiii, 217–25, *219*, 336–37
　Black Death, 219, 222–24, *222*, *223*
　malaria, 178, 202, 226, 227, 306–7, *307*, 337
　sickle-cell anemia, 306–7
　tuberculosis, 211–13, *211*, *212*, 215–16, 333
　water supply and, *225*, 226–27
Dixon, Andrew, 246
DNA (deoxyribonucleic acid), 73, 76, 77, 81, 94, 215, 217, 218, 224, 248
　antigens and, 93
　of chimpanzees, 300
　cultural evolution and, 305–8
　of Darwin's finches, 88, *89*
　of fish, 232, *233*
　"junk," 327
　mitochondrial, 296–99
　mutation of, *see* mutations
　Neanderthal, *296*, 299–301
　tree of life and, 102–9, *102*, 111–14, *114*
　of whales, 140
Dobzhansky, Theodosius, 80–85, 320
dogs, 159, 165, 167, 174, 227
dolphins, 135, 159
Doolittle, Russell, 329

Doolittle, W. Ford, 112–13
Dorudon, 140
Down House, 37, *38*, 44, 48, 192, 341, 342, 343
Dpp, *123*, 124
dragonflies, 79, 155, 241
Draper, John William, 52–53, 54
Drosophila melanogaster, 80–81
Drosophila pseudoobscura, 81
drought, 87–88, 151, 185
drugs:
for AIDS and HIV, 218–19
see also antibiotic resistance
ducks, 173, 175, 245
dugongs, 165
Dunbar, Robin, 270, 272, 273, 285–86, 290, 291
Dusenbery, David, 234
dust, interplanetary, 104, 160–61

Earth, 9–12
　age of, 9, 58–63, *59*, *60*, 70, 315, 322
　creation of, 9, 12, 318
　history of, life's history compared with, 70–71, 109–10
　magnetic field of, 160, 323
　orbit of, 167–68
　plate tectonics and, 53–54, 162
earthquakes, 17, 20, 21, 31, 164
Echinogammarus, 181
ecological niches, 127–28, 156, 165
ecosystems, 149, 150, 164, 165, 185
　biological invaders and, 179–83
　of Krakatau, *152*, *153*, 154
　Permian-Triassic extinction and, 154–55
　of South America, 167
Ecuador, 88, 226
Ediacarans, 67, *67*, 68, 69, 71, 122
Edinburgh, 5–6, 14, 342
Edinburgh Review, 50–51
Edinburgh University, 5–6
eggs, 73, 77, 91, 229, 230, 232–35, 239–45, 248
Egypt, 323–24
　Napoleon's invasion of, 15
　whales with legs in, *137*, 140
Ehrlich, Paul, 197–98, 199, 202
Einstein, Albert, xvi, 323
elephant birds, 170, 175
elephants, 84, 159, 165, 246
　living vs. fossil, 10
embryos:
　development of, 30, 31, 39, 49–50, 117, 120, *123*–24, 131, 230, 238, 244
　fossils of, 68, *68*, 69
Emlen, Stephen, 247–48, 250–51, 275

England:
　Darwin's return to, 24–25, 42
　evolution denounced in, 15, 30, 39–40, 51–54
　social Darwinism in, 316–17
Enterococcus, 216
Environmental Protection Agency (EPA), 204
enzymes, 105, 108, 194, 218, 305
　digestive, 327–30
Epperson, Susan, 321
Escherichia coli, 108, 216
Essay on the Principle of Population, An (Malthus), 34–35
Ethiopia:
　baboons in, 285
　hominids found in, 262–63, *264*
eucalyptus trees, 171
eugenicists, xvi, 316
eukaryotes, 66
　mitochondria of, 111
　in tree of life, 101, *102*, 103, *107*, *109*, 111, 114, *114*
Europe, 170, 183, 298, 300, 301, 304, 305
　CCR5 mutation in, *222*, 223–24
　pesticide use in, 200–201
eurypterids, 149
evolution, theory of, xi–xiv
　categories of evidence in, x–xi
　Chambers's views on, 38–40, 52
　as dangerous idea, xvi–xvii, 33, 39
　Darwin's discovery of, 4
　Darwin's secrecy about, 33, 37–41, 43
　Erasmus Darwin's views on, 13, 14, 27
　Geoffroy's theory of, 28–31, *29*
　Lamarckian view of, 14, 28, 29, 34, 74, 80, 85, 309, 316, 318
　mechanisms of, *see* catastrophic mass extinctions; natural selection
　Owen's attack on, 30–31, 50–53
　as scientific revolution, xi, 25, 54–55
　social Darwinism and, 316–18, 342
　social implications of, 52, 317, 318, 342
　teaching of, 316–22, 325, 333–36, 338
evolutionary trees, *89*, *138*, 262, *263*, 298, 300, 324–25
　see also tree of life
Evolution (TV show), xv–xvii
Ewald, Paul, 225–26, *225*
exaptation (preadaptation), 134, 308–9
extinctions, xvi, *11*, 28, 48, 143–86
　acceleration of, 175–78
　alien invasion and, 178–81
　"background," 146
　climate change and, 170

coevolution and, 207–9
contemporary threat of, 144, *168*,
    181–84
curve of, 144–46, *145*
Darwin's views on, 143–44, 146
by fragmentation, 176–78
at Mahaulepu, 171–75, *172*
mass, *see* catastrophic mass extinc-
    tions
eyes, 49, 118, 125, 127–31, 309
Pax-6 gene and, 121
shortcomings of, 128–31, *130*
social intelligence and, 269–70
see also vision

faces, attractiveness of, 277, 279
Faed, John, *280*
Farrell, Brian, 198–99
Feldman, Marcus, 81
fibrinogen, 329–30
finches, *see* Darwin's finches
Finding Darwin's God (Miller), 338–39
fire, 171, 179–80, 262
fish, 30, 101, *327*
    with double eyes, 131
    in Lake Victoria, 88–92, *90*
    legs and toes on, 118
    origin of tetrapods from, 132–34
    sexual vs. asexual, 231–33
Fisher, Ronald, 78–79, 80, 237
FitzRoy, Robert, 7–9, *7*, 16, 17, 20, 32
    Christian devotion of, 7, 16, 19, 24,
        54, 340
Flannery, Tim, 170–71
flatworms, 49, 67
fleas, 127
Fleming, Alexander, 214–15, *214*
flies, 34, 281
flood, Noah's, 9, 11, 54, *321*, 322
flowers, 282
    coevolution and, 189–94, *189*, *191*
forests, 69, *70*, 154, *155*, *168*, *169*, 176,
        178, 183, 184–85, 307
    in Australia, 171
    of Costa Rica, 207–9
    fragmentation of, 176–78, 182
    of Pacific Northwest, 195–97
    petrified, 21
    tropical, 166, 176, 181, *182*
Formica ants, 249
Fortas, Abe, 321
fossils, fossil record, 67–71, *67*, *68*, *69*,
        169, *296*, 326, 336, 338
    Chambers's views on, 38–39
    Cuvier's views on, 10–11, *11*, 14–15
    Darwin's discovery of, 18–19, *19*, 21,
        24, 31–32, *31*, 33
    Darwin's explanation of, 48, 49,
        64–65, 71, 143–44

dating of, 64–71, *70*
evidence of evolution in, x, 49, 324
formation of, *3*, 9
history of extinctions in, 143–46,
        171–75
of human remains, 261–67, *264*,
        265, 297–98, 300, 304
Lamarck's views and, 14–15
of land animals, 57
in Mahaulepu, 171–75
Owen's views on, 31–32, *31*, 44, 51
transitional forms in, x, 49, 324–25
tree of life and, 101–2
of whales, 136–41
Fouches family, 146
fovea, 131
Frair, Wayne, 324
France, 113
    cave paintings in, 293–95, *294*, 297
    evolutionary theories in, 14–15,
        28–31, *29*
    fossils studied in, 10–11, *10*
frogs, 70, 121, 124, *128*, 185
fruit flies, 80–82, 85, 87, 337
    in Hawaii, 173
    master-control genes in, 121, 122
    mutations in, 119–21, *119*
    nervous system of, 124
    sex of, 240–41, 243–44
fundamentalism, Christian, 317–20
fungi, 66, *102*, 194–95, 209
fungus-growing ants, 204–7, *206*, *207*
fusion evolution, 111–15

Gabon, 221
Galápagos Islands, 21–23, *22*
    species unique to, *23*
    see also Darwin's finches
Galileo, xi
gametes, 233–34
    see also eggs; sperm
Garden of Eden (Brueghel), *313*
gazelles, 266, 267
geese, 173
gemmules, 73
gender control, reproductive success
        and, 245
genes, genetics, 74–85
    altruism and, 248–49
    antifreeze, 327–28, 330
    body-building, 117–27
    behind Cambrian explosion, 122–24
    of chloroplasts, 112–13
    decision making influenced by, 275
    defined, 76
    duplication of, 110–11, 124–26, 331
    Hox, 120–21, 125–26, 128
    of humans vs. apes, 262
    master-control, 121–22, 134

maternal vs. paternal, 244
Mendel's research on, 74–77, *75*
mitochondrial, 113, 114–15
modern synthesis and, 74, 80–85
of peacocks, 250
race and, 81
silencing of, 331–32
transfer of, 107–8, *107*
tree of life and, 101–3, 107–15
see also chromosomes; DNA;
        genomes; RNA
Genesis Flood (Morris), 321–22
genetically altered crops, 203–4
Genetics and the Origin of Species
        (Dobzhansky), 80, 82, 83, 85
genomes, 108, 111, 114, 222
    of HIV, 217–18
    of humans, 81–82
    lack of diversity in, 307–8
    sequencing of, 337
Genyornis, 169
Geoffroy Saint-Hilaire, Etienne, 28–31,
        29, 123
geology, geologists, 27, 55, 149,
        160–64
    in dating of life, 57–66, 71
    Earth as viewed by, 9–12
    history of extinctions as viewed by,
        144–45
    Lyell's views on, 16–17, *17*, 19, 20,
        21, 54, 59
    see also rocks
Georgia, Republic of, 267
Geospiza fortis, *see* Darwin's finches
Gerhart, John, 124
Germany:
    eugenics in, xvi, 316
    Neanderthals in, 261
Gingerich, Philip, 136–37, *137*, 140
gingko trees, 198–99
giraffes, neck of, 14
glaciers, 90, 126, 167–70, 183
Glires, 159
global warming, 126, 150, 183–86
goats, 174, 176, 178
gobies, round, 181
God, 6, 40–41, 54, 313–44, *339*
    benevolence of, 13, 30
    as creator, 9, 50, 315
    as designer, 12–13, 28, 30, 39, 59
    divine evolution and, 44, 51
Goethe, Johann Wolfgang von, 28
gonorrhea, 216
gorillas, 51, *52*, 261, 262
Gould, James, 32
Gould, Stephen Jay, 283, 284, 308–9
Gouverneur-Général Loudon, 151–52
Grant, Peter, 87–88, 89
Grant, Robert, 6, 14, 15

Grant, Rosemary, 87–88, 89
grasses, 166, 180
grassquits, 88, 89
Graves, Bill, 335
gravity, Newton's theory of, 322, 323
Gray, Asa, 314–15, 338, 342, 343
Great American Interchange, 166–67
Great Chain of Being, 12, 14, 48
Great Lakes, 180–81
Greeks, ancient, 200, 308
greenhouse effect, 150, 166, 183
Greenland, 109, 149
    rocks of, 65–66, 295
    tetrapods found in, 133–34
grooming, 285–86, 291
Guam, 179
Gubbio rocks, 159–62
Gulf of Mexico, 162–63
gulls, 83, 84
guppies, evolution of, 86–87, 86, 325
gymnosperms, 154, 198–99

habitat, destruction of, 175–79, 181,
    184–85
Hadrothemis defecta, 241
Hadza, 267
Haeckel, Ernst, 316, 342
Hahn, Beatrice, 221, 224
Haig, David, 244
Haiti, K-T rocks in, 162, 163
Haldane, J. B. S., 198
half-life:
    extinction and, 177–78, 182
    radioactive, 61, 62, 64
Halkieria, 69
halteres, 119
Hamilton, William, 232, 248–50
Hare, Brian, 271
Hawaii:
    biological invasions of, 174, 178–80
    Burney's finds in, 171–75
    fire in, 179–80
Hawkes, Kristen, 267, 268, 274
Heinrich, Bernd, 209
Hellabrun, bombing of, 254
hemoglobin, 306, 331
hens, 237, 240
Henslow, John Stevens, 6, 8, 24
herbivores, 148, 161, 165, 170–71
heredity, 33–35, 46, 50, 73–97
    acquired traits and, 14, 29, 34, 46
    Darwin's failed understanding of, 71,
        73–74, 101
    natural selection linked to, 71, 74,
        78–80, 82, 84–94
    pangenesis and, 73–74
    sexual displays and, 237
    see also chromosomes; DNA; genes,
        genetics; genomes; mutations; RNA

Hillaire, Christian, 293–94
Himalayas, 164, 166
hippocampus minor, 51, 52
hippopotamuses, 137, 140
hip-to-waist ratio, 277, 278
Hirsch, Vanessa, 220–21
History of Creation (Haeckel), 316
HIV (human immunodeficiency virus),
    217–25, 330, 333
HIV-1, 220–22, 220
HIV-2, 220–21
Hoffman, Paul, 126
holdovers, evolutionary, xi
Holloway, Linda, 335
homeosis, 119, 120
hominids, 262–68, 264, 265
    evolutionary tree of, 262, 263
    see also specific hominids
Homo, 266–67
Homo erectus, 282, 285, 296, 297, 298,
    304
    extinction of, 301, 304, 305
Homo ergaster, 267, 267
homologies, 41, 42, 50
Homo neanderthalensis, see Nean-
    derthals
Homo sapiens, xi, 14, 33, 52, 71, 296,
    297, 298, 300, 304
    as agent of extinction, 144, 169–72,
        174–86
    eyes of, 121, 128
    genome of, 81–82
    impact of culture on evolution of,
        305–11
    Morris's views on, 323–24
    new kind of mind of, 301–5
    superiority of, xii, xiii–xiv, 209
honeybees, 192, 209, 248
honeycreepers, 173
Hooker, Joseph, 37–38, 43, 45, 53, 54
horses, 84, 85, 136, 159, 166, 167, 169,
    174, 208, 209
Hox genes, 120–21, 120, 125–26, 128
Hrdy, Sarah Blaffer, 284
human evolution, social roots of,
    259–91
    alliances and, 268–70
    conflicts of interest and, 278–81
    five great transitions in, 262
    language and, 285–91
    Pleistocene passions in, 272–78
    "theory of mind" and, 270–73
Humboldt, Alexander von, 6
Hunt, Kevin, 264
hunter-gatherers, 267, 275–76, 285,
    302, 304
hunting, 124, 126, 127, 136
    by humans, 167, 169, 170, 175, 176,
        181, 301

Hutterites, 285
Hutton, James, 9–10, 16–17
Huxley, Thomas Henry, xii, 43–44, 44,
    52, 261, 343
    as Darwin's ally, 44, 44, 50–54, 59
    Owen's disagreements with, 44,
        51–53
hydroids, 181

ibis, sacred, 15
ice ages, 90, 126, 167–70, 183, 208–9
Ichthyostega, 133, 133, 134
IFG2 (insulin-like growth factor 2), 244
iguanas, 21, 23
immune system, 92–94, 214, 218, 219,
    222, 224, 233, 238, 279, 331
inclusive fitness, 249–51, 274
Index, The, 342
India, 136, 137, 162, 166, 305
Indian Ocean, 245
    coral reefs in, 23–24
Indonesia, 45, 46, 132, 305
Industrial Revolution, xv
infanticide, 247, 253, 253, 256
infidelity, 240, 279–80
Insectivora, 159
insects, 29, 30, 67, 69, 70, 123, 159,
    177, 337
    bacteria compared with, 215
    birds' eating of, 88–89
    coevolution and, 189–93, 191, 192,
        194, 197–209, 199, 203, 206, 207
    extinctions of, 149, 150, 155
    eyes of, 128
    hometoic genes in, 119
    Hox genes in, 120
Institute for Creation Research, 322
insulin-like growth factor 2 (IGF2), 244
intelligence, 316–17
    fluid, 303
    social, 268–70
Intelligent Design, 325–26, 330–32,
    336
invertebrates, 29, 49, 69
    algae eating, 127
    Geoffroy's views on, 29
Iraq, Shanidar Cave in, 301
iridium, 160–63
islands, equilibrium of diversity of,
    153–54
isotopes, 60–65, 149, 172
Italy:
    genes in, 308
    Gubbio rocks from, 159–62

jacanas, 247–48
Janzen, Daniel, 207–9
jealousy, 279–80, 280
jellyfish, 67, 122

Jenny (orangutan), 33
Jenyns, Leonard, 8
jicaro, 209
Johanson, Donald, 264
John Paul II, Pope, *333*
Johnson, Phillip, 332–33
Journal of Researches (Darwin), *32*, 36, 46
Joyce, Gerald, 106

Kahoolawe, 174
kangaroos, 127, 158, *158*, 169, 170, 171
Kansas, creationist crusade in, 334–36, 338
Kanzi (bonobo), 266
Karroo Desert, 146–49, *147*, 295
Kasting, James, 185–86
Kauai, Mahaulepu on, 171–75, *172*, 178–79, 181–82
Keeling Islands (Cocos Islands), 24
Kegl, Judy, 287–88
Kelvin, Lord (William Thomson), 58–61, *59*, 64
Kentucky, creationist crusade in, 318
Kenya, 177, 263–64
Kipling, Rudyard, 282
Klein, Richard, 298, 302–4
Knoll, Andrew, 150–51
koalas, 158, 171
Komdeur, Jan, 245
Koza, John, 96
Krakatau, 151–54, *152*, 161, 176
Kreisworth, Barry, 216
Krings, Matthias, 300
K-T boundary, 162–65, 167
Kyte, Frank, 163–64

Lacalli, Thurston, 130
lactase, 305–6
lactose, 305
Lamarck, Jean-Baptiste Pierre Antoine de Monet, chevalier de, 14–15, 28, 34, 74, 80, 309, 316, 317, 318
    Geoffroy compared with, 28, 29
lancelets, *117*, 124–26
    eyespot of, *125*, 130–31, *130*
language, 285–91, 303, 304, 308
    Nowak's model of, 288–90
    sign, 287–88, *287*
La Niña, 87
Lankester, Ray, 135
Larkin, Chris, xv
larynx, 290–91
Latin America, cane toads in, 179
Lawson, Nicholas, 22
lead, 61–66, *62*
Leakey, Maeve, 263–64
legs:
    evolution of, 132–34, *133*
    whales with, 136–41, *137*, *138*, 324–25

lemurs, 49, 170, 270
lentiviruses, 220, 224–25
Leopold, Nathan, 319
Lewontin, Richard, 283–84
life:
    artificial, 94–97, 311
    dating of, 57–71, 70–71
    search for origins of, 104–7, *104*, *105*, 295
    tree of, *see* tree of life
limestone:
    from Bottaccione Gorge, 159–62
    extinctions measured in, 149, 159–60
Linnaeus, Carl, 75, 135
Linnean Society, 45
lions, 246–48, *247*, 253, 273
Lipson, Hod, 96
lizards, 169, 179
    whiptail, 230, 231
lobe-fins, 132–33, 134
lobsters, 29, *29*, 40, 118
Loeb, Richard, 319
London, 5, 27, 35, 36, 37
Lootsberg Pass, 148
Louisiana, 336
Lucy (*Australopithecus afarensis*), 264, *265*, 285
Lull, Richard, 132
lungfish, 30, 132, 133, 326, *326*
lungs, 326, *326*
lycopsids, 185
Lyell, Charles, 17, 21, 31, 33, 45, 54, 342
    on corals, 23
    Darwin influenced by, 16–17, 19, 20, 24–25, 40, 59
lymph nodes, 93
Lystrosaurus, 148, 149

MacArthur, Robert, 153–54
Madagascar, 170, 175, 184, 189, 193
Madeira, 16
Mahaulepu, 171–75, *172*, 178–79, 181–82
malaria, 178, 202, 226, 227, 306–7, *307*, 337
malathion, 202
Malawi, 128
Malthus, Thomas, 34–35, *34*, 40, 45, 46, 48
mammals, x, 30, 49, 71, *71*, 101, 156–60, 165–69
    coevolution and, 208, 209
    descent of whales from, 135–41, *138*
    explosion of, 141, 165–69
    extinctions of, 10–11, 18, 21, 166–67, *167*
    metabolism of, 156–57, 159
    placental, 158–59, 165, 244, 331

synapsids compared with, 148
    voice box of, 290–91
mammoths, 10, 11, 169
manatees, 41, *42*, 50, 165
Man's Mastery of Malaria (Russell), 202
Man's Place in Nature (Huxley), 52
marine animals, extinctions of, 141, *145*, 146, 149, 161, 165
marsupials, 158, *158*, 159, 167, 169, 171, 179
master-control genes, 121–22, 134
mastodons, 169
Mauritius, biological invaders in, 181
mayflies, 246
Mayr, Ernst, 82–84, 320
medicine, 307, 337, 341–42
    Darwin's rejection of career in, 4, 5–6
medicine, evolutionary, 211–27
Meishan, 149–50
Mel (chacma baboon), 268
Melander, A. L., 201, 202
Melinis minutiflora, 180
memes, 309
Mendel, Gregor, 74–77, *75*, 80
mesonychids, 136–40, *138*
Mesozoic era, 145, 146, 157
metabolism of mammals, 156–57, 159
Meteor Crater, 63, *63*
meteorites, 63, *63*, 104, 160
Mexico, 162–64, *163*, 175–76
mice, 121, 208, 244
Microsoft, 95–96
milk drinking, 305–6
Miller, Keith, 238
Miller, Kenneth, 338–39
missionaries, 7, 19, 340
Mithen, Stephen, 303
mitochondria, 111–15, *114*
"Mitochondrial Eve," 297
moa, 170
Moby Dick (Melville), 135
mockingbirds, 32
modern life, 293–311
    cave paintings and, 293–95, 297, 302
    cultural evolution and, 305–11, *310*
    Neanderthal DNA and, *296*, 299–301
    new kind of mind and, 301–5, *302*
    old view of, 295–96
    "Out of Africa" hypothesis of, 297–99, *299*
modern synthesis, 74, 80–85, 102, 107, 114
molasses grass, 180
moles, *42*
mollusks, 17, 30, 40, 69
Monism (Monist League), 316, 318

monkeys, 159, 167
  social life of, 268
  sooty mangabey, 220–21
  vervet, 268, 288–89
monotremes, 158, *158*
monsoons, 166
monsters, 30, 118–21
Monte Verde, 169
Morgan, Thomas Hunt, 80–81
Morris, Henry, 321–24
Moschorinus, 148
mosquitoes, 178, 226, 227, 306
moths:
  coevolution and, 189–90, 193
  wings of, x, 119
Mount, the, 5, 6, 8, 14, 25
mountains, formation of, *17, 20, 35*
Mueller, Ulrich, 205–6, *206*
Müller, Miklos, 114
Muller, Paul, 201
mullet, 173, 179
Murray, John, 46
mussels, zebra, 180–81, 184
mutations, 78–82, 85, 86, 107, 108,
  114, 128, 231–33, 330
  of bacteria, 215
  CCR5, *222,* 223–24
  to duplicated genes, 111
  in fruit flies, 119–21, *119*
  of HIV, 218
  lactase switch and, 305–6
  language and, 288–89
  neocortex and, 270
  pesticide resistance and, 201, 202–3
Myanmar (formerly Burma), 175
Mycobacterium tuberculosis, 211–13, *211,*
  215
myoglobin, 331

Napoleon I, Emperor of France, 113,
  342
National Museum of Paris, 14, 28
National Research Council, 334
Native Americans, 7, 19, 306
Natural History of Rape, A (Thornhill
  and Palmer), 281–82
natural selection, x, xii, 37–38, 42–51,
  146, 155, *283,* 316, 334–36
  in artificial life, 94–97
  disease and, 222, 224
  fighting colds with, 92–94, *93*
  heredity linked to, 71, 74, 78–80, 82,
    84–94
  human evolution and, 260–62,
    273–78
  Malthus's population theory and,
    34–35, *34,* 46, 48
  pigeon breeding compared with, 46,
    46, 48

sexual selection vs., 236
social Darwinism and, 317
tree of life and, 101–15
nature:
  design of, 12–13, 28, 30, 35, 39,
    325–32
  laws of, 28, 31, 39, 332
  meaning of life and, xiii
Nature, 282
Nazi Germany, xvi
Neanderthals, 261, 296, 297–301, 303,
  304, *304*
  DNA of, *296, 299*–301
  extinction of, 301, 303, *304*
neocortex, 156–57, 270, 285
neo-Lamarckians, 74, 85
nerve cord, *117,* 122–25, *123,* 130
  of lancelets, *117, 125*
  see also spinal cord
nervous fluid, Lamarckian view of, 14
New Guinea, 285, 305
  Binumarien people of, 274, 283
Newton, Isaac, xvi, 55, 322, 323, 344
newts, rough-skinned, 195–97, *196*
New York Times, 273–74
New Zealand, 170
Nicaraguan Sign Language, 287–88,
  287, 290
Nile perch, 91–92
Noah's flood, 9, 11, 54, *321,* 322
North America, 166–67, 169, 173, 183
  humans in, 170
North Sea, 83–84
notochords, 125
notothenioids, 326–28
Novacek, Michael, 158, 167
Nowak, Martin, 288–90
Nuer people, 274

O'Brien, Stephen, 222–24
oceans, 9, *17,* 66, *70,* 144, 154, 155, 162,
  166
  Cambrian explosion in, 126–27
  in Mahaulepu, 172, 173
  Permian-Triassic extinction and, 149,
    150–51
octopi, 16, *128*
Of Pandas and People (book about Intel-
  ligent Design), 325
Ofria, Charles, 94
Ogbourne, Eleanor Martha, xv
oil and minerals, search for, 336
Oken, Lorenz, 30
Oklahoma, 320, 336
O'Leary, Maureen, 139, 140
"On the Intellectual Development of
  Europe" (Draper), 53
*On the Origin of Species by Means of Nat-
  ural Selection* (Darwin), xiv, *38,* 59,

  74, 136, 143, 261, 314, 324, 338,
    343, 344
  argument of, 46–50
  coevolution in, 192
  as defensive book, 48–49
  Huxley as champion of, xii, 44, 59
  irony of, 55
  origin of, 27–46
  Oxford meeting debate over, 52–54
  publication of, 46, 58
  reviews of, 50–51, 53, 314–15
Opabinia, 69
opossums, 158, 167
optical illusions, 269, *269*
optic nerve, 129, 131
orangutans, 33, 45, 51, *52,* 265
orchids, 189–90, *191,* 192–93, 208
ornaments, 301–2, *302*
Orr, H. Allen, 331
Osborn, Henry Fairfield, 85
Osco-Umbro-Sabellian civilizations,
  308
Osorno, Mount, 19, 20
ostriches, 28
"Out of Africa" hypothesis, 297–99,
  299
overpopulation, dangers of, 34, *34,* 45,
  48
Overton, William, 322, 323, 325
Ovid, 136
ovulation, 234, 279
Owen, Fanny, 35
Owen, Richard, 30–32, *31,* 68
  archetypes as viewed by, 40–41, 44,
    50, 51
  on divine evolution, 44, 51
  evolution attacked by, 30–31, 50–53
  Huxley's disagreements with, 44,
    51–53
owls, 173, 174
Oxford meeting (June 1860), 52–54
oxygen, 62, 111, 114, *114,* 164, 306, 326
  in ocean, 126, 155
  photosynthesis and, *101*

Pääbo, Svante, 299–300
Pacific Northwest, forests of, 195–97
Pacific Ocean, 87, 163, 164
Pakicetus, 137, 138–39, *138*
Pakistan, 136, 137, *137,* 140
paleoanthropologists, 262, 297, 298,
  301–3
Paleocene epoch, 71, 160
paleoecologists, 171–75
paleontologists, x, 55, 64–65, 67–69,
  71, 127, 143–44, 170, 171, 178, 324,
  333
  Cambrian explosion as viewed by,
    122

mammals as viewed by, 157–59
mass extinctions as viewed by, 146–49
modern synthesis and, 74, 80, 84, 85
and origin of tetrapods, 132–34
and origin of whales, 136–41
Paleozoic era, 145, 146
Paley, William, 12–13, 15, 25, 49, 53, 94, 325–26
Palmer, Craig, 281–82
palp, 241–42
Panama, Isthmus of, 166
pancreas, digestive enzymes in, 327–28
Pangaea, 126, 169
pangenesis, 73–74
Pan troglodytes troglodytes, 221
papaya, 208
parasites:
    sexual reproduction and, 231–33, 237
    triumph of, 213–14
paris green, 200, 201
Patagonia, 18, 21, 33
Patterson, Claire, 63
Paul (chacma baboon), 268
Pax-6 gene, 121
peacocks:
    inclusive fitness and, 250
    tails of, 229, 230, 235–37, *236*, 238, 239
peahens, 229, 236–37, *236*, 239, 250, 277
peas, Mendel's research on, 75–77, 80
penicillin, 214–15, *214*, 216
Penton-Voak, Ian, 279
perch, Nile, 91–92
Permian period, *71*, 147, 155, 162
Permian-Triassic extinction, 146–51, 155, 157, 162
Perrett, David, 277
Peru, cholera in, 226
pesticides, coevolution and, 200–203, 209
Petrie, Marion, 235, 237, 238, 250
pharmaceutical companies, 214, 216
phenylalinine, 307
phenylketonuria, 307
pheromones, 234, 241
Philippines, 184
Phillips, John, 145
photoreceptors, 128–29, *130*, 131, 309
photosynthesis, 23, *101*, 111–12, 126, 149, 194, 197
pigeon breeding, 43, 44, 46, *46*, 48, 78, 236
pigs, 174, 175, 178
Pimm, Stuart, 177, 182–83
Pinker, Steven, 287
pipefish, sex of, 238–39
placental mammals, 158–59, 165, 244, 331

plankton, 160, 161, 164, 165, 168, 336
plants, 66, 69, *70–71*, 101, *101*, *102*, 107, 149, 154, 166, 168
    breeding of, 35
    with canals of latex or resin, 198
    chloroplasts in, 111–12, *112*
    coevolution and, 189–95, *189*, *191*, 197–209
    Goethe's views on, 28
    in Hawaii, 171–74
    Mendel's research on, 75–77, 80
plasmids, 107
Plasmodium, 227
Plasmodium falciparum, 306, 307
plate tectonics, 53–54, 162
platypuses, 30, 158, *158*, 159
Pleistocene epoch, *71*, 208–9
Plymouth, 3, 9, 15
Poland, 176
Pollack, Jordan, 96
Polynesians, 174, 178
polyps, 23
Pope, Alexander, 12
porpoises, 135
Povinelli, Daniel, 271
preadaptation (exaptation), 134, 308–9
Precambrian era, 64, 65, 68, 69, *70–71*, 126
pregnancy, 234–35, 244
primates, 71, 131, 159, 165, 166, 175
    color vision of, 237–38
    lentiviruses in, 220, 221, 224–25
    social life of, 268–69
Principles of Geology (Lyell), 16–17, *17*, 23
prokaryotes, 103
proteins, 76, 77, 78, 117, 203, 214, 215
    Bmp-4, 123–24, *123*
    Bt, 203–4
    chordin, *123*, 124
    Dpp, *123*, 124
    fibrinogen, 329–30
    tree of life and, 103–6, 108, 110, 112
Protocetus, 325
protozoans, 66, 69, 209, 337
    asexual reproduction of, 230
pseudogenes, 331–32
psychology, evolutionary, 276–84
pterodactyl fossil, *57*
pterosaurs, 161
Public Health Research Institute, 216
punctuated equilibrium, xiii
Punta Alta, 18
pythons, 268

Quarterly Review, 50
quartz, shock waves in, 162, *164*
quillworts, 154, 155, 185
quinolones, 217

rabbits, 159
racism, xiii, 317, 318
radioactivity, 59, 60–65, *60*, 177, 322
radiometric dating, 336
Ramsay, Marmaduke, 6
rape, 281–82
rats, 174, 175, 178, 224
Rattlesnake, HMS, 44
Raven, Peter, 197–98, 199, 202
reciprocal altruism, 251–53, 273
Red Queen hypothesis, 231–33, *232*
Red Sea, 305
reed buntings, 246
Reid, Alastair, xv
relativity theory, 323
religion:
    science vs., 9–12, 36, 39, 53, 54, 59, 315–16, 320–26, 330–44
    see also creationism
reptiles, x, 30, 71, 101, 141, 155
retina, 128–31
    detached, 129, 131
retinal pigment epithelium, 129, 131
Reznick, David, 86–87
rhinoceroses, 159, 165, 169, 208
rhinoviruses, 225–26
Rice, William, 243–44
Rickettsia prowazekii, 113–14
ring species, 83–84, *83*
RNA (ribonucleic acid), 76, 77, 94, 217, 327
    tree of life and, 103, 105–9
"RNA world," 106, 213
robots, 96–97
rocks, 10, 66, 335
    Acasta, 57–58, *58*, 62–64
    age gap of, 63–64
    of Greenland, 65–66, 295
    Gubbio, 159–62
    of Karroo Desert, 146–49, *147*, 295
Rodhocetus, 140
Rome, ancient, 200
Romer, Alfred, 133, 134
roosters, 240
    combs of, 237, 238, *238*
Russell, Dale, 161
Russell, Paul, 202
Russia:
    burials in, 303
    Napoleon's invasion of, 113
    tuberculosis in, 211–13, *212*, 215–16
Rutherford, Ernest, 61–62

sago palms, 198
Sahara Desert, 305
Saint Jago, 16, *17*
Saint Mark's basilica, 283–84, *283*
salamanders, 70, 176
Salmonella, 217

San Jose scale, 201
Sauroposeidon, 157
sawflies, 119
Schizachyrium condensatum, 180
Schultz, Ted, 205, *206*
science:
    "creation," 320–22, 325
    limits of, 332–34
    religion vs., 9–12, 36, 39, 53, 54, 59,
        315–16, 320–26, 330–44
    test of, 322–25
    theories in, ix, x, 322–23, 335
    twofold task of, ix
    see also creationism
scientific method, 332
Scopes, John, 318–20
Scotland, rocks of, 10
Scott, Eugenie, 336
sea cucumbers, 330
seals, elephant, 235
sea pens, 14, 67
Sedgwick, Adam, 6, 11, 12, 30, 39–40
seeds, 87, 89, 197
    distribution of, 43, 178, 207–8
    in Mahaulepu, 171, 173
Selfish Gene, The (Dawkins), 309
Sepkoski, John, 145–46, 169
sex, sexual reproduction, 73, 77, 82, 91,
    229–56
    altruism and, 248–53
    asexual reproduction vs., 230, 231
    attraction and, 277–78, *277,* 279
    battle of sperm in, 239–42
    of bonobos, 254–56
    coevolution and, 190–93
    Darwinian family life and, 246–48,
        247
    disadvantages of, 230–31
    as evolutionary adaptation, 230
    female choice and, 235–40, 277, 278,
        279, 281
    maternal investments and, 245
    Red Queen hypothesis and, 231–33,
        232
    success of, 231–33
sexes:
    chemical warfare of, 242–44
    conflict between, 230, 252–53, *253,*
        254, 278–81
sexual selection, 235–39, *236, 238, 239,*
    261, 273
Seychelles warblers, 245
Shackleton, Nicholas, 168
shale, Australian, 66
Shanidar Cave, 301
sharks, 127, 132, 213
sheep, 174, *176,* 229
shrimp, 29, 40, 127
Siberia, 150, 162

sickle-cell anemia, 306–7
sign language, 287–88, *287,* 290
Simpson, George Gaylord, 85, 87, 320
Singh, Devendra, 278
SIVs (simian immunodeficiency
    viruses), 220, 221
slime molds, *102*
sloths, 159, 167, *167, 169,* 208
smallpox, 226
snails, 173, 174, 178
    shells of, 284
snakes, 118, 127, 156, 179
    red-sided garter, 195–97
    vestigial pelvic bones in, xi
social Darwinism, 316–18, 342
social intelligence, 268–70
sociobiology, 273–84
    critics of, 274, 281–84, *283*
    evolutionary psychology and,
        276–84
Sociobiology (Wilson), 273–74, *283*
sog, *123,* 124
songs, genelike properties of, 309
South Africa, 193, 265, 340
    Karroo Desert in, 146–49, *147,* 295
South America, 31–32, 33, 166–67,
    169, 185
    Beagle in, 17–21
    cholera in, 226
Soviet Union, science in, 320
Spaniards, 209
species:
    adaptation of, 14
    design of, 12
    divergence of, 30, 48
    diversity of, *see* biodiversity
    Dobzhansky's views on, 82, 83
    evolution of, 88–92; *see also* evolu-
        tion, theory of
    extinctions of, *see* extinctions
    Lamarck's view of, 14
Spencer, Herbert, 316–17
sperm, 73, 77, 229, 230, 232–35, 248
    battle of, 239–42
spiders, 185
    Australian redback, 229, 230, 231,
        241–42
spinal cord, 29, *29,* 123–24
    of humans vs. apes, 262–63
sponges, 14, 122, 154
Sputnik satellite, 320
squid, eyes of, 129–30
squirrels, 49, 208
starfish, 67, 121
stepparents, 280–81
sterols, 197
Steubing, Elizabeth, 313–14
Stoneking, Mark, 300
Streptococcus, 216

Streptomyces, 206–7
Stringer, Christopher, 296, 297
stromatolites, *110*
subspecies label, 83
Sudan, Nuer people of, 274
suicide, 7
    of male redback spider, 229, 230,
        231, 241–42
sulfate (SO_4), 150
sulfur, 200, 201
Sumerians, 200
Sumner, William Graham, 317
Sundstrom, Liselotte, 249
sunlight, 66, 11, 168
supercontinent, breakup of, 126, 169
supernova, 161
Supreme Court, U.S., 321
swallows, barn, 238, *239, 240,* 246
Sweden, 223
swordtail fish, 237
symbiosis, 111–14, 337
    cultural, 310
synapsids, *143,* 148, 149, 155, 156, 157
Syngnathus typhle, 238–39

Tanganyika, 128
Tanzania, hominids found in, 264
technology, 176, 259, 260, 303, 309–11
    artificial life and, 94–97
teeth, 208
    of elephants, 84
    of horses, 166, 208, 209
    of humans, 262, 263
    of plants, 197
    of whales, 139
Temple of Nature, The (Erasmus Dar-
    win), 13
Temple of Serapis, 17, *17*
Tennessee, 318–20
    Supreme Court of, 320
Tertiary period, 160
    K-T boundary and, 162–65, *167*
testosterone, 238, 279
tetrapods, 148, 326
    origin of, 132–35
Texas, 162
textbooks, 316, 317, 320, 336
Thewissen, Hans, 137
Thomson, William, *see* Kelvin, Lord
Thornhill, Randy, 281–82
Through the Looking Glass (Carroll), 231,
    232
Tibetan plateau, 166
Tierra del Fuego, 7, 19, 21
toads, 174, 213
    cane, 179
Tomsk Prison, tuberculosis in, 211–13,
    216
Tooby, John, 276

tools, toolmaking, 260, *260*, 261, 262, 265–68, *266*, 273, 285, 291, 301
modern, *293*, 295, 296, 301–4
topminnows, 233
tortoises, 21, 22
Toth, Nicholas, 265–66
transitional forms, x, 30, 49, 324–25
transmutation, 14
tree of life, *27*, 101–15, 331
branches of, *102*, 108, *109*, 111
as hypothesis, 102–3
as mangrove, 107–9, *107*
see also evolutionary trees
Triassic period, *71*, 147, 154, 156
see also Permian-Triassic extinction
trilobites, 149
Trinidad, evolution of guppies in, 86–87, *86*, 325
triploblasts, 67–69, 122
Trivers, Robert, 251–52, 284
tsunamis, 161, 162, 164, 174
tuberculosis, 211–13, *211*, *212*, 215–16, 333
Turkana, Lake, 263–64
Turkey, 301
turtles, 148, 155
typhus, 113

Ungulata, 159
United States:
arrival of Darwinism in, 314–16
biological invasions in, 179, 180–81
birds in, 176–77, *177*
evolution battles in, 316–22, 325–26, 330–38
global warming in, 183
pesticide use in, 201, 202
social Darwinism in, 316–17
uranium, 61–66, *62*, 149
Uspallata Pass, 20–21
Ussher, James, 9, 315

vaccines, 336–37
Valdivia, 20
vancomycin, 216

Various Contrivances by Which British and Foreign Orchids Are Fertilized by Insects, The (Darwin), 193
vertebrates, 28–31, 118, 120, 160, 184
archetype of, 40–41
arrival of, on land, *71*, 132–34
dawn of, 124–26, 132
eyes of, 125, 127–31, *128*, *130*
fossils of, 69–71, *70*
Geoffroy's views on, 28–29, *30*
lancelets compared with, 125, 126, 130–31
nervous system of, 123–24, *123*
Vestiges of the Natural History of Creation (Chambers), 38–40, 45, 50, 52
Vibrio cholerae, 225
Victoria, Lake:
as ecological niche, 127–28
speciation in, 89–92
viruses, 107–8, 179, 213, 216
cold, 219, 225–26
human immunodeficiency, *see* HIV
vision:
brain and, 269, *269*
color, 237–38
volcanoes, 9, 17, 21, 126, 144, 165, 173
crater hypothesis and, 23
on Krakatau, 151–52, *152*, 154, 161
mass extinctions and, 149, 150–51, *152*, 162
of Osorno, 19, *20*
Vrijenhoek, Robert, 230, 233

Waal, Frans de, 255
Waksmann, Selman, 211–12, 214
Wallace, Alfred Russel, 45, 46, 59–60, 236, 261
warblers, Seychelles, 245
Ward, Peter, 146–49, 151, 157, 165, 185
Washington, pesticide use in, 201
Wason Test, 276
wasps, 197, 213
water supply, disease and, *225*, 226–27
Waynforth, David, 277
weapons, 303, 304

Wedgwood, Emma, *see* Darwin, Emma Wedgwood
Wedgwood, Joseph, 8
weevils, 198
Westminster Abbey, 343–44
whales, x, xi, 159, 165
holdovers in, xi
origins of, 135–41, *137*, *138*, *139*, 324–25
Wheaton College, 313–14
Whipsnade Park, 250
White, Randall, 301, 302
Whiten, Andrew, 268–69, 272
Wilberforce, Samuel, 52–54, 314
Williams, George, 129
Williams syndrome, 269–70
Wilson, Allan, 296–97, 299
Wilson, Edward O., 153–54, 273–74, 283, 339–40, *339*
Wilson, Woodrow, 317
•Woese, Carl, 103, 108–9, 112–13
Wollya Bay, 19
wolves, 169, 227
wombats, 169, 171
woodpeckers, ivory-billed, 177, *177*
World War I, 316
World War II, 214, 254
worms, 121, 213, 337
Wrangham, Richard, 253, *253*, 255, 256, 271
Wright, Sewall, 78–79, 80
Wyoming, 165

Xanthopan morgani praedicta, 193

Yellowstone National Park, *105*
Yersinia pestis, 224
Yucatán Peninsula, 162–64

Zaire, HIV in, 219
zircons, 62, *62*, 64, 65, 149
zoology, zoologists, 6, 43–44, 74, 80, 85, 182
Zoology of the Voyage of HMS Beagle (Darwin), 19

Where more than one credit exists on a page, the images are listed in clockwise order starting from the top.

ABBREVIATIONS

AA — Animals Animals
AMNH — Courtesy Dept. of Library Services, American Museum of Natural History
AP/WW — AP/Wide World Photos
BAL — Bridgeman Art Library
BCI — Bruce Coleman Inc. New York
CP — Culver Pictures
FMC — French Ministry of Culture and Communication, Regional Direction for Cultural Affairs—Rhone-Alpes region—Regional department of archaeology
GC — The Granger Collection, New York
GH — Runk/Schoenberger/Grant Heilman Photography, Inc.
MF — Masterfile
MEPL — Mary Evans Picture Library
NGS — National Geographic Society Image Collection
NHM — The Natural History Museum
PA — Peter Arnold, Inc.
PR — Photo Researchers
SPL/PR — Science Photo Library/Photo Researchers

FRONTISPIECE: Background: Laura Varacchi/WGBH. Left to right: Sam Abell/NGS; Geological Survey of Canada/SPL/PR; Ernest A. Janes/BCI; FMC
PART ONE: p. 1: Sam Abell/NGS; Geological Survey of Canada/SPL/PR; Ernest A. Janes/BCI; FMC
CHAPTER 1: p. 2: Sinclair Stammers/SPL/PR; 3, 5: MEPL; 7: Royal Naval College, Greenwich, London, UK/BAL; 11, 13, 15, 16: MEPL; 17: Neg. No. 338957 AMNH; 19: Neg. No. 326796 AMNH; 22: Sam Abell/NGS; 23: Schafer & Hill/PA; Kelvin Aitken/PA
CHAPTER 2: p. 26: Courtesy Oxford University Press; 27: CP; 29: Neg. No. 338680 Photo by Beckett AMNH; 31: Hulton Getty/Liaison Agency; 32: MEPL; 34: GC; 36, 37: George Richmond (1809-96) Down House, Downe

Kent, UK/BAL; 38: MEPL; 41: Ron Sefton/BCI; 44: MEPL; 47: CP; 52: Neg. No. 338956 AMNH
CHAPTER 3: p. 56: NHM; 57: Transparency No. K10275(4) AMNH; 58: NHM; 59: GC; 60: MEPL; 62: Mark A. Schneider/PR; 63: Pekka Parviainen/SPL/PR; 67: John Dawson/NGS; 68: Harvard University via AP/WW; 69: Transparency No. K10275(4) AMNH; 70: Neg. No. 992(1)/AMNH.
CHAPTER 4: p. 72: Tripos Associates/PA; 73: Hans Reinhard/BCI; 75: Bettmann/CORBIS; 86: Photograph by Benoni Seghers; Courtesy of David Reznick; 90–91: Hans Reinhard/BCI; 93: Manfred Kege/PA
PART TWO: p. 99: Geological Survey of Canada/SPL/PR; Ernest A. Janes/BCI; FMC; Sam Abell/NGS
CHAPTER 5: p. 100: GH; 101: Science Pictures Limited/CORBIS; 104: Ronald E. Royer/PR; 105: Raymond Gehman/CORBIS; 110: John Reader/SPL/PR; 112: Science Pictures Limited/CORBIS; 114: Professors P. Motta & T. Naguro/SPL/PR.
CHAPTER 6: p. 116: GH; 117: Kjell Sandved/PR; 119: David M. Phillips/PR; All others Oliver Meckes/PR; 129: Adam Hart-Davis/SPL/PR; AA/Bruce Watkins; James Robinson/PR; Kjell Sandved/PR; Gregory G. Dimijian/PR; 133: J. P. Sylvestre/BIOS/PA; 137: Chris Schmidt/WGBH; 139: Courtesy of Carl Buell
CHAPTER 7: p. 142: Neg. No. 035338 Photo by Thomson/AMNH; 143: GC; 147: Kate Churchill/WGBH; 152: GC; 157: Neg. No. 338591 AMNH; 158: Tom McHugh/PR; Gary Bell-TCL/MF; 163: Geological Survey of Canada/SPL/PR; 164: Dr. David Kring/SPL/PR; 167: Transparency No. 2554(4) AMNH; Tom McHugh/PR; 172: Lida Pigott Burney; 177: Transparency No. 6287(2) Photo by Beckett AMNH;
PART THREE: p. 187: Ernest A. Janes/BCI; FMC; Sam Abell/NGS; Geological Survey of Canada/SPL/PR
CHAPTER 8: p. 188: Dr. Jeremy Burgess/SPL/PR; 189: AA/Leen Van Der Slik; 191: Neg. No. 338955 AMNH; 192: AA/Leen Van Der Slik; 196: Lee Rentz/BCI; Stephanie Ito/WGBH;

199: AA/Breck P. Kent; 203: Richard R. Hansen/PR; 206: Jill Shinefield/WGBH; 207: J.P. Varin/Jacana/PR
CHAPTER 9: p. 210: Oliver Meckes/Eye of Science/PR; 211: Jean-Loup Charmet/SPL/PR; 212: Antoine Gyori/ Corbis Sygma; 214: Hans Oswald Wild/TimePix; Science Museum/ Science Society Picture Library; 219: Chris Steele-Perkins/Magnum Photos, Inc.; 223: Private Collection/ BAL; 225: Oliver Meckes/Gelderblom/ PR; Jill Shinefield/WGBH; 226: Jean-Loup Charmet/SPL/PR
CHAPTER 10: p. 228: AA/Stouffer Prod.; 229: Jonathan Scott–TCL/MF; 232: GC; 236: Ernest A. Janes/BCI; 238: Gloria H. Chomica/MF; 239: Louis Quitt/PR; 241: M. Fogden/ BCI; 247: Jonathan Scott–TCL/MF; 253: Norman Tomalin/BCI; John Heminway/WGBH; 254: John Guistine/BCI
PART FOUR: p. 257: FMC; Sam Abell/ NGS; Geological Survey of Canada/ SPL/PR; Ernest A. Janes/BCI
CHAPTER 11: p. 258: D. Roberts/SPL/PR; 259: Neg. No. 4744(5) Photo by D. Finnin/C. Chesek AMNH; 260: John Reader/SPL/PR; 265: John Reader/ SPL/ PR; Neg. No. 4744(5) Photo by D. Finnin/C. Chesek AMNH; 266: Peter Davey/BCI; 267: Transparency No. K19057 AMNH; 275: M.P. Kahl/PR; 277: Photograph Courtesy Devendra Singh; 280: Bury Art Gallery and Museum, Lancashire, UK/BAL; 283: Historical Picture Archive/ CORBIS; 287: Susan Meiselas/ Magnum Photos, Inc.
CHAPTER 12: p. 292: Archivo Iconografico, S.A./CORBIS; 293: Meckes/ Ottawa/PR; 294: FMC; 296: John Reader/SPL/PR; 302: Archivo Iconografico, S.A./CORBIS; Neg. No. 338394 AMNH; 304: John Reader/SPL/PR; 307: Meckes/Ottawa/PR; 310: Bettmann/ CORBIS
CHAPTER 13: p. 312: Scala/Art Resource, NY; 313: GC; 317: Private Collection/ BAL; 319: AP/Wide World Photos; 321: Philadelphia Museum of Art/ CORBIS; 326: Tom McHugh/PR; 327: Courtesy of C.-H. Chris Cheng; 329: CNRI/SPL/PR; 333: AP/Wide World Photos; 339: Ira Wyman; 343: GC

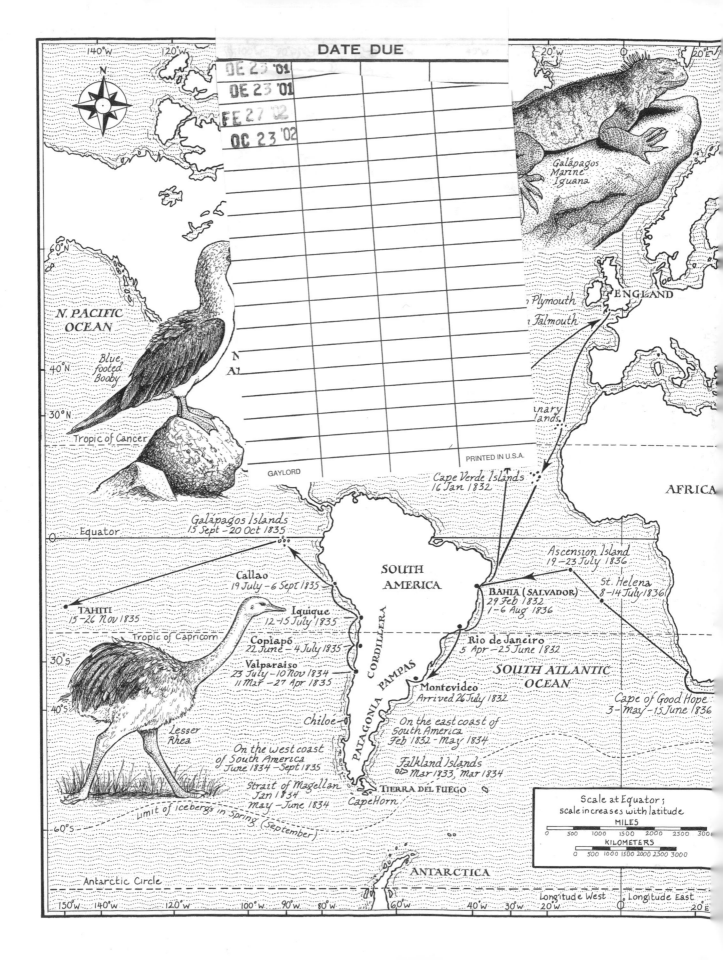

N. PACIFIC OCEAN

Blue-footed Booby

Tropic of Cancer

Galápagos Marine Iguana

Plymouth
ENGLAND
Falmouth

Canary Islands.

Cape Verde Islands
16 Jan 1832

AFRICA

Galápagos Islands
15 Sept – 20 Oct 1835

Equator

SOUTH AMERICA

Callao
19 July – 6 Sept 1835

Ascension Island
19 – 23 July 1836

St. Helena
8 – 14 July 1836

TAHITI
15 – 26 Nov 1835

Iquique
12 – 15 July 1835

BAHIA (SALVADOR)
29 Feb 1832
1 – 6 Aug 1836

Copiapó
22 June – 4 July 1835

Tropic of Capricorn

Rio de Janeiro
5 Apr – 25 June 1832

Valparaiso
23 July – 10 Nov 1834
11 Mar – 27 Apr 1835

CORDILLERA

PAMPAS

SOUTH ATLANTIC OCEAN

Montevideo
Arrived 26 July 1832

Cape of Good Hope
3 May – 15 June 1836

Chiloé

PATAGONIA

On the east coast of South America
Feb 1832 – May 1834

Lesser Rhea

On the west coast of South America
June 1834 – Sept 1835

Falkland Islands
Mar 1833, Mar 1834

Strait of Magellan
Jan 1834
May – June 1834

TIERRA DEL FUEGO
Cape Horn

Limit of icebergs in spring (September)

Scale at Equator;
scale increases with latitude

MILES
0 500 1000 1500 2000 2500 3000

KILOMETERS
0 500 1000 1500 2000 2500 3000

Antarctic Circle

ANTARCTICA

Longitude West   Longitude East

150°W 140°W 120°W 100°W 90°W 80°W 60°W 40°W 30°W 20°W 0 20°E